The Science of Horticulture

The Science of Horticulture
VOL. 1

Edited by :

K. V. PETER

Director

World Noni Research Foundation, 12, Rajiv Gandhi Road
Srinivasa Nagar, Perungudi, Chennai - 600 096
Tamil Nadu

NEW INDIA PUBLISHING AGENCY
Pitam Pura, New Delhi-110 034

Published by
Sumit Pal Jain *for*
New India Publishing Agency
101, Vikas Surya Plaza, CU Block, L.S.C. Mkt.,
Pitam Pura, New Delhi- 110 088, (India)
Phone: 011-27341717, Fax: 011-27341616
E-mail: newindiapublishingagency@gmail.com
Web: www.bookfactoryindia.com

ISBN : 978-93-80235-47-9

For any clarification, Contact Professor K.V. Peter at : peter.kv@gmail.com

Composed and Designed by NIPA

Contents

Fruit Science

Visit us at : www.ouat.ac.in
Tel: 0674-2397700 (O), 2561606 (R) Fax : 0674-2397780
E-mail : ouat_dproy@yahoo.co.in/ouat_dproy@rediffmail.com

ORISSA UNIVERSITY OF AGRICULTURE & TECHNOLOGY
BHUBANESWAR - 751 003, ORISSA

Prof. D.P. Ray
Vice-Chancellor

Foreword

Horticulture Science is getting special attention due to the realization that nutritional security and freedom from hunger are possible with more production and availability of fruits, vegetables, medicinal and aromatic plants, spices, tuber crops and plantation crops. The Indian Council of Agricultural Research (ICAR) has now revised the post-graduate syllabi in Horticulture with more content of Science in them. Vegetable Science, Fruit Science, Floriculture and Landscape Planning and Plantation Crops, Spices and Medicinal Plants are the four scientific disciplines proposed. Use of biotechnology in improvement of crops is more pronounced in Horticultural crops. The WTO regime has necessitated appellation of Geographical Indications (GI) and Intellectual Property Rights (IPR) and it is necessary to expose students and teachers to these developments. Pesticide "residue free" horticultural products are much sought after. The present book 'The Science of Horticulture' is a compilation of 16 chapters contributed by 27 eminent men of Science, Teachers and Lawyers who have worked on the subjects under review for years. The plethora of chapters reflect the wide spectrum of Horticulture Science. Biotechnology especially transgenic technology has been effectively utilized in vegetables, ornamentals and tuber crops.

I congratulate all the contributors for contributing the chapters in academic style and format for the benefit of post-graduate students. I compliment Dr. K.V. Peter, basically an established Professor of Horticulture who edited this much valued book.

I complement the New India Publishing Agency, New Delhi for the publication of the book in an excellent way.

D.P. RAY

Preface

'State of Agriculture'released by National Academy of Agricultural Sciences, New Delhi reveals critical information on Horticulture, its past, present and future. India is the second largest producer of fruits and vegetables with 11% and 12% respective shares in world production. Varied agro-climatic conditions favour cultivation of a large number of horticultural crops (fruits, vegetables, root and stem tubers, aromatic and medicinal plants and spices and plantation crops (coconut, arecanut, cashew, cocoa, palmyrah, tea, coffee, cardamom, cinchona, oil palm, rubber and wattle). In India, emphasis on horticultural development is focused through various programmes: National Horticultural Mission, Technology Mission for Integrated Development of Horticulture in Northern hilly states, Development of commercial Horticulture and Capital Investment Subsidy Scheme of the National Horticulture Board (NHB) and Integrated Development of Coconut including Technology Mission on Coconut.

'The Science of Horticulture' exposes all the stake holders-students, scientists, extension officers, farmers, policy planners to recent scientific research findings in Horticulture. Marketing of Horticulture produces has assumed prime importance in a Global Market with establishment of World Trade Organisation (WTO). Many trade related agreements were signed among over 110 countries. The Agreement on Trade Related Aspects of Intellectual Property Rights (TRIPS agreement) specifically represents the resolve of all member countries to use intellectual property as one of the means of achieving economic balance and diminishing the trade barriers. The legislative history elaborates the genesis of TRIPS agreement, General Agreement on Tariffs and Trade (GATT), Madrid agreement, Chairman's draft, Dunkel draft, general protection under article 22 and Certain aspects of TRIPS definition. Malabar Pepper, Alleppy Green Cardamom, Assam Tea, Darjeeling Tea, Nilgiri Tea and Coorge Orange are a few horticultural crops which got registered under Gi appellation. The Chapter I is contributed by Latha S. Nair, Intellectual Property Firm, Gurgaon.

Hi-Tech Horticulture is based on the principles of minimizing the uses of inputs, maximizing the efficiency of inputs including space, water and energy and optimizing the output-the economic product per unit of time, space, energy and labour-. Hi-tech horticulture is discussed under micro propagation, protected cultivation, micro irrigation and fertigation and high density planting. Merits of micro propagation are mass multiplication of plants, disease free planting materials, year round production, uniform size of plants and easy transportation and storage. Factors affecting micro propagation are gene type, contamination, culture media and hardening. Protected cultivation structures-Quonset Green House, Curved Roof Green House, Goble Roof Green House-, covering materials for green house, green house design and design of environmental control are dealt in Chapter II. Patil *et al.*, Hi-tech Horticulture Project, UAS Dharwad explain in detail the priciples of Hi-tech Horticulture.

Radio tracer studies are done for genetic improvement of crop plants, soil-plant-water and soil fertility research, root activity studies in fruit crops, fertilizer management in vegetable crops and identification of nutrient efficient varieties of scion and rootstocks in fruit crops. Other uses of radiotracer and radiation are determination of volume of fruit trees, detection of unorthodox movement of nutrients, monitoring of internal dynamics of water, pesticide biodegradation and management of insect pests.

Biodiversity is the wealth of any nation for sustenance and biosecurity. Survey, collection, description, documentation and maintenance of plant germplasm are pre requisites for any long term measures for food security for man, animals, birds and other fauna. The horticulture crops constitute 13.08% of total cropped area and contribute about 28% to Agricultural Gross Domestic Product. Over 15000 species of higher plants occur in India, of which 33% are endemic. These 4900 endemic species belong to 141 genera and 41 families and are mainly distributed in the Himalayas, Southern Peninsular Region and North Eastern region. Today only about 150 plant species are important in meeting the food needs of mankind worldwide. Among horticulture crops, the widely grown fruits comprise of bananas, orange, apple, mango and melons. The vegetable crops are tomato, cabbage, onion, squash, potato, sweet potato and beet root. Germplasm of above crops are essentially to be conserved. The conservation methods are *in situ* conservation, *Ex situ* conservation, field gene banks, seed conservation, sub-zero temperature storage, modified atmosphere storage of seeds, botanic gardens, cryopreservation, pollen storage and DNA storage.

Chapter V deals advances in male sterility in vegetable crops. Male sterility is classified into two major groups-Genetic male sterility and non genetic. On phenotypic basis, genetic male sterility is classified into three classes *i.e* Sporogenous, Structural and Functional. Non genetic male sterility is classified as chemical, physiological and ecological. In chapter V, male sterility mechanisms in bell pepper, cabbage, cauliflower, broccoli, carrot, onion and muskmelon are elucidated. At global level, CMS and CGMS are the most widely utilized systems in most of the vegetables. Research on transgenic male sterility is in progress.

Polyembryony is considered a desirable character in citrus, mango, jamun and almond to obtain true to type planting materials. Formation of embryo with asexual means is apomixis. In polyembryonate species,extra embryos are formed by nucellar cells (2n), integuments (2n), synergids (n), antipodal cells (n) etc. Polyembryony is a type of apomixis that may occur by cleavage of pro-embryos, formation of embryos by cells of embryo sac other than egg, development of more than one embryosac within the same ovule or multiple polyembryony and formation of embryo by some sporophytic/maternal tissues of ovule. Topics of discussion in chapter VI are polyembryonic crops, factors affecting polyembryony, causes of polyembryony, inheritance of nucellar embryony, organogenesis, degree of polyembryony, identification of nucellar seedlings and significance of polyembryony.

Post harvest losses of horticultural crops are estimated 20-40% and the reduction to an achievable 10% would go a long way to reach targets of requirement. Food irradiation involves exposure to short wave energy to achieve extension in shelf life, insect disinfestations and elimination of food-borne pathogens and parasites. Compared to heat or chemical treatments, irradiation is more effective and an appropriate technology to destroy food-borne pathogens without much loss to the quality of food. Today, more than 42 countries worldwide including India have approved use of irradiation for over 100 food items. South Africa has cleared more than 40 food items. Benefits of food irradiation are reduction of pathogenic micro organisms, decontamination, extension of shelf life, disinfestations and inhibition of sprouting. Chapter VII deals with principles of food irradiation.

A perspective on plant pathology of Horticultural crops is vital to understand the major events in the Science of Plant Pathology. Democritus (470BC) noted plant blights and described a way to control them using olive ground cake left after extraction of olive oil. The ophrastus (300BC), Robert Hook (1670), Pier Antonio Mitcheli (1729), Carl Von Linne (1735), Needham (1743), Anton de Bary (1861-63), Adolph Mayer (1886), Ivanovski (1892), Beijernick (1898), Stanley (1935), Bawden (1936), Gierrer and Schramm (1956), Vermeulen (1963), Doi (1967) and Diemer (1971) are a few outstanding plant pathologists who worked on diseases and their management. Chapter VIII elaborates history of plant pathology of horticultural crops. Chapter IX deals with history of plant breeding with special emphasis on vegetables. Hooke (1665) described the cell. Millington (1676) reported that anthers function as male organs. These early historical findings would pave way to the rapid development and use of science in vegetable improvement. Basics in vegetable seed technology are described in chapter X. Seed is the basic, the cheapest, critical but the most effective and critical agricultural input for attaining higher productivity. Categories of seed- nucleus, breeder, foundation, certified-ensure quality specifications. Seed, field and then seed certifications are done for all categories other than nucleus and breeder seed. Field inspections are done at three stages-vegetative or pre-flowering, flowering and post harvest. Principles of seed production are elucidated for standards and maintenance of quality of annual crops, biennial crops, root crops and vegetatively propagated vegetables.

Plant Growth Promoting Rhizobacteria (PGPR) are getting attention for their beneficial effects by root colonization. They stimulate plants growth as being done by plant growth promoters. The important PGPR used in vegetables are *Azospirillum, Azotobacter, Pseudomonas fluorescens*, Phosphate Solubilising Bacteria (PSB) and *Bacillus subtilis*.Effect of PGPR on chilli, tomato, cucumber, lettuce, potato, okra, knol khol, cabbage, onion, cauliflower, brinjal, watermelon, capsicum, French bean, bitter gourd, broccoli and peas are reported for enhanced nutrient content and value of produces. Effect of PGPR on pest and disease suppression is reported. The plants with PGPR were less injured in relation to plants without PGPR.

Genetically Modified (GM) horticultural crop varieties are being developed for higher yield, better nutritional values and quality and resistance to biotic and abiotic stresses. Gene sequencing in plants is a pre-requisite for genomics and subsequent identification of useful genes (Chapter XII).

Chapter XIII deals with molecular tools for characterization of horticultural crops. Barcoding of genotypes has become relevant in the context of registration of crop varieties and consequent benefit and right to breeders and farmers. Chapter XIV describes transgenic ornamentals with reduced ethylene production and perception. Prolonging shelf life of cut flowers like carnations and chrysanthemums by interfering with ethylene production is attempted. Chapter XV elaborates biotechnology of tuber crops especially cassava,sweet potato and yams. Transgenic research on tuber crops is also reviewed. Chapter XVI deals with viticulture in the Napa Valley, California USA. Napa valley is known for its vineyards. Issues of site selection, planting material, vineyard layout, infrastructure and cultural practices are discussed. A century and a half of grape growing and wine making are narrated. Today Napa valley houses around 400 Wineries. Richard Mendelson the author, is a lawyer specializing in vineyards and Wine law with the law firm of Dickenson, Peatman and Fogarty in Napa, California.

The Science of Horticulture is thus a compilation of 16 chapters on various facets of Horticulture contributed by 27 scientists from India, Japan and USA.

List of Contributors

PATIL A.A.
Hi-Tech Horticulture Project
University of Agricultural Sciences
Dharwad-580 005, Karnataka
Email:smmantur@gmail.com

GEETA BASSI
Department of vegetable crops
Section on seeds, Punjab Agricultural University, Ludhiana, Punjab-141004
Email:jagwant.kanwar@yahoo.com

MANNIKERI I.M.
Hi-Tech Horticulture Project
University of Agricultural Sciences
Dharwad-580 005, Karnataka
Email:smmantur@gmail.com

KANWAR J.S.
Department of vegetable crops
Section on seeds, Punjab Agricultural University, Ludhiana, Punjab-141004
Email:jagwant.kanwar@yahoo.com

RAJMOHAN K.
Department of Biotechnology
Kerala Agricultural University
P.O. Vellayani, Trivandrum,
Email:rajmohan33@yahoo.com

PETER K.V.
Kerala Agricultural University
P.O. KAU-680656
Email:kvptr@yahoo.com

KUNDAN KISHORE
ICAR Research Complex for NEH Region
Sikkim Centre, Tadong, Gangtok-737102
Email-kkhort2003@yahoo.com

LATHA R. NAIR
Intellectual Property Law Firm
K & S Partners, Gurgaon
Email:latha@knspartners.com

SHEELA M.N.
Department of Biotechnology
Kerala Agricultural University
P.O. Vellayani, Trivandrum
Email:rajmohan33@yahoo.com

BIRADAR M.S.
Hi-Tech Horticulture Project
University of Agricultural Sciences
Dharwad-580 005, Karnataka
Email:smmantur@gmail.com

TRIPATHI P.C.
Principal Scientist (Horticulture)
National Research Centre for Women in Agriculture
Bhubaneswar, Orissa -751003
Email:prakaashtripathii2000@yahoo.co.in

SADHAN KUMAR P.G.
Professor, Kerala Agricultural University
KAU, P.O.680656, Thrissur, Kerala
E mail: pgsadhankumar@yahoo.com

DHALL R.K.
Department of Vegetable Crops
Punjab Agricultural University
Ludhiana-141 004
E-mail: dhallrk2007@rediffmail.com

RICHARD MENDELSON
Dickenson, Peatman and Fogatry in
Napa, California USA
Email:mendelson@dpf-law.com

ROBERT STEINHAUER
Dickenson, Peatman and Fogatry in
Napa, California USA
Email:mendelson@dpf-law.com

NIRMALA DEVI S.
Professor, Kerala Agricultural
University, KAU
P.O., Thrissur-680 656, Kerala
E mail: pgsadhankumar@yahoo.com

KOTUR S.C.
Isotope Laboratory
Division of Soil Science & Agricultural
Chemistry, Indian Institute of
Horticultural Research, Hessaraghatta
Lake Post, Bangalore-560 089
Karnataka, Email:sckotur@gmail.com

DOIJODE S.D.
Division of Genetic Resources
Indian Institute of Horticultural Research
Hassaraghatta Lake Post
Bangalore-560 089
Email:sddoijode@hotmail.com

MANTUR S.M.
Hi-Tech Horticulture Project
University of Agricultural Sciences
Dharwad-580 005, Karnataka
Email:smmantur@gmail.com

SABINA ISLAM
Division of Vegetable science
IARI, New Delhi 110012
Email:tusar@rediffmail.com

SHIGERU SATOH
Graduate School of Life and
Environmental Sciences
Kyoto Prefectural University Sakyo-ku
Kyoto 606-8522, Japan

PRADEEP KUMAR T.
Kerala Agricultural University
P.O. KAU-680 656
Email:kvptr@yahoo.com

BEHERA T.K.
Division of Vegetable science
IARI, New Delhi 110012
Email:tusar@rediffmail.com

TAKAKO NARUMI
Department of Horticulture
Faculty of Agriculture, Kagawa University
Miki-cho, Kagawa Prefecture 761-0795
Japan

JAYALAKSHMY V.G.
Associate Professor
Department of Plant Breeding and
Genetics, Kerala Agricultural University
P.O. Vellayani, Trivandrum 695522
Email:Jayavgj@Yahoo.com

VEENA ASHOK
Senior Scientist
Indian Institute of Horticultural Research
P.O.Hessarghatta Lake, Bangalore-560089
Email:veenaashok@iihr.ernet.in

YUSUKE KOSUGI
Prefectural Institute of Agricultural
Biotechnology, Seika-cho
Kyoto Prefecture 619-0244
Japan (*ssatoh@kpu. ac.jp)

❑❑❑

Acknowledgement

I am grateful to Prof. Dr. P.I. Peter, Chairman, Noni Biotech., Chennai for support and facilities. It is his foresightedness and generosity which enthuse me to contribute to Science Journalism. I am grateful to all the 27 eminent scientists who contributed chapters to this book. I acknowledge the publisher New India Publishing Agency, New Delhi for timely publication. I appreciate Dr. P.G. Sadhan Kumar, Professor, AICVIP, Kerala Agricultural University for Logistics. I am gratefull to Prof. D.P. Ray Hon. Vice-Chancellor, OUAT, Bhubaneswar and a renouned Horticulturist for the Foreword to the book.

K.V.PETER

Devotion

The book 'The Science of Horticulture' is devoted to my brother Prof. K.V. Thomas and sister-in-law Mrs. Shirley Thomas who demonstrated how a woman can be supportive and be a pillar of strength to a successful husband. Prof. K.V.Thomas is the Minister of State (Agriculture, Co-operation, Consumer Affairs, Public Distribution) Government of India (2009 –).

General Horticulture Science

The Science of Horticulture : Vol. 1, 2011
© *K.V. Peter (ed.), pp. 1-22*
New India Publishing Agency, New Delhi (India)
E-mail : newindiapublishingagency@gmail.com
Web: www.bookfactoryindia.com

Chapter 1 Geographical Indications and TRIPS Agreement

LATHA R. NAIR

World Trade Organisation

The World Trade Organisation (WTO) was the result of years of painstaking efforts on the part of governments and other bodies to establish a globally acceptable basis for a new economic order. It was established to correct the imbalances and disparities in international trade and bring about an evenly balanced world trade order that equally benefits the developed, developing and least developed nations. While there are several agreements under the WTO, the Agreement on Trade Related Aspects of Intellectual Property Rights (TRIPS Agreement) specifically represents the resolve of all member countries to use intellectual property as one of the means for achieving economic balance and diminishing the trade barriers.

---

1 Latha R Nair is a partner with the Gurgaon based Intellectual Property Law firm, K&S Partners. She has been advising the Government of India in the protection of two well-known geographical indications namely, Basmati and Darjeeling and also represents the interests of a couple of well-known wine geographical indications outside India.

Legislative History

While the term 'geographical indication' has been in use for some decades, the first multi-lateral agreement dealing with the term is the TRIPS Agreement.[2] Intellectual property rights, recognized as 'private rights' in the preamble to the TRIPS Agreement, are copyrights, trademarks, geographical indications, industrial designs, patents, lay-out designs of integrated circuits and trade secrets. By time the WTO was established, well-oiled and more or less uniform machineries of protection in respect of these rights were already established by in its member states. As regards the concept of geographical indications, active protection at that time was offered only by France and some European countries. Since the Agreement left it open to the member countries to determine the appropriate method of implementing its provisions within their own legal system and practice[3], member countries have sought to fulfill the obligations in respect of geographical indications under TRIPS through the means available under their domestic laws. Needless to say, these means differ in nature and content, thereby resulting in uneven levels of protection.

The European Union's interest in affording protection to geographical indications at the international level was evident during the Uruguay Round negotiations going back to 1987 when they proposed guidelines for an intellectual property component in the next trade round of the General Agreement on Tariffs and Trade (GATT). The European Communities subsequently presented their first text of proposals to this effect, which triggered the process towards the current text of the TRIPS Agreement.[4] Section C of the said text, comprising Articles 19 to 21, was devoted to geographical indications. The genesis of the language in the current Article 23 of the TRIPS Agreement for 'additional protection' for geographical indications pertaining to wines and spirits was found in this text, although it did not discriminate between wines and spirits and other goods.[5]

The United States text also proposed a text and section 18 of the same provided that geographical indications be protected by providing for their registration as certification or collective marks. Section 19 of the proposal

---

2 Daniel Gervais, *The TRIPS Agreement – Drafting History and Analysis*, second edn, Sweet and Maxwell.

3 Article 1 of the TRIPS Agreement.

4 See Document MTN/GNG/NG11/W/68 of March 29, 1990. Source: Daniel Gervais, *The TRIPS Agreement – Drafting History and Analysis*, second edn, Sweet and Maxwell.

5 Sergio Escudero, T.R.A.D.E Working Papers, 10, International Protection of Geographical Indications and Developing Countries, South Centre, July 2001. Also see "Intellectual Property and International Trade" by Carlos M. Correa &Abdulquawi A Yusuf, Kluwer Law International, 1998, Page 441 (Communication received from Argentina, Brazil, Chile, China, Colombia, Cuba, Egypt, India, Nigeria, Peru, Tanzania, Uruguay and Pakistan)

further provided that use of non-generic appellations of origin for wine should be prohibited when such use misled the public as to true geographic origin of the wine.[6]

A proposal made by a group of developing countries required the members to provide protection for geographical indications including appellations of origin against any use which is likely to confuse or mislead the public as to the true origin of the product. A footnote of this proposal defined geographical indications as 'any designation, expression or sign which aims at indicating that a product originates from a country, region or locality'.[7]

A proposal from Switzerland[8] also provided for protection of 'geographical indications' and 'appellations of origin'. Under Article 220 of the proposal, both these terms were defined as below:

A geographical indication is any direct or indirect reference to the geographical origin of a product, including characteristics or qualities which are related to that origin.

...an appellation of origin is a qualified geographical indication denominating a country, region or locality indicating that a product is originating therefrom and has qualities, reputation or other characteristics which are essentially or exclusively attributable to the geographical origin, including human or natural factors. A denomination which has acquired a geographical character in relation to a product which has such qualities, reputation or characteristics is also deemed to be appellation of origin.

Apart from defining 'geographical indications', the Swiss proposal also provided for protection of geographical indications related to services.

The Japanese text of the proposal[9] merely proposed to protect geographical indications 'by complying with the provisions under the Madrid Agreement for the repression of False Deceptive Indications of Source of Goods of 1891, as last revised in 1967'.

6 Sergio Escudero, T.R.A.D.E Working Papers, 10, International Protection of Geographical Indications and Developing Countries, South Centre, July 2001.

7 Ibid. Appellation of origin would refer to geographical names only within the meaning of the Lisbon Agreement and non-geographical names were not covered therein. The proposal of the developing countries essentially sought equal protection for geographical indications and appellations of origin.

8 Document MTN.GNG/NG11/W/73. Source: Daniel Gervais, *The TRIPS Agreement – Drafting History and Analysis*, second edn, Sweet and Maxwell.

9 See Document MTN.GNG/NG11/W/74, dated May 15 1990. Source: Daniel Gervais, *The TRIPS Agreement – Drafting History and Analysis*, second edn, Sweet and Maxwell.

In addition, the Chairman of the Negotiating Group on Trade Related Aspects of Intellectual Property Rights, including Trade in Counterfeit Goods prepared a report on the progress of the negotiations that had taken place so far. This report is also referred to as 'Chairman's Draft' or 'Composite Text' and section 3 of the same referred to the protection of geographical indications. As per the Chairman's draft, member countries were to protect geographical indications from:

...any usurpation imitation or evocation, even where the true origin of the product is indicated or the appellation or designation is used in translation or accompanied by expressions such as 'kind', 'type', 'style', 'imitation' or the like.[10]

Under the said Draft, the members were obliged to co-operate with a view to establishing an international register for protected geographical indications, which would facilitate the protection of geographical indications including appellations of origin. The Draft did not discriminate between geographical indications for wines and those for other products.

The Draft Final Act dated December 20, 1991 was circulated by Mr. Arthur Dunkel, the then Director General of GATT, after negotiations and approval of all the negotiating partners. This draft of the TRIPS Agreement called the 'Dunkel Draft'[11] dealt with geographical indications under section 3 thereof. Eventually, this draft was approved at the Marrakesh Ministerial Meeting as the current text of the TRIPS Agreement.

Section 3 of the TRIPS Agreement

Section 3 of TRIPS Agreement deals with geographical indications and has three Articles, namely, Articles 22, 23 and 24. Article 22 defines what geographical indications are and provides for a general framework of protection for geographical indications relating to all products, whereas, Article 23 provides for an additional level of protection for geographical indications relating to wines and spirits. Article 24 contains certain exceptions to the protection of geographical indications.

General Protection Under Article 22

Article 22 which contains the basic definition of 'geographical indication' applies to all products and describes the general level of protection available in respect of the same.

---

10 Document MTN.GNG/NG11/W/76. Daniel Gervais, *The TRIPS Agreement – Drafting History and Analysis*, second edn, Sweet and Maxwell.

11 This document was circulated under the reference MTN.TNC/W/FA and was presented on December 20, 1991. Source: Daniel Gervais, *The TRIPS Agreement – Drafting History and Analysis*, second edn, Sweet and Maxwell.

Article 22.1

The text of Article 22.1 reads as follows:

Geographical indications are, for the purposes of this Agreement, indications which identify a good as originating in the territory of a member, or a region or locality in that territory, where a given quality, reputation or other characteristic of the good is essentially attributable to its geographical origin.

The definition does not envisage that the indication be a proper geographical name. It could include traditional non-geographical terms such as Feta, Alphonso, Basmati etc, which have acquired connotations of nexus with a specific geographical region. Such indications should identify the good as originating in the geographical area. To illustrate, Swiss knives and Darjeeling tea respectively indicate that the knives originate from Switzerland and the tea originates from Darjeeling, India. Both the designations 'Swiss' and 'Darjeeling' are proper geographical names. On the other hand, the non-geographical names listed above namely, 'Feta' denotes a cheese from Greece, 'Basmati' denotes a rice from a region in the Indian sub-continent and 'Alphonso' denotes mangoes from a region in the State of Maharashtra. These traditional names have acquired geographical connotations through customary usage over the years.

The definition also envisages that the good possesses a certain quality, reputation or other characteristics. In the Swiss knife and Darjeeling tea examples, Swiss knife is known for its excellent craftsmanship and Darjeeling tea is known for its renowned and sought-after flavour and fineness. It may be noted that the definition is flexible enough to cover even region-evocative reputation possessed by a good as a sufficient basis for its recognition as a geographical indication.

Further, under the definition, such quality, reputation or other characteristics must be essentially attributable to its geographical origin. To illustrate, Swiss chocolates are descriptive of chocolates made in Switzerland and denote to the public a definable group of products of distinctive reputation.

Certain Aspects of the TRIPS Definition

The definition under TRIPS for geographical indications does not seem to include services. However, member countries are free to extend such protection to services in their legislation. Countries like Liechtenstein, Peru and Switzerland do have laws which protect services as geographical indications. In fact, some countries or regions thereof have acquired reputation / acclaim for services such as medical services, traditional healing methods, herbal massages etc, which would prima facie qualify as geographical indications. The traditional Ayurvedic massages and medical services from the state of Kerala in India could qualify for such protection if such services are regulated within the state and are practiced according to prescribed and laid down standards and structured procedures to maintain and preserve their reputation and integrity.

Yet another aspect not addressed by the TRIPS definition is when more than one country share a geographical indication, how would it be protected. As the definition stands today, it limits the geographical area to the 'territory of a member, or a region or locality in that territory'. Hence, if a given geographical indication identifies a region which falls in more than one country, then it would not fall within the TRIPS definition. One regional law that provides for such a situation is the European Community Regulation 510 / 2006. Under this Regulation, where more than one country may have a shared claim in a given geographical indication, all the countries concerned may reach an overall solution and file a joint application in respect of the geographical indication concerned.

An example of such cross border ownership in a geographical indication would be Himalayan natural mineral waters[12] shared by India and some of its neighbouring countries.

Article 22.2

The text of Article 22.2 reads as follows:

In respect of geographical indications, Members shall provide for legal means for interested parties to prevent:

(a) The use of any means in the designation or presentation of a good that indicates or suggests that the good in question originates in a geographical area other than the true place of origin in a manner which misleads the public as to the geographical origin of the good.

(b) Any use which constitutes an act of unfair competition within the meaning of Article 10 bis of the Paris convention.

When this provision is read in conjunction with the last sentence of Article 1.1 of TRIPS it is clear that a member country is free to choose the legal

12 Currently, there is a brand of natural mineral water being marketed under the trademark 'HIMALAYA' and claimed to be sourced from the Himalayas. If 'Himalayan waters' could be considered a protectable geographical indication owing to the origin and the special properties of the natural mineral waters derived from the region, any appropriation of the same as a trademark would appear to be in conflict with the geographical indication 'HIMALAYAN WATERS'.

means within its own legal system and practice to implement its obligations. Article 1.1, reads as follows:

Members shall give effect to the provisions of this Agreement. Members may, but shall not be obliged to implement in their law more extensive protection than is required by this Agreement, provided that such protection does not contravene the provisions of this Agreement. Members shall be free to determine the appropriate method of implementing the provisions of this Agreement within their own legal system and practice

As explained at the beginning of this chapter, driven as they were by private enterprises, the means and form of protection of intellectual property rights other than geographical indications were able to achieve a certain level of uniformity of structure even before TRIPS came into force. On the other hand, geographical indications, lacking in the incentive of private enterprise (being a matter of collective community rights and thus, considered a function of state responsibility) and spread through the spectrum of natural resources evenly shared between the industrial and non-industrial countries, could not achieve any similar uniformity of structure.

Taking advantage of the flexible 'legal means', countries like the United States and United Kingdom protect geographical indications through certification trademarks, others like Japan and Germany protect these by way of registration as a collective mark and yet another group of countries like India and France protect geographical indications through *sui generis* legislations. Then there are countries like Canada and Australia which have geographical indication laws only for the protection of geographical indications relating to wines and spirits. For geographical indications in respect of products other than wines and spirits, these countries offer protection through registration as certification trademark.

This diverse approach has resulted in an inchoate and unsatisfactory means for protection of geographical indications. For example, India protects geographical indications through the certification trademark regime as well as *sui generis* legislation. A *sui generis* legislation primarily mandates a public register of defined specifications and an automatic right of use in favour of all those who are legitimately involved in the production, distribution and sale of the product which conforms to these specifications. On the other hand, the certification trademark regime operates through a system of licensing arrangements between the non-trading proprietor and all other persons involved in the production and sale of the product concerned. Although the enactment by India of the *sui generis* legislation may be considered a TRIPS compliant step, any registration obtained under the same may not ensure uniform protection throughout the world since certain countries protect geographical indications only through certification trademarks. Thus, overseas protection of a geographical indication in such countries, validly registered under the *sui generis* legislation in India would require the proprietor to create a licensing mechanism for administration of rights in these countries. On the other hand, in other countries which provide similar *sui generis* protection, the proprietor would not be required to establish any similar licensing system. Hence, global protection of geographical indications would differ from country to country in form and structure, burdening the administering authorities of these geographical indications not only with substantial expenditure of resources but also affecting the legal adequacy of the protection available.

Article 22.2(a)

Although the plain language of Article 22.2(a) seems to limit an actionable use to deceptive labeling or advertisements as to origin of the good, it may include other means of deceptive use of expressions which are understood in the public mind as a description of the manufacturing process, characteristic features etc, if such means are liable to mislead the public. For instance, 'Blended with Scotch' a term proprietary to the Scotch Whisky trade cannot be used in respect of blends of whiskies other than those blends each of which is a Scotch Whisky. In the event such a description is used, the Scotch Whisky Association (SWA) would have an actionable claim provided they are able to show that it is likely to mislead the public as to the geographical origin of the whisky produced.

This provision lays emphasis on proof of public deception while stipulating a requirement that a misleading use can be prevented. Such proof would include evidence of public perception by way of affidavits/ public or market surveys etc, all bearing on the question of public recognition of the geographical indication in the country of disputed use as well as likelihood of confusion in the public mind inherent in the use complained of.

Article 22.2(b)

This Article provides that 'legal means' must be offered by members to prevent any use of a geographical indication which constitutes an act of unfair competition within the meaning of Article 10bis of the Paris Convention.

Article 10bis states as follows:
[Unfair competition]

(1) The countries of the Union are bound to assure to nationals of such countries effective protection against unfair competition.

(2) Any act of competition contrary to honest practices in industrial or commercial matters constitutes an act of unfair competition.

(3) The following in particular shall be prohibited:

(a) all acts of such a nature as to create confusion by any means whatever with the establishment, the goods, or the industrial or commercial activities, of a competitor;

(b) false allegations in the course of trade of such a nature as to discredit the establishment, the goods, or the industrial or commercial activities, of a competitor;

(c) indications or allegations, the use of which in the course of trade is liable to mislead the public as to the nature, the manufacturing process, the characteristics, the suitability for their purpose, or the quantity of the goods.

While the acts of unfair competition are very broadly defined, there is a lot of emphasis on the element of confusion, falsehood or misleading of the public. This requires the proprietor of a geographical indication to demonstrate by evidence that the geographical indication concerned is well recognised among members of the public or the trade in the country of disputed use. The scope of 'unfair competition' may be interpreted to include any use of a geographical indication that could result in its dilution and erosion of its reputation. The concept of dilution as traditionally understood has proceeded on a plane which does not require proof of confusion or deception. It is only necessary to prove that the mark concerned is a well known mark and its use on or in relation to dissimilar goods or services would result in dilution of its source identifying function.

Indian courts have fashioned out newer and proactive approaches to the acts of unfair competition and extended the common law remedy of passing off to restrain use of a well known mark by unscrupulous traders in respect of dissimilar goods. The classic case to date is the *Mercedes Benz* case[13] decided by the High Court of Delhi. In this case, at the instance of the owners of the well-known car makers, the court restrained the use by the defendant on its packaging cartons, of a device which consisted of a three pointed human figure wearing an underwear within an illuminated ring with the words 'BENZ' and the description 'German Perfection - Great Mileage in Comfort'.

While granting the interim injunction, the court noted that a mark like BENZ was not up for grabs and could not be diluted through its use on such goods. This decision has been consistently followed by other High Courts in India and it is a settled law in India that the owner of a well known mark, whether overseas or Indian, can restrain local defendants from using the same in respect of even dissimilar goods or services.

13 AIR 1994 Del 239.

However, the scope of 'unfair competition' is not uniformly or consistently understood or applied in all jurisdictions, thereby creating an environment of legal uncertainty and difficulties of enforcement for owners of geographical indications.

Article 22.3

The text of Article 22.3 reads as follows:

A member shall, ex officio if its legislation so permits or at the request of an interested party, refuse or invalidate the registration of a trademark which contains or consists of a geographical indication with respect to goods not originating in the territory indicated, if use of the indication in the trademark for such goods in that member is of such a nature as to mislead the public as to the true place of origin.

Trademarks and geographical indications are two distinct and different concepts with trademark being a private monopoly right and geographical indications being a public/ collective right. Consequently, a registered proprietor of a trademark gets an exclusive proprietary right in the mark and is entitled to prevent others from using the same without authority. Geographical indications, being collective community rights cannot, therefore, be protected as trademarks. Most countries have provisions in their trademark legislations excluding geographical names or marks which indicate quality or origin as matters excluded from registrability.

Article 22.3, relates to the obligation of a member country to refuse or invalidate a trademark which contains or consists of a geographical indication with respect to goods not originating in the territory indicated, if use of such indication misleads the public as to true origin. Hence, if a geographical indication of country X is being used by a national of country *Y* as a trademark or as part of a trademark, country Y, if requested by country X, is obliged to refuse such registration or invalidate it only if such use is a misleading use. Consistently with the tenor of Article 22.2(a), country *X* will have to adduce evidence to the effect that the use of the impugned trademark is indeed misleading. In other words, it would be necessary to prove that the use of the impunged mark is likely to mislead. This Article is subject to Article 24.5 which exempts from preventive action, trademarks used in good faith before TRIPS.

Under the Article, country *Y* shall, *ex officio,* if its legislation permits, refuse or invalidate such trademark. However, for exercising such discretion, the relevant Trademark Office or the concerned official should have the knowledge that the trademark or part of it is indeed a geographical indication and should, therefore, be refused or invalidated.

Hence, it would appear that *ex officio* actions are subject to subjective opinions and the perceptions or the lack thereof on part of the examining officers.

Article 22.4

The text of Article 22.4 reads as follows:

The protection under paragraphs 1, 2, & 3 shall be applicable against a geographical indication which, although literally true as to the territory, region or locality in which the goods originate, falsely represents to the public that the goods originate in another country.

This Article prevents the use of homonymous indications, i.e., when two countries or places have the same name and one of them is famous for a geographical indication for a particular product, the use by the other of the same name in respect of its corresponding product may be prevented if it falsely represents the origin.

To give a hypothetical illustration, both India and Pakistan have 'Hyderabad' as a city in their respective countries. Pearls and pearl jewelry from Hyderabad, India is well-known. Assuming 'Hyderabadi pearls' of India to be a geographical indication, any use by those in Hyderabad of Pakistan of the expression 'Hyderabadi pearls', although literally true as to the region in which it originates, may falsely represent to the public that it originates in Hyderabad of India. In such event, such use would be actionable since it would be violative of the rights of 'Hyderabadi pearl' traders in India.

In summary, the scope of protection provided to geographical indications under Article 22 is against consumer deception and acts of unfair competition. This is a general level of protection extended to all geographical indications.

Higher Level of Protection for Wines and Spirits

Article 23 offers a higher level of protection to geographical indications relating to wines and spirits. While Article 22 deals with protection of geographical indications at a general level and relating to all products, Article 23 deals with those geographical indications relating to wines and spirits only. This Article was incorporated at the insistence of several wine producing countries of the European Union who wanted a higher level of protection for wines and spirits.

Article 23.1

The text of Article 23.1 reads as follows:

Each Member shall provide the legal means for interested parties to prevent use of a geographical indication identifying wines for wines not originating in the place indicated by the geographical indication in question or identifying spirits for spirits not originating in the place indicated by the geographical indication in question, even where the true origin of the goods is indicated or the geographical indication is used in translation or accompanied by expressions such as kind, type, style, imitation or the like[4)?.

4. [TRIPS Agreement note] Notwithstanding the first sentence of Article 42, Members may, with respect to these obligations, instead provide for enforcement by administrative action.

Unlike Article 22 which emphasizes on public confusion and deception, Article 23 provides to geographical indications in respect of wines or spirits a higher level of protection even in the absence of confusion or deception as to origin and even where a defendant has qualified its use of the protected geographical indication with an indication of the true origin of the goods or with de-localising terms such as 'kind', 'type', 'style', 'imitation' or the like.

To illustrate further, under Article 23.1, if a whisky is made in India by employing an identical method used for distilling Scotch Whisky originating from Scotland and which is identical in taste and other qualities to that of Scotch Whisky, it cannot be described as any of the following:

(i) Indian Scotch
(ii) Scotch from India
(iii) Scotch style Indian whisky
(iv) Even if only a blend of Scotch and Indian whisky is used, it cannot be described as « Blended with Scotch »

Since all the above descriptions clearly indicate the origin, none of them could be labeled as 'misleading' as to origin of the goods. However, Article 23.1 does not take into account the 'absence of the misleading factor as to origin'. Under this article, even if the true origin of the wine is indicated while using a geographical indication relating to a wine or spirit product, members are obliged to provide legal means to prevent such use.

On the other hand, under Article 22.2(a), even if a geographical indication is incorrectly used, it is sufficient if the true origin is specified. For instance, if a potato replicating the taste and quality of Idaho potatoes from the United States is grown and produced in India, it may be described as any of the following:

(i) Indian Idaho Potatoes
(ii) Idaho Potatoes from India
(iii) Idaho type Potatoes from India
(iv) Idaho Potatoes grown in India

Absent any reputation of Idaho Potatoes in India as a geographical indication, the above descriptions might be regarded sufficient to prevent confusion and deception as to origin under Article 22.2(a).

Further, under the Article, an aggrieved rights holder need not waste resources to adduce evidence as to misleading nature of the geographical indication unlike that of a situation under Article 22.2(a) where the burden of proving the misleading nature of the geographical indication lies with the rights holder.

Article 23.2

The text of Article 23.2 reads as follows:

The registration of a trademark for wines which contains or consists of a geographical indication identifying wines or for spirits which contains or consists of a geographical indication identifying spirits shall be refused or invalidated, *ex officio* if a Member's legislation so permits or at the request of an interested party, with respect to such wines or spirits not having this origin.

The scope of protection provided to trademarks, containing or consisting of geographical indications for wines and spirits under Article 23.2, is wider than that under the corresponding Article 22.3 in respect of geographical indications relating to non-alcoholic goods. Under Article 22.3, a Member may refuse or invalidate such a trademark only if it is of such a nature as to mislead the public as to the true place of origin. However, there is no such condition for refusal or invalidation under Article 23.2. In other words, mere incorrect or inaccurate use of a geographical indication is sufficient to refuse invalidation under Article 23.2.

Barring the vested rights under Article 24.5 which exempts from preventive action pre-TRIPS trademarks acquired and/or used in good faith, essentially, Article 23.1 and 23.2 have the effect of protecting geographical indications relating to wines and spirits against any risk of genericide through use of de-localising terms.

Article 23.3

The text of the Article reads as follows:

In the case of homonymous geographical indications for wines, protection shall be accorded to each indication, subject to the provisions of paragraph 4 of Article 22. Each member shall determine the practical conditions under which the homonymous indications in question will be differentiated from each other, taking into account the need to ensure equitable treatment of the producers concerned and that consumers are not misled.

The counter part of this Article is Article 22.4. This Article deals with similar indications for wines and spirits whose use is not misleading or deceptive under Article 22.4. Wherever the indications concerned are not misleading, both may be protected. Further, the Members are to determine the conditions necessary to differentiate both the indications.

Article 23.4

The text of this Article reads as follows:

In order to facilitate the protection of geographical indication for wines, negotiations shall be undertaken in the Council for TRIPS concerning the establishment of a multilateral system of notification and registration of geographical indications for wines eligible for protection in those Members participating in the system.

There is no corresponding paragraph to this Article under Article 22. This Article grants one more privilege to the rights holders for protection in respect of geographical indications relating to wines and spirits. This is by providing for negotiations to be undertaken at the Council for TRIPS towards the establishment of a multilateral system for notification and registration of geographical indications relating to wines and spirits.

Article 24

Article 24 has nine paragraphs out of which six deal with the exceptions to the protection of geographical indications. While Article 24.1 and 24.2 deal with the obligations of the Members of the WTO on continuing negotiations, Article 24.3 makes it obligatory for a member not to diminish the existing protection for geographical indications on the date of entry into force of the WTO Agreement.

Article 24 should be read in conjunction with Article 23. While Article 23 is understood to be providing a higher level of protection to geographical indications in respect of wines and spirits, this so-called absolute protection is substantially cut down in Article 24. This is the result of a consensus between the proponents of Article 23 protection spearheaded by the European Union and the opponents who feared that such a wide protection might be detrimental to the prior acquired rights in certain geographical indications. In a way, this Article balances the obligations and privileges under Article 23.

Article 24.1

The text of the Article reads as follows:

Members agree to enter into negotiations aimed at increasing the protection of individual geographical indications under Article 23. The provisions of paragraphs 4 through 8 below shall not be used by a Member to refuse to conduct negotiations or to conclude bilateral or multilateral agreements. In the context of such negotiations, Members shall be willing to consider the continued applicability of these provisions to individual geographical indications whose use was the subject of such negotiations.

Reading Article 23.4 in conjunction with Article 24.1, a clear framework for further multilateral negotiations regarding the protection of geographical indications for wines and spirits emerges. There are two specific objectives set forth by these two Articles: first, Article 23.4 contemplates negotiation over establishment of a multilateral system of notification and registration of geographical indications for wines. These negotiations have been taking place in the TRIPS Council for some time now. Secondly, Article 24.1 requires members to 'enter into negotiations aimed at increasing the protection of individual geographical indications under Article 23', specifically for wines and spirits. The Article further stipulates that even if a geographical indication is subject to the exceptions stipulated in Article 24.2 through Article 24.8, this factor should not be used by a Member to refuse to conduct negotiations.

This Article represents a departure from the Most Favoured Nation (MFN) prohibition contained in Article 4 of TRIPS Agreement which mandates a member country to accord immediately and unconditionally to the nationals of all other members any advantage, favour, privilege or immunity granted to the nationals of any other country. Taking advantage of this Article, member countries are allowed to conclude bilateral or multi-lateral agreements aimed at increasing the protection of individual geographical indications in respect of wines and spirits and any limitations imposed by virtue of Article 24 cannot stand in the way of any negotiations in this behalf.

To illustrate, the appellation of origin 'Champagne' is considered semi-generic by the United States. However, if France wants to restore the original significance of 'Champagne' in the United States and wishes to conduct negotiations in future, it is not open for the United States to refuse such negotiations on the ground that nationals of the United States have been using the name 'Champagne' in respect of wines produced in that country for more than 10 years preceding 15 April 1994 (exception under Article 24.4).

In 2002, the European Union concluded its negotiations with South Africa wherein South Africa agreed to refrain from using the terms 'port' and 'sherry' in respect of wines by phasing them out in its export markets within five years and domestic markets within twelve years. Similarly, it was agreed to phase out the terms 'grappa' and 'ouzo' over a period of five years.[14]

14 See http://europa.eu.in/comm/agriculture/publi/newsletter/41/41_en.pdf

If these names were phased out at the end of these terms, the European Union would give 5,000,000 Euros to South Africa for the development and restructuring of the local wine and spirit industry.[15] Critics feel that this is a blatant example of commercial upper-hand by the European Union. Whether the European Union would have been able to successfully conclude an agreement with a country which is on a higher or equal bargaining plane is a relevant question in this context.

Article 24.2

The text of Article 24.2 reads as follows:

The Council for TRIPS shall keep under review the application of the provisions of this Section; the first such review shall take place within two years of the entry into force of the WTO Agreement. Any matter affecting the compliance with the obligations under these provisions may be drawn to the attention of the Council, which, at the request of a Member, shall consult any Member or Members in respect of such matter in respect of which it has not been possible to find a satisfactory solution through bilateral or plurilateral consultations between the Members concerned. The Council shall take such action as may be agreed to facilitate the operation and further the objectives of this Section.

This Article supplements the scope of Articles 23.4 and 24.1 by providing for the TRIPS Council to keep under review the application of the provisions of 'this Section'. The reference to 'this Section' and not 'this Article' obviously means that all the three Articles, namely, 22–24, are covered within its scope. The role of the Council might be to ensure that interested WTO members negotiate in good faith.[16]

Article 24.3

The text of Article 24.3 reads as follows:

In implementing this Section, a Member shall not diminish the protection of geographical indications that existed in that Member immediately prior to the date of entry into force of the WTO Agreement.

This Article ensures that no member is going to diminish the level of protection existing on the date of entry into force of the WTO Agreement in that member country. This section reiterates the essence of Article 1 of the TRIPS Agreement which says that members may implement in their laws

---

15 Faan Wolvaart, "Last Port of call for Sherry?" Bowman Gilifilan Inc Newsletter, 2002

16 Daniel Gervais, *The TRIPS Agreement – Drafting History and Analysis*, Second edn, Sweet & Maxwell

more extensive protection than is required under TRIPS. Hence, TRIPS-plus provisions, whether existing prior to the date of entry of TRIPS or even afterwards, are encouraged provided such provisions do not contravene TRIPS provisions.

Article 24.4

The text of Article 24.4 reads as follows:

Nothing in this Section shall require a Member to prevent continued and similar use of a particular geographical indication of another Member identifying wines or spirits in connection with goods or services by any of its nationals or domiciliaries who have used that geographical indication in a continuous manner with regard to the same or related goods or services in the territory of that Member either (a) for at least 10 years preceeding 15 April 1994 or (b) in good faith preceeding that date.

The exceptions to the protection of geographical indications commence from Article 24.4 onwards. As explained before, these exceptions cannot be used by members to refuse to conduct negotiations under Article 24.1.

Article 24.4 protects what are known as 'prior rights' in respect of a geographical indication identifying wines or spirits. For instance, if a geographical indication belonging to country *A* is being used by a national of country *B* as a geographical indication in respect of same or related goods or services either (i) for at least 10 years prior to 15 April 1994 or (ii) in good faith preceeding that date, country *B* is not obliged to prevent such use.

The Article raises the following comments:

(i) The language of the Article refers to the 'Section' thereby referring to Articles 22 and 23, although the scope of its application seems to be limited to wines and spirits. Hence, effectively, the exception excludes geographical indications in relation to products other than wines and spirits.

(ii) There are two alternate conditions under the Article wherein a member is not obliged to protect another member's geographical indication. In the first condition of prior continuous use of 10 years before 15 April 1994, proof of bad faith is not a pre-condition. Instead, it has been stipulated as the second alternate condition. The Article, therefore, raises the issue as to what would be the status of a geographical indication adopted in bad faith but used for more than 10 years prior to 15 April 1994. Even under trademark law, adoption of a trademark tainted by bad faith is liable to be refused or prevented. Since mere proof of bad faith is a matter of evidence, it would be an arduous task for a rights holder to establish this as per the second condition. comparatively, the first condition is easier to establish. Ideally, both these conditions should have been made concomitant.

(iii) The use must be in relation to the 'same or related goods' or services of the WTO member in question. The Article is silent about the use amounting to dilution and tarnishing of a geographical indication by use on dissimilar goods.

(iv) Lastly, the Article excludes use of terms which are deceptively or confusingly similar to the geographical indication in question. For instance 'CALOGNIAC' for a Brandy produced in California to evoke 'COGNAC'.

Article 24.5

The text of Article 24.5 reads as follows:

Where a trademark has been applied for or registered in good faith, or where rights to a trademark have been acquired through use in good faith either:

(a) Before the date of application of these provisions in that Member as defined in Part VI; or

(b) Before the geographical indication is protected in its country of origin, measures adopted to implement this Section shall not prejudice the eligibility for or the validity of the registration of a trademark, or the right to use a trademark, on the basis that such a trademark is identical with, or similar to, a geographical indication.

While Article 24.4 protects prior rights in respect of geographical indications relating to wines and spirits, Article 24.5 deals with prior rights in respect of trademarks vis-à-vis geographical indications. The expression 'deals with' is used in the previous sentence rather than 'protects', because Article 24.5 is not clear as to which would prevail. In other words, in a situation where a later geographical indication is found to be in conflict with a prior trademark, the article does not say whether one would prevail over the other or whether both will co-exist.

To illustrate, if a trademark has been applied for/registered or has been acquired through use in good faith in a country *A*, either before the date of coming into force of TRIPS in that member or before the date of protection of the geographical indication in its country of origin, measures adopted to implement Articles 22, 23 and 24 of TRIPS 'shall not prejudice the eligibility for or the validity of the registration of such trademark, on the ground that such trademark is identical with or similar to the geographical indication. Neither does the Article clearly state that the geographical indication will prevail over the trademark or vice versa nor does it state that both the geographical indication and the trademark will co-exist. The Article seems to apply to all geographical indications and not merely to those relating to wines and spirits.

Article 24.6

The text of Article 24.6 reads as follows:

Nothing in this Section shall require a Member to apply its provisions in respect of a geographical indication of any other Member with respect to goods or services for which the relevant indication is identical with the term customary in common language as the common name for such goods or services in the territory of that Member. Nothing in this Section shall require a Member to apply its provisions in respect of a geographical indication of any other Member with respect to products of the vine for which the relevant indication is identical with the customary name of a grape variety existing in the territory of that Member as of the date of entry into force of the WTO Agreement.

The above article consists of two parts, one part containing a general principle and the other, its specific application to products of the vine. According to the general principle, if a term is customary in common language as the common name for certain goods or services in a country, then Articles 22–24 do not apply to such goods and services. An Indian example of such situation is the use of the term '*Madira*', which in several Indian languages means 'alcohol' or 'wine'. However, in Portugal, a wine called 'Madeira' is produced in the Madeira Island and is protected as an appellation of origin. It is debatable if Madeira may be considered an exception to protection under this Article.

Under the second part, the principle is applied to products of the vine for which the relevant indication is identical with the customary name of a grape variety existing in the WTO Member. To illustrate, if country *A* has geographical indication *X* in respect of a wine produced from *grapes* and in country *B*, if *X* is the customary name for a grape variety, country *B* is not obliged to protect the geographical indication *X* belonging to country *A* because in country *B*, *X* is the customary name for a grape variety.

This part basically addresses the problem that wines are sometimes named after the grapes used in making them. Hence, over a period of time, these names become indirect geographical indications such as the 'Riesling' wine from Germany which uses 'Riesling' grape for its production. In such cases, if the grape variety existed as on the date of entry into force of the WTO Agreement, use of the name of the grape variety on the wine is permitted.

In the first part of the Article, the expression used is 'the term customary in common language as the common name for such goods or services'. Hence, if the Madeira producers of Portugal want to protect the name 'Madeira' in India, they would need to prove that it is not a generic term in India in respect of wines or alcohol. For this, they would need to produce evidence of the kind required in a similar situation in trademark law. Hence, evidence such as reference to the wine in the print and visual media, public surveys, affidavits etc. would have to be resorted to by such rights holders. Procuring this kind of evidence is no mean task.

However, in the second part, the expression used is 'the customary name of a grape variety'. Hence in the previous illustration concerning geographical indication *X* from country *A*, all that country *B* needs to prove for refusing protection is that there is a fairly widespread use of the term as a grape variety in that country and, therefore, protection cannot be extended.

The Article raises the following comments:

(i) Considering the fact that 'wine' is produced from fruits and rice as well, the reference in the Article is interestingly limited to wines produced from grapes alone. Feni is an alcoholic drink produced in Goa, India from the apple of Cashew nuts. Obviously, in the hypothesis given in the previous paragraph, *Feni* would not fit in because it is not a product of 'vine'! Also, there are a number of countries which produce wines from fruits other than grapes as well as from rice. These are clearly excluded from the applicability of the Article because they are not products of the vine.

(ii) The first part of the Article refers to geographical indications in general (perhaps, to that extent the example of 'Madeira'is not apt) and the second part relates to geographical indications with respect to 'products of the vine' The test under the second part of the Article seems to be easier to establish than that in the first part as explained before.

Article 24.7

The text of Article 24.7 reads as follows:

A member may provide that any request made under this Section in connection with the use or registration of a trademark must be presented within five years after the adverse use of the protected indication has become generally known in that Member or after the date of registration of the trademark in that Member provided that the trademark had been published by that date, if such date is earlier than the date on which the adverse use became generally known in that Member, provided that the geographical indication is not used or registered in good faith.

Drafted quite inelegantly and complexly, this Article is described as the procedural counterpart of Article 24.5 by some authors.[17] The Article seeks to impose a time frame of five years within which any measures to prevent the use or the cancellation of a trademark containing or consisting of a geographical indication may be initiated.

Article 24.8

The text of Article 24.8 reads as follows:

The provisions of this Section shall in no way prejudice the right of any person to use, in the course of trade, that person's name or the name of that person's predecessor in business, except where such name is used in such a manner as to mislead the public.

The purpose behind this Article is to preserve and recognise a person's right to use his or her own name or that of a predecessor in business in the course of trade. The only exception to such use is when the concerned person is using it in such a manner as to mislead the public.

Article 24.9

The text of Article 24.9 reads as follows:

There shall be no obligation under this Agreement to protect geographical indications which are not or cease to be protected in their country of origin, or which have fallen into disuse in that country.

This Article removes any obligation on a member country to protect geographical indications that are not or which cease to be protected in their country of origin, or which have fallen into disuse in the country of origin.

Controversies Surrounding Section 3 of TRIPS

The preamble to the TRIPS Agreement states as follows:

Members,

Desiring to reduce the distortions and impediments to international trade, and taking into account the need to promote effective and adequate protection of intellectual property rights, and to ensure that measures and procedures to enforce intellectual property rights do not themselves become barriers to legitimate trade...

17 Daniel Gervais, "The TRIPS Agreement – Drafting history and Analysis", second edn, Sweet & Maxwell

However, Section 3 of the TRIPS Agreement dealing with geographical indications appears to be an anathema to the preamble. It has been the centre of controversy for several years for its preferential and discriminatory treatment to geographical indications relating to wines and spirits under Article 23, which has been a cause for discontentment among Members.

With discussions for and against extension of Article 23, the Intellectual property community seems to have turned bipolar with one end taking a view that Article 23 must be extended to geographical indications relating to products other than wines and spirits as well and the other end taking the view that there is no need for such an extension and that the status quo should be maintained.

Reference

Latha R Nair and Rajendra Kumar 2005.Geographical Indications-A search for Identity.LexisNexis Butterworths, New Delhi, 359p.

The Science of Horticulture : Vol. 1, 2011
© K.V. Peter (ed.), pp. 23-63
New India Publishing Agency, New Delhi (India)
E-mail : newindiapublishingagency@gmail.com
Web: www.bookfactoryindia.com

Chapter 2 Principles of Hi-Tech Horticulture

PATIL, A.A., MANTUR, S.M., BIRADAR, M.S. AND MANNIKERI, I.M.

Growth of horticulture is the cornerstone for augmenting employment opportunities in agriculture sector both directly and indirectly. Of late, there is a marked shift in the cropping pattern towards diversification into high value commercial crops in the country. To make horticulture viable and profitable, the thrust should be on improving the resource use efficiency, so that the relative profitability could be enhanced. Success in horticulture is more technology driven compared to other crops. The time bound removal of quantitative curbs on imports and other barriers to access domestic market under WTO, of which India is a signatory, will require Indian Horticultural produce / products to become competitive both in the domestic and international markets. This would call for use of Hi-tech horticulture technologies. The term Hi-Tech Horticulture was coined in recent years. There are no standard definitions for this. The Department of Agriculture and Co-operation,GOI in 1999 defined Hi-tech Horticulture as "any technology which is modern, less environment dependent, capital intensive and has the capacity to improve the productivity and quality of horticultural crops". The technologies encapsulated in Hi-Tech Horticulture are; genetically modified horticulture crop varieties derived from bio technology and genetic engineering, micro propagation, protected cultivation, micro irrigation and fertigation, high density planting, organic farming in horticulture, integrated nutrient, pest and disease management and post harvest management. Hi-tech Horticulture technologies are discussed in detail under following heads:

I. Micro propagation
II. Protected cultivation
III. Micro irrigation and fertigation
IV. High density planting

I. Micropropagation

The term micro propagation is used specifically to refer to application of "tissue culture" techniques to the propagation of plants starting with very small plant parts growing aseptically in a test tube or other containers. Tissue culture is based first on the principle of totipotency, i.e. the concept that every living cell has the genetic potential to reproduce entire organism. The technique of *in-vitro* cultivation of plant cell, tissue or organs is primarily denoted to solve two basic problems, first to keep plant cells free from microbes and second to ensure the desired tissue development by providing suitable nutrient media and other environmental conditions. The first problem can be eliminated using modern equipments and careful handling necessary for maintaining aseptic environmental conditions. While second problem remains an area of active research mainly on manipulation of culture media, especially growth regulators and to a much lesser extent on environmental conditions.

Merits of Micro Propagation

1. *Mass multiplication of plants*: Only one plant can be produced from a cutting/ seed, whereas one explant can produce more number of new plants. Very a few stock plants are required to provide the explants needed to produce thousands of new plants. This is the most useful for multiplication of new hybrids, mutants, unique seedlings and or disease free stock plants.

2. *Disease free planting materials*: The diseases transmitted from parent to offsprings often can be eliminated through micro propagation. External contamination can be removed by surface sterilization of explants. Many internal viruses can be eliminated by using the apical meristem which will be free from viruses.

3. *Year round production*: Tissue culture is not limited by the season or the weather. Working condition in the laboratory is ideal for year round production.

4. *Uniform size of plants*: Uniform size of plants helps in establishment of orchard uniformly, which helps in carrying out cultural operations and harvesting.

5. *Easy Transportation and storage*: Easy exchange of plant materials, since many plantlets can be multiplied in a container. *In-vitro* plantlets can be stored for several years and hence can be conserved for long duration.

Factors Affecting Micro Propagation

1. *Genotype and choice of explants*: Regeneration capacity varies with species, cultivars and cultural conditions. In general, those species, the easiest to propagate conventionally, have also been the easiest to micro propagate. Most herbaceous plants can be easily propagated by conventional or micro propagation, woody perennials are much more difficult to regenerate and new principles of culture have to be developed. Success in the most woody perennial species is related to their juvenility status. Shrubs are generally more easily handled than trees. In micro propagation, selection of proper explants is a major requirement. Explants selected should be potential and are to be from active growing phase.

2. *Contamination*: Fungal and bacterial contamination with the explants is one of the major hurdles in developing sterile cultures. Overcoming this problem is a big challenge especially with the perennial horticultural species. Disinfecting explants, tools and working area to remove contaminants from the surface is utmost important.

3. *Culture media*: A basal medium must be selected or established through systematic analysis. The most widely used is the MS medium (Murashige and Skoog). Commercial micro propagation requires optimization of not only the inorganic elements but also of other factors of the medium. This can be done by systematically testing a range of concentration of individual elements combined in various ways with other elements.

4. *Hardening*: Plantlets multiplied through micro propagation need to be hardened carefully before taking them to main field. Acclimatization of regenerated plantlets is therefore very essential. In ornamentals and other annuals, the field establishment is not much problematic. In most cases the regenerated plantlets have to be gradually acclimatized by hardening under greenhouse and shade house.

Requirements of Tissue Culture Laboratory

1. *Washing room*: It is essential for washing, drying and storage of vessels. It should be furnished with wash basin.

2. *Media room*: It is required for preparation, sterilization and storage of culture media. This room should be equipped with wash basin, double distillation unit, hot water bath (i.e. steamer), hot plates with magnetic

stirrers, electronic weighing balances, pH meter, refrigerator, cupboard for storage of chemicals and media, filter sterilization unit, centrifuges, vacuum pump, etc.

3. *Sterilization room*: Sterilization room should be adjacent to media room but in a separate corner taking all care of safety. An autoclave unit should be placed in this room.
4. *Incubation room*: Laminar air flow hoods are fitted in this room for aseptic manipulation. A small table having stereoscopic microscope or inverted microscope may be adequate for culture observations. It should be equipped with air conditioners and air curtains at doors.
5. *Culture room*: Culture room should have two door facility, controlled temperatures (usually 25±2°C with air conditioners) and humidity (60-80% RH with humidifiers), culture racks fitted with light (generally 1000 lux light generated by fluorescent tubes with automatic timers to regulate photoperiod) and a horizontal shaker for agitation of liquid cultures. It is desirable to have a generator set to the culture room in the event of electricity failures.
6. *Polyhouse*: Two types of polyhouses are desirable. A fully automatic polyhouse having fan and pad cooling system along with foggers and drip irrigation and mini sprinkler systems depending on sensors are necessary to provide controlled temperatures (25±2°C) and humidity (80-100% RH) required for development of fresh *in vitro* plants. The plantlets are further hardened in semi-automatic polyhouses where natural ventilation is provided. In these polyhouses, plantlets are repotted in bigger containers supplemented with organic and inorganic manures. After 2-4 months period, tissue cultured plantlets are transferred to shadehouses and finally to the field with initial shade cover or in congenial atmospheric condition. Polyhouses should be connected with a generator set for general safety.

Preparation for Culture

1. *Choice of explants*: For the success of micro propagation, the kind of explants and from where they are collected are very much critical. They may vary with purpose of the culture, the species and cultivars. In the absence of definite information on a given species or cultivar, one can use available information on a closely related species or cultivar or carry out experimentation with the particular plant. While collecting shoot tip, plant should be relatively in active growth phase. The size of the explants varies from 1-5 mm of meristem tip for meristem culture. Smaller the explants, the more effective is the elimination of pathogens. On the

other hand, smaller the explants, the more difficult to establish and the lower survival rate. In general explants of 0.25 to 1 mm can be used. Virus status must be confirmed by indexing.

2. *Sterilization*: All the materials like vessels, instruments, medium, plant material, etc. used in incubation and culture rooms must be free from microbes. This is achieved by one of the following approaches.

   Dry heat: Glass wares, plastic wares and instruments may be sterilized by dry heat in an oven.

   Wiping with alcohol: The surfaces that cannot be sterilized by other techniques e.g. platform of laminar flow cabinet, hands of operator, etc. are sterilized by wiping them thoroughly with 70 % alcohol and the alcohol is allowed to dry. The use of mouth mask is also advocated to avoid contamination.

   Flame sterilization: Instruments like forceps, scalpel, needles, scissors, etc. are ordinarily sterilized by dripping them in 80-90 % alcohol (ethanol) followed by flaming. These instruments are repeatedly sterilized during the operation to avoid contamination. It is customary to flame the mouth of culturing vessels prior to inoculation or subculture.

   Surface sterilization: All *in-vitro* plant materials to be used for culture are treated with an appropriate sterilizing agent to inactivate the microbes called 'surface sterilization'. The protocol is dependent on the surface and type of explants which determine the contamination load and tolerance to the sterilizing agent. The sterilizing agents used for surface disinfection are calcium hypochlorite (9-10%), sodium hypochlorite (2%), mercuric chloride (0.1-1.0%), silver nitrate (1%), bromine water (1-2%), hydrogen peroxide (H_2O_2,10-12%) and antibiotics (4-50 mg/l). Among these calcium or sodium hypochlorite and mercuric chloride are the most commonly used for 10-30 minutes followed by 4-6 rinses with sterilized distilled water.

   Autoclaving: Culture vessels (both empty and containing media) are generally sterilized in an autoclave or pressure cooker at 121°C at 15 p.s.i. or 1.06 kg/cm^2 for 15-40 minutes. Care should be taken to properly stopper all the vessels and open the autoclave only when its pressure gauge indicates zero pressure. Use of special thick and long handgloves are recommended while handling the hot autoclaved material.

   Filter sterilization: Some growth regulators e.g. GA_3, zeatin, ABA, certain vitamins and enzymes are heat sensitive. Such compounds are filter sterilized by passing their solution through a membrane filter of 0.45 micron or lower size. Then, filter sterilized heat liable compounds are added to cooled autoclaved media (40°C). The laminar air flow cabinet creates aseptic working space by blowing filter sterilized air through filter at velocity of 1.8 km / hr.

3. *Culture indexing*: Culture indexing test for contamination by specific pathogens in containers is done as follows. Cut off a piece of plant tissue, dice it and place the pieces into 15 ml of nutrients broth optimum for bacterial and fungal growth. Positive results are shown in cloudiness of the medium and should appear within 4-5 days, if made at initial explants stage.

Stages in Growing the Cultures

General patterns of treatments and handling procedures are well established, many of the requirements of individual stages must be developed for different species and even cultivars. The process can be described as optimizing the variables that affect each stage of the process. These include subculture interval, concentration of ingredients in the medium, relative cytokinin / auxin concentrations, size of propagules, light, temperatures, etc. Micro propagation may include a preliminary pretreatment culture step in which the explants are initially placed in a petri dish on the surface of an agar medium of basal salts and sugar but without hormones. Using a liquid medium in the initial stages or frequent transfers in agar medium for several days may leach out the toxic material.

Stage I: Establishment and stabilization: The function of this stage is to establish the explants in culture and to stabilize the culture for multiple shoot development. This may involve stimulation of auxiliary shoots, initiation of adventitious shoots on excised shoots and initiation of callus from the cut surfaces. The medium selected may vary with the species, cultivars and kind of explants used. Control of development is the most often achieved by manipulating levels of auxin and cytokinin. Usually 4-6 weeks are required to complete stage I and produce explants ready to be transplanted to stage II. Some woody plant species can take upto a year for this to occur.

Stage II: Shoot multiplication: In the multiplication stage each ex-plant has expanded into a cluster of micro shoots arising from a basal mass of callus like tissue. This structure is divided into separate micro shoots, which are transplanted into new culture medium. The kind of medium used depends on the species, cultivar and type of culture. Usually it is essentially the same as that used in stage I, but often the cytokinin and mineral supplement level are increased. Adjustments may need to be determined by experimentation. Different species vary in the optimum size of the micro shoot and method of cutting apart. There is generally a minimum mass of tissue necessary to produce uniform rapid multiplication in the next transfer. Division may be made by vertically cutting the tissue mass into sections, keeping some of the base with each piece. Number of micro shoots produced ranges 5-25 depending upon the species and conditions of culture. Multiplication may be repeated several times to increase the supply of material to a predetermined level for subsequent rooting and transplanting. During multiplication, off type propagules sometimes appear, depending on the kind of plant and method of regeneration. In some plants restricting the multiplication phase to only three propagation is recommended to avoid development of off-type micro shoots. Other procedures which can reduce the potential for variability include reducing hormone concentration and avoiding basal callus which may give rise to adventitious shoots in sub culturing.

Stage III: Pre-transplant and *in vitro* rooting: The purpose of the third stage is to prepare the micro plants for transplanting from the artificial heterotrophic environment of the test tube to an autotrophic free living existence in the greenhouse and on to their ultimate location. This preparation may involve rooting, but it also may involve conditioning of the micro plant to increase its potential for acclimation and survival during transplanting. For example, the agar and/or sucrose concentration might be increased. Light intensity is sometimes increased during this period. For *in vitro* rooting, individual micro shoots are sub cultured into new containers in a sterile medium with reduced or omitted cytokinin, an increased auxin concentration and often a reduced inorganic salt content. Selection for uniformity and rouging of obviously abnormal, aberrant or diseased micro shoots should be made prior to stage III.

Stage IV: Transplanting, acclimation and *ex vitro* rooting: Rooted or unrooted micro shoots are removed from the culture vessels, agar is washed away completely to remove a potential source of contamination and the micro plants are transplanted into a standard pasteurized rooting or soil mix in small pots or cells in a more or less conventional manner. Initially, micro plants should be protected from desiccation in a shaded, high humidity tent or under mist or fog. Several days may be required for new functional roots to form. Once the micro plants are established in the rooting medium, they should be gradually exposed to a lower relative humidity and higher light intensity for hardening.

II. Protected Cultivation

India is endowed with diverse agro-climatic conditions, which help to grow all kinds of horticulture crops, almost throughout the year, in one part of the country or the other, but the quality of produces under open field conditions particularly in case of high value flowers and vegetables are not suitable for international as well as standard domestic markets. Sometimes there is no guarantee of consistent production from open cultivation as the crop is exposed to a number of natural environment factors which change frequently.

The need of the hour is to increase the productivity and quality of produce to meet the demand of quality conscious consumers. A break through in production technology that integrates market driven quality parameters with the production system, besides ensuring a vertical growth in productivity is required. One such technology is "Protected cultivation", or generally called greenhouse technology. It is the technique of providing favourble environmental or growth conditions to the crop. It is rather used to protect plants from the adverse climatic conditions by providing optimum conditions of light, temperature, humidity, carbon dioxide etc. for the best growth of plants to achieve maximum yield and the best quality. A greenhouse is a covered structure which protects plants from wind, precipitation, excess solar radiation, temperature extremes and considerable pests and diseases attack.

Cultivation of crops under greenhouse in the world at present is practiced in more than 50 countries. China adopted plastic greenhouse technology to a greater extent. The smaller land holdings and short growing seasons in the Netherlands and Japan are the reasons for adopting greenhouse technology. Water conservation is one of the reasons for greenhouse crop production in Israel and Middle-East countries. India has greenhouses, mainly glass houses for the last 4-5 decades for research and off season protection of plant materials, but the interest in commercial application is recent. The choice of crop to be raised in a greenhouse is made on the basis of physical size of the structure and economics of crop production. As a result, high value horticulture crops, particularly cut flowers and vegetables are more popular for cultivation in greenhouse.

Structures for protected cultivation: A greenhouse is a quasi-permanent structure, covered with a transparent and translucent material ranging from simple home made designs to sophisticated pre-fabricated structures wherein environment could be modified suitable for propagation or growing plants. Materials used to construct a greenhouse frame may be wood, bamboo, stone pillar and steel whereas covering materials may be glass, various rigid or flexible plastic materials. Depending on covering materials, different terminologies are used in the contest of greenhouse structures.

Glasshouse : A greenhouse with glass as the covering material is called glasshouse.

Polyhouse : It is a greenhouse with polyethylene as the covering material

Type of structures: The greenhouse design and cost range from a simple plastic walkin tunnel costing Rs.200 / m^2 to a climate controlled, saw tooth greenhouse with automatic heating and cooling system costing more than Rs. 3000/m^2. The selection of greenhouse design should be determined by the grower's expectations, need, experience and above all its cost effectiveness and available market price for the produce. Greenhouses are classified depending on shapes, which also determine their cost, climate control and use in terms of crop production. Commonly used structural designs are briefly described below.

1. *Quonset greenhouse*: The roof of this type of greenhouse will be semi round. The roof and sides can be covered with poly ethylene, fiber glass or polycarbonate sheet. This type of structure requires relatively lesser cladding material. In this, multispan saw-tooth type structure can be installed with top ventilation.

2. *Curved roof greenhouse*: The semi-circular tunnel greenhouse structure appeared with introduction of poly ethylene film as covering material. These structures are simple and easy to construct. But these structures have the disadvantage of poor ventilation because the curved roof is not amendable to provide top ventilation.

3. *Gable roof greenhouse*: This is the most basic structure similar to a hut like structure and is perhaps the first version of a greenhouse with glass as the covering material. The structure allows large opening in the side walls and at the ridge for high rates of natural ventilation. Modern gable shaped greenhouses are multi span units with bay width of 6-12 m.

Covering materials for greenhouses: The covering materials are also called as cladding or glazing materials. The purpose of a greenhouse covering is to allow sunlight to pass through it, so that the energy is retained inside. Glass was the main covering material in the early greenhouses. With the introduction of plastic materials, there are now several alternatives available for greenhouse coverings. A brief description of greenhouse covering materials is given below:

Glass: A clean and transparent material provides the maximum light transmittance to the extent of 90 %. Since it is heavier in weight, it requires elaborate structures for adequate support. It is fragile and can break with minimum shock or vibrations resulting in high maintenance cost.

Acrylic: This material has long service life, good light transmittance (80%), moderate impact resistance, but prone to scratches. Being inflammable and costly, it is not a preferred martial.

Polycarbonate: It is available in single or double wall sheets of different thickness. A new polycarbonate sheet has good light transmittance of about 78 %, but reduces with age. It has excellent impact resistance and low inflammability. High cost limits its use on large scale.

Fiberglass reinforced plastic panels (FRP): These plastics consist of polyester resins, glass fibers, stabilizers etc. It has an initial light transmittance of about 80 % and has high impact resistance with a service life ranging 6-12 years. Good quality FRP material for greenhouse covering is not quite assured.

Polyethylene: A clear, new polyethylene sheet has about 85% light transmittance. Its higher strength and low cost have made it the most popular replacement to glass. An ultra- violet (UV) stabilized plastic sheet can have a service life of minimum 3 years. These sheets are generally available in 7 and 9 m. widths with 200 micron (0.2 mm) thickness. Now a days, plastic film with specific characteristics, like, anti-drip, anti-dust, sulphur resistant, high diffusivity etc are available and should be used depending on the requirement.

Concept of Greenhouse Design: The basic concept of designing of greenhouse includes light transmission, crops to be grown, environment control requirement, frame material and cladding material. Under Indian conditions, quonset type, multi span greenhouse is the most suitable because of its low cost and ease of fabrication. An ideal glazing material should transmit the visible portion of sunlight and filter the ultraviolet light. Ultraviolet resistant low density polyethylene (UVLDPE) single film cladding of 200 micron thickness is sufficient for Naturally Ventilated (NV) greenhouse and fan and pad (FP) greenhouses. Double inflated film of UVLDPE is required for heated greenhouses. The UVLDPE film is economical, 90% transparency for light with 3 years usable life. The size of individual pieces of film should not be more than 100 m^2 and this should be properly fastened to frame. The use of silpaulin film should be avoided due to its low transparency and less life. Naturally ventilated greenhouses are the most suited for mild agro- climatic conditions as in Bangalore , Pune and coastal regions. Greenhouses with FP system of cooling can be used in dry and warm seasons of Northern plains and they are the most suitable for nursery raising and rooting of cuttings due to high relative humidity inside. Greenhouses with heating system are generally required during winter in northern hilly regions and at nights in plains for better growth of plants. The greenhouse can have more than one kind of environmental control system based on the variation in the climate at a location during the year.

Design of Structure: The structure has to carry one or more loads of dead load, live load, wind load and or snow load depending upon the locality and it has to be designed accordingly. Sometimes, the greenhouse should be able to take the dead load plus live load or dead load plus wind load plus half live load. The greenhouse is the best fabricated out of galvanized iron pipes and the foundation should be able to take designed load.

Design of Environmental control: The design of environmental control system depends upon the meteorological data of location (solar insulation, light intensity, wind speed, wind direction, ambient temperature and relative humidity) and the temperature and RH to be maintained in the greenhouse.

1. *Naturally ventilated (NV) greenhouse*: The temperature in these greenhouses is above the ambient temperature. The basic purpose is not to have excess temperature in mild climates. The NV greenhouse can also be used to increase temperature in cold climates. The ventilation in the greenhouse takes place by (a) thermal buoyancy and (b) wind velocity. More is the ventilation in the greenhouse, lesser the temperature built up. The calculation of ventilation area required is very scientific, however a greenhouse should have approximately 25% of floor area as opening area inside walls and an equal area of opening at ridge. The wall and ridge opening have to be throughout length and provided with 40 mesh nylon insect proof net. The temperature build up inside the greenhouse can be controlled by controlling the size of wall opening or roof opening by using adjustable opening by rolling or unrolling LDPE film over wall opening or ridge opening using crank mechanism. The irrigation also reduces the temperature inside the NV greenhouse by 2 to 3°C. The size of NV greenhouse has to be less than 50 m x 50 m to avoid excess temperature build up. The shade nets when used inside the greenhouse reduce the heat load on plants.

2. *Fan and pad greenhouse*: These greenhouses consist of the wet, cross fluted cellulose pad on one end and the exhaust fans on the other end. These greenhouses are not suitable for high humidity areas. The pad should be provided through out width, the height of pad can be 5% of the length of the greenhouse. The 10% of the circulating water on pad should be wasted to avoid build up of salts on pad. The water flow rate should be 9 litres / minute / m length of pad. The length of greenhouse should be restricted to 60 m.

3. *Heated greenhouse*: The heat loss through film is reduced to 60-65% when double inflated film is used. The total heat loss of the greenhouse is determined by multiplying the exposed area of the glazing walls, wind factor, tightness of the greenhouse and the temperature to be maintained inside the greenhouse in comparison to ambient temperature. In centrally heated GH, the boiler of required capacity can be installed and steam at

102°C or water at 85°C can be circulated through GI pipes installed above plants and around the side of the greenhouse.

The unit heaters in which fuel burns inside the combustion chamber, hot fumes pass through the heat exchanger and the fan is used to blow air over heat exchangers are also used in small greenhouses. Infra-red heaters consisting of exhaust fume pipes and reflectors are also used. The solar heater can also be used for heating the water and circulating hot water inside the greenhouse through pipes at night.

4. *Insect proof net house*: Insect–proof nets are available in different intensities of perforations, ranging 25 to 60 mesh (perforations/cm^2). Nets of 40 or higher mesh are effective means to control entry of most flying insects and save crop from diseases. These structures reduce the risk of vectors causing viral diseases. Higher mesh size., however, reduces the air exchange of the structure, thereby, enhancing temperature and humidity inside the net house. For most flower crops, 30 and 40 mesh nets are used. Now a days, nets with characteristics like photo-selectivity, better UV stabilization and higher mechanical strength are available.
5. *Shade net house:* Shade nets are perforated plastic materials used to cut down the solar radiation and prevent scorching or wilting of leaves caused by marked temperature increases within the leaf tissue from strong sunlight. These nets are available in different shading intensities ranging from 25 to 90%. Leaf vegetables and ornamental greens are recommended to be grown under shade nets whose growth rates are significantly enhanced compared to unshaded plants when sunlight is strong. These nets are also very useful in managing the temperature inside the greenhouses. There are specialized nets having reflective surfaces which are generally called as thermal screens. These nets are effective in saving of energy required for temperature control inside the greenhouses.

Temperature control in greenhouse: Thermostats can be used to control exhaust fans in fan and pad greenhouse or controlling fuel combustion in centrally heated greenhouse. In naturally ventilated greenhouse ,it can be controlled by side and top ventilation.

Relative humidity control: Humidistat can be used to control the water circulating pump for pad in FP greenhouse. The foggers or misters can be controlled through humidistat to maintain RH in NV greenhouse and centrally heated greenhouse. High humidity can be maintained in GH used for nursery raising or for rooting of cuttings by misting or fogging. This type of GH is also known as mist chamber.

Light intensity control: The shade nets of required cut off capacity can be used to control light intensity. The HDPE net or the LDPE film with reflective type stripes when used in NV greenhouse, increases inside temperature due to restriction in ventilation. These shade nets decrease the temperature inside the FP greenhouse.

Protected Cultivation of Flowers and Vegetables

As a result of globalization of trade and liberalization of Indian economy, there is immense scope for export of high value flowers and vegetables from India, besides, meeting the increased demand in domestic market. To get higher productivity of superior quality produce, the crop should be grown under protected environment conditions adopting modern techniques like micro irrigation, fertigation etc.

India with a diversified climate grows most of the high value flowers and vegetables under partially environment controlled protected structures or naturally ventilated polyhouses across the county. High value and low volume flower crops like rose, gerbera, carnation, chrysanthemums, anthuriums, orchids, liliums etc., and vegetables like capsicum, tomato, cucumber and exotic vegetables like broccoli, Chinese cabbage, red cabbage, lettuce, cherry tomato, zucchini, pockchoy, celery, brussels sprout, asparagus and leeks can be grown under protected environments, besides, healthy seedling production under cover. Production technologies for protected cultivation of important crops are given below.

Rose

Rose is a leading cut flower grown commercially all over the world. It ranks first in global cut flower trade. This flower has a worldwide consumption of more than 40 billion stems annually. Its cultivation demands special care and attention, so that plants bloom to their maximum potential. To meet the perpetual demand, cultivation of roses in greenhouse has become very commercial.

Soil: For successful rose cultivation, soil should be well drained and a depth of 50 cm with high organic matter is preferred. It requires slightly acidic soil with pH 5.5 to 6.0.

Climate: The temperature between 15 to 27°C is optimum for good growth and temperature below 15°C hinders plant growth. If temperature goes more than 28°C, it should be brought down by providing ground level misting. Humidity of 60 to 65% is optimum for its growth and this can be maintained by operating foggers/ misters for about 1-2 minutes / hr during winter and 1-2 minutes for every half an hour during summer.

Varieties: While selecting the varieties, one should know about their yield, shelf life and demand in the market. Rose varieties are classified into three groups based on the stem length. Long stemmed varieties are Skyline (yellow), Polo (white), First Red (red), Grand Gala (red) etc. Medium stem ones are Passion (red), Lambada (salmon), Nobles (pink), Naranga (orange) and Miracle (orange) etc., and short steamed varieties are Golden Gate (yellow), Print stone (peach), Spinks (yellow) and Sweetness (pink).

Propagation: Good quality budded plants are commercially propagated through budding . Plants budded over *Rosa indica or Rosa mannetii* root stocks are better. The budded plants of 4 to 8 months old are used for planting.

Land preparation and planting: Land should be ploughed thoroughly and brought to a fine tilth after removing unwanted materials. Add well decomposed farm yard manure (10 kg/m^2). Addition of coir pith (2 kg/m^2), vermicompost (1 kg/m^2) and neem cake (200 g/ m^2) improves soil structure, besides nutritional status of growing media. Before preparation of the beds ,media should be sterilized with formaldehyde (5%). Raised beds of 25-30 cm height and width of 90-100 cm all along the greenhouse length are prepared. Path of 50-60 cm in between two beds is taken. Planting is done on the bed with paired row system in zig zag fashion by giving spacing of 40-45 cm between two rows and 20-25 cm between the plants. With this 5-6 plants / m^2 of green house area could be accommodated. While planting, care should be taken that budded portion should be above the ground and soil around plant is pressed gently to remove the air around root system. Irrigation is given immediately after planting with rose head can or hose pipe with rose head and also for initial 15 days. Higher humidity is maintained initially for better establishment.

Canopy Management Practices

i. *Pruning and bending*: Crop canopy management is very important in commercial production of roses. Young plants, if allowed to flower immediately, the entire structural frame work of plant is disturbed. Hence, developing appropriate plant architecture is very essential by adopting pruning and bedding. To develop basel shoots, pruning is carried out after two months of planting and it is done just above the budded portion leaving two to three leaves. Recently bending is followed instead of pruning. Here blind and weak shoots are bent towards the path leaving two buds at base. While bending, flower buds and sprouts are removed. Shoots coming after pruning or bending play very important role in flower yield. Hence remove all the shoots by pruning or bending retaining 3 to 4 productive shoots.

ii. Pinching side shoots: Remove the shoots developed on root stocks at a regular interval. The side shoots from flowering stem are also removed regularly to get quality flower.

Irrigation: Water requirement of rose plants depends on the type of soil, temperature, humidity and stage of the crop. Generally 7-8 litres/m^2/day are required. Irrigation is done by drip irrigation system, having emitters at 30 cm apart on the lateral line. Ground level misters are also adopted to increase humidity.

Nutrition: Total nutrient requirement of rose is N-125g, P and K-75g each, Ca-40g and Mg-8g/m^2/year This dose of nutrient is spilt equally and provided on daily bases through fertigation system. Nutritional dose is altered based on nutrient status of the soil which is analysed at regular intervals. Soil EC and pH are also analysed regularly and corrected, if necessary.

Harvesting and yield: Flowers can be harvested 30-40 days after pruning. Flower stems are harvested at tight bud stage when one or two petals start opening. Harvesting stage varies with the variety, market distance, temperature and buyers/consumers need. For local and domestic market, flowers are harvested at little open stage, while for distant market or export at tight bud stage. Harvesting should be done either in the morning or evening hours. Depending on the varieties, 200-300 flowers can be harvested/m^2/year.

Post harvest handling: Immediately after harvest, flowers should be stored at 3-5^{0}C temperature in cold storage to remove field heat, reduce rate of respiration and water loss. Based on the stem length, flowers are graded and made into bundles of 20 flowers which are covered with soft corrugated paper by putting rubber bands. These bundles are packed in corrugated boxes and sent to the market. Flowers can be stored in cold house at 2-3°C temperature for about 5 to 7 days.

Grading: After cooling, the stems are shifted to air-conditioned grading room. In the grading room, defective stems with blemishes or symptoms of diseases and pest infestation are removed. The stems are sorted out in different grades of length by colour coded board or automatic grading machines. The quality parameters for grading are :

1. Long stemmed varieties are graded from 40 m onwards with the difference of 10 cm whereas short-stemmed varieties are graded from 40-65cm with the difference of 5 cm.
2. Leaves should be dark green and healthy. There should not be any yellowing due to micro-nutrient deficiency. Leaves should also be free

from dust, chemical residue, chemical injury, spider mites or powdery mildew.

3. The flowers should not be bullhead, too open, too tight, with bent neck, damaged by thrips or botrytis etc.
4. Bud size should be representative to the variety.
5. The length of the neck should not be too long.

Plant Protection

Insect pests: Thrips, aphids, red spider mite, bud borer and stem borer. Sucking pests like thrips and aphids can be controlled by spraying oxydemeton methyl, monocrotophos or dimethoate (2ml/l). Bud and stem borers can be controlled by spaying of quinolphos (2 ml/l). Red spider mite is a serious pest which can be controlled by spraying dicofol (2.5 ml/l) or wettable sulphur (3g/l). In case of severity ,it can be effectively controlled by spray of abamectin (0.5 ml/l).

Diseases: Powdery mildew, black and *alternaria* leaf spots.

Powdery mildew can be controlled by spraying wettable sulphur (3g/l), black and *alternaria* leaf spots can be controlled by spraying carbendazim (1g/l) or mancozeb (2g/l).

Gerbera

Gerbera (*Gerbera jamesoni*) is one of the important cut flowers commercially grown in several countries of the world. It is commonly known as transvalls daisy or African daisy. There is great demand for gerbera in recent years as it is used in different flower arrangements, bouquets and flower vases. Flowers are in good demand locally during marriage and festival seasons.

Soil: Gerbera requires well drained soil with high organic matter having acidic pH of 5.5-6.0. Deep soil is suitable as gerbera roots grow up to 50-60 cm. Before selecting the land for cultivation, it is better to test the physical and chemical properties of the soil. Soil with EC of less the 1 dS / m is better.

Climate: Though gerbera crop can be cultivated in moderately warmer areas in open sunny situations or in shaded conditions, but to get higher productivity of superior quality flowers, cultivation under greenhouse is feasible. Optimum temperature should be maintained, as it influences the leaf number, stem length and flower yield. Temperature of 23°C is required for flower initiation and about 25 to 27°C is optimum for leaf growth. Temperature below 12°C and above 35°C affect the flower production.

Varieties: Different varieties of attractive colours are available and the demand varies with the consumer preference in day to day market. Based on the consumer demand, breeders release several varieties every year in the market. Hence, it is difficult to name specific varieties. The varieties in demand at present are Savannah, Pink elegance, Danaellin, Topic Blend, Goliath, Sunway, Rosalin, Dalma, Winter Queen, Amulet, etc,. Generally varieties with light coloured petals having dark center are in good demand.

Land and bed preparation: Land should be ploughed thoroughly and brought to fine tilth. After removing unwanted materials, add well decomposed farm yard manure (10 kg/m^2). Addition of coir pith (2 kg/m^2)/ vermicompost (1 kg/m^2)/ neem cake (200 g/m^2) will improve soil structure, besides nutritional status of growing media.

Before preparation of the beds, media should be sterilized with formaldehyde (5%). Raised bed of 45 cm height and 70 cm width of greenhouse length are prepared. Path of 30 cm in between two beds is given. Planting is done on the bed with paired row system in zig zag way by giving the spacing of 35 cm between the two rows and 25-30 cm between the plants. With this 6-8 plants/ m^2 of greenhouse area could be accommodated. While planting, care should be taken that the crown portion should be just above (1 to 2 cm) the ground level and root ball should not be disturbed while planting. Soil above the bed should not be disturbed while planting. Top soil of the bed should be racked every fortnight to a depth of 3-5 cm and it helps in better root aeration. Humidity of about 80-90% has to be maintained immediately after planting and also for about 4-6 days. There should not be moisture stress at this stage.

Irrigation and fertigation: The pH of irrigation water should be between 6.5 to 7.0 and EC 1 dS/m. It is better to analyse water before use. Irrigation is given immediately after planting with rose head can or hose pipe and also for about 10-12 days till establishment of plants. High humidity is to be maintained thorough misting or fogging for initial four weeks, irrigation through drip system is given after 15 days of planting. Water requirement is about 5-6 litres/day /m^2. Moderate moisture should be maintained and based on the season, irrigation can be altered. Excess moisture invites root or crown rot and leads to death of plant.

Nutrients are supplied through fertigation with very low concentration on daily or alternate basis. The quality of nutrients available in the growing media or supply of nutritional dose has direct influence on growth and productivity. Unlike in open conditions, scheduling of fertilizer application inside the greenhouse is tricky and has to be monitored carefully. Fertilizer application based on growth stages in different ratio of NPK such as 2:3:1 at early vegetative stage (15-20 days), 3:2:1 at rapid vegetative stage (up to 45-50 days), 1:2:2 at bud formation stage (50-60 days) and 1:1:3 at the beginning of the flowering (60-90 days) and also after flowering regularly. The nutrients applied through irrigation should not be more than 1g / litre of water. Micro nutrients are applied through foliar sprays regularly and also when deficiency of particular element is noticed.

Harvesting and yield: Total crop duration is about 24-30 months. Based on the climatic conditions, flowering starts 8 to 12 weeks after planting. Peak flowering is observed 7 months after planting. Flowers are harvested when outer 2-3 rows of the disc-florets are fully developed. Flowers are harvested usually in the morning or late in the evening by side ward pulling of the stalk. The heel of the stock should be cut about 2-3 cm above the base and flower stalk should be kept in clean water. Flower yield is about 150-200 / m^2 /year in standard varieties.

Post harvest handling: Flowers should be kept in fresh water immediately after harvest for about 4 hours in cold house at 7 to 8^{0}C. Harvesting stage, stem length, firmness of stalk, etc. decide the flower quality. Individual flower head is put in thin plastic sleeve and a bundle containing 10 flowers are made, about 30-40 such bundles are packed in corrugated boxes.

Grading

Grade	Stem length (cm)	Flower diameter (cm)
I	>50	>10
II	<50	<10

Plant protection: Among insect pets, red spider mite and thrips are important. To control mites, spray 2.5 ml dicofol or abamectin 0.5 ml and for thrips oxydemeton methyl 2.0 ml / litre of water is effective. Root rot and crown rot caused by *Pythium, Rhizoctonia, Phytophthora, sclerotium* are important and to control these, drenching with metaloxyl 3.0 g / litre can be used. Powdery mildew is another disease which occurs when humidity increases. Spraying of wettable sulphur 3.0 g / litre is advised.

Carnation

Carnation stands next to rose in world flower trade. It is available in different colours, remains fresh for longer period in vase and the best suited for export as the flowers are light in weight and have very good rehydration capacity. It is grown in greenhouse condition in recent years due to its higher demand.

Soil: Sandy loam soil with a pH 6-7 is suitable for its cultivation. Add soil, sand and farm yard manure in 2:1:1 ratio and addition of coir pith (2 kg/m^2) will improve the drainage. Media should be sterilized by using 5% formaldehyde (10 l/m^2). After putting formaldehyde to media, completely cover with plastic sheet for about 4-5 days. Then media is flushed with water to drain out the left over chemical.

Climate: Carnation requires day temperature of 20-25°C and night temperature of 10-15°C . It requires higher sunshine but during summer light should be cut down by providing 30-50% shading net inside the greenhouse. Humidity of about 50-60% is optimum for plant growth and flower development. Fluctuation in day and night temperature results in calyx splitting and malformation of flower.

Bed preparation: Raised bed of 75-80 cm width and 30 cm height are prepared along with the greenhouse length. Leave 50-60 cm path in between two beds. Add 100 g calcium nitrate, 600g super phosphate, 50g potassium chloride, 50 g magnesium sulphate / m^2 area and mix it thoroughly in the soil.

Varieties: There are two types in carnation viz, standard (sim) and spray. Standard varieties bear a single bold size flower on the stem while in spray type, small flowers are produced on the branches of flower stem. A few important varieties of both the types with different coloures are given below.

Flower colour	**Standard type**	**Spray type**
Red	Domingo, Master, Grana, Imphala, Aicardi, Goudina	Alister, Darling, Red Arrow, Durango
Pink	Cherry solar, Bolagro, Petra, Pink dover, Pinkshiva, Buemonde	Happyness, Resca, Rosini, Cherry Bag
White	Dovar, Sansura, Nordica, Mistral, Viking, Alibaba	Vergo, Closeup, Crystal, Amaliya
Yellow	Yellow solar, Tahiti, Solaroro, Kiro Diana, Yellow Firato	Sonia, Ureka, Castillo, Koreno
Orange	Solar, Orange, Pinto,Dali, Firato	Indira, Nikhi, Macarena
Purple	Lavender, Jargia, Biasis	Purple chopin, Rosetti
Variegated (mixed)	Pikaso, Rodrick	Jakobin, Sapora, Challenger.

Planting: Planting can be done thoughtout the year under greenhouse conditions. However, mid winter is suitable for planting. Rooted tip cuttings are used for planting. Planting net of 7.5x7.5 cm size is stretched on the bed at 12 cm height, then planting is done on a bed in 4 rows at a spacing of

15x15cm for standard type and 15x20cm or 20x20cm for spray type. In one square meter greenhouse area, 15 to 20 plants can be accommodated. Planting should not be done too deep to avoid stem rotting as it is highly susceptible to *Fusarium* wilt and *Rhizoctonia* root rot.

Irrigation and fertigation: For a few weeks after planting, it is better to provide irrigation through misting for better establishment. Drip irrigation can be started 3 weeks after planting. Water requirement is about 4 to 5 litres/ day/ m^2. Nutrients should be given at all stages of plant growth regularly. The following doses of nutrients are given through fertigation at different stages of growth.

Stage	**Fertilizers (g/m²/week)**				
	N	**P**	**K**	**Ca**	**Mg**
Planting to end of pinching	4.64	1.62	3.62	2.55	0.55
End of Pinching to until harvesting	4.69	1.50	5.58	2.72	0.51
During harvesting	4.88	1.50	7.03	3.09	0.50

Netting for support: Netting is an important practice in carnation cultivation. To avoid bending and breaking of long stem, netting made of wire or nylon is essential. Nets of different sizes are tied to iron poles provided at 3 m distance on the edge of either side of the bed in 4-5 layers. First layer of size 7.5 x 7.5 cm is provided before planting then second one is of 12.5x12.5 cm, third layer is of 15x15 cm and fourth is of 20x20 cm size , provided at a distance of 10-15 cm between layers.

Pinching: Pinching is an important cultural operation in the successful production of top quality carnation flowers. After planting, the plant continues to grow a main stem. If left unpinched, this main stem produces a crown flower. Pinching means to remove the tip of the main stem. This allows the side shoots to develop and these side shoots produce the first flush. Pinch the plants leaving 6 pairs of leaves (internodes). This is called single pinching. The plants are usually pinched for the first time about 3 weeks after planting. Second pinching can also be done 3 weeks after first pinching. All the side shoots produced after first pinching can be pinched leaving 4 pairs of leaves. By this, flowering will be delayed. This is called double pinching. For staggered flowering 50 % of side shoots produced after first pinching will be pinched leaving 4 pairs of leaves. This is called single and half pinching. If temperature is more, second pinching is not preferred.

Disbudding: From standard carnations, the side buds may be removed for main flower to develop. In case of spray carnations, only the central terminal bud is removed to encourage lateral flower buds to develop. Care must be taken to avoid any injury to the main stem.

Harvesting: Flower harvesting starts from 15 to 20 weeks after planting. Harvesting in standard type is done at paint brush stage (just opening of petals in the flower bud). In case of spray type, flowers are harvested when two buds are opened and remaining buds at paint brush stage. Stems are harvested with sharp knife leaving 15 cm stalk at the base and should be kept in fresh water immediately after harvest. While harvesting, care should be taken to see that, side shoots and supporting net should not be damaged. Carnation yield depends on the varieties, growing region, climate, planting time and type of pinching. Every year, yield/m^2. of green house area will be about 200-250 stems.

Post harvest handling: After harvesting, standard carnations are classified based on the stem length and made into a bundle containing 20-25 stems. Then it is wrapped with soft corrugated sheet by putting a rubber band and such 20 bundled are kept in corrugated box of size 122 x 50 x 30 cm and sent to market. Spray carnations are packed in a bundle containing 5 stems. The flowers after packing are pre-cooled at 0°C temperature and can be stored for 3 weeks at 3-4°C storage temperature.

Grading: Before packing, flowers should be graded. The grading is based on the degree of bud opening, minimum flower diameter and stem length. In the United States, the Society of American Florists recommended following grades for the standard carnations in U.S.A.

Grade		Minimum flower diameter (mm)	Minimum stem length (cm)
Blue (fancy)	Tight	50	55
	Fairly tight	62	
	Open	75	
Red (standard)	Tight	44	43
	Fairly tight	56	
	Open	69	
Green (short)	Tight	None	30
	Fairly tight	None	

Plant protection: Red spider mite is an important pest of carnation. To control this pest, maintain temperature at optimum level and spray 2.5 ml dicofol or abamectin 0.5 ml or wetteble sulphur 3.0 g / litre of water. Wilt, stem rot and foot rot are important diseases in carnation. To avoid spread of these diseases, provide proper drainage. Drench beds with copper oxychloride 4.0 g or carbandizim 2.0 g / litre of water.

Anthurium

Among tropical cut flowers, anthurium occupies the place of pride. This is a popular flower among flower arrangers. The cut flowers of anthurium are valued for myriad of colours, showy spathe and long vase life. It is mainly used for different flower arrangements and grown as potted plant to keep as indoor plant. There is great demand for quality anthuriums in international market. Recently commercial cultivation is being taken up in Kerala, Karnataka and North Eastern states adopting suitable structures based on the existing agro climatic conditions.

Media and Bed preparation: Great attention should be given while selecting the media as the crop once planted remains for about 6-7 years. Several materials can be used as media like charcoal, coconut fiber, coconut husk, tree bark, paddy husk, coco peat, brick pieces, rock wool, coarse sand, etc. In India coir pith and coconut husk are used as growing media. Media should be well drained, should hold sufficient amount of moisture and nutrients. It should not be easily decayed and cause diseases to the plants. Media containing well sterilized soil, sand and well decomposed farm yard manure in 3:1:1 ratio can be used. Recently farmers cultivating anthurium in polyhouses are using coconut husk which is giving good results. The pH of media (5 to 6.5) is optimum for growth.

Beds are prepared including irrigation systems. The beds of 90 to 120 cm width and 45 cm height with convenient length depending upon greenhouse are prepared. The pathway between two beds is 45-60 cm. It is very important to have good drainage system in the beds. A perforated PVC pipe is passed through the center of bed through which excess irrigation water is collected and drained out.

Climate: Anthurium grows naturally in a shaded environment, well protected from adverse weather conditions. When anthurium is cultivated, it must be protected from several weather conditions like wind, rain, cold, heat, sunlight, CO_2, temperature and relative humidity. Plant growth and flower production depend on the temperature. Day temperature of 26 to 30°C and night about 18 to 20°C are ideal for the crop. Night temperature should not go below 15°C, it delays the flowering. High temperature of 35°C and relative humidity up to 80-85% are not problems for the crop. Temperature more than 35°C and a very low relative humidity of 20-30% are not suitable for the crop. Anthurium is a shade loving plant, higher temperature affects the flower and also leads to leaf scorching. Shade requirement varies with varieties and the existing climate. Generally 75% shading net is used.

Varieties: Anthurium varieties are classified into 3 groups based on the spathe character. Varieties with heart shaped spathe are called 'standard' (commonly grown); 'obake'- type has bi-coloured spathe and 'tulip' type has cup shaped spathe. However, Hawaiian, Holland and other types of anthurium varieties are listed below based on colour of spathe.

Hawaiian: *Red:* Osaki, Kosohara, Koumana, Micke mouse, Hawaiian Red, Crinkle Red

Orange: Nitta, Sunburst, Sunset orange, Diamond Jubilee.

White: Manamist, Morocco, Suchiro, Chameleon, Hidden Treasure,

Pink: Blush, Candy Strip, Candy Queen

Holland: *Red:* Tropical, Avo Rosette, , Fla Red, Arizona, Vio letta, Avo netta

Orange: Fla Orange, Avogino, Casino.

White: Acropolis, Uranus, Fla Exotica, Bianco

Pink: Anneke, Surprise, Cheers, Magic Pink, Snata

Others: *Red:*Mauritius Red, Tanaka, Calipso, Modonna, Butterfly, Sweet Heart

Orange: Mauritius Orange, Farao

White: Mauritius White, Lime, Meringo, Ivory, Grace, Lambada

Pink: Agnihotri, Lady Gem, Paradise Pink, Passion, Sultan

Potted varieties: Mia Diablada, Condor, Bonina, Titicaca, Florida, Falicita,

Propagation: It is through suckers . Commercially, tissue culture plants are used for planting.

Planting: Planting can be done throughout the year but care should be taken to avoid planting in extremely hot or cold periods. After bed preparation, planting is done at a distance of 30x30cm, accommodating 6-7 plants / square meter. Close planting is to be avoided for better aeration and uniform spreading of plants.

Irrigation and fertigation: Irrigation is done daily twice once in the early morning and again in the noon and one more extra irrigation during sunny days is required. Irrigation should not be done late in the evening. Ideal pH of water is 5.7 to 6.2 and EC 1.2 to 1.3 dS/m

Fertilizers are given daily at very low concentration through irrigation system under greenhouse condition. Anthuriums are fertigated with a solution containing nitrogen 14 ppm, potassium 176 ppm, calcium 60 ppm, magnesium 24 ppm, nitrate 91 ppm, sulphate 48 ppm and phosphate 31 ppm, besides trace elements.

Other Cultural Operations

i. *Thinning of leaves*: The old leaves of anthuriums grow at the expense of production of flowers. Too many leaves may even harm the flowers. Anthurium plant has to be thus pruned to retain adequate number of leaves, without any adverse effect on flower production and quality. This would also improve health of plants in view of better air circulation. The leaves are to be cut as frequently as possible in order not to lose the equilibrium of plants. Pruned leaves that are healthy can also be sold as cut foliage. The number of leaves retained and the frequency of cutting also depend upon the size and orientation of leaves. On an average 3-5 leaves are sufficient on a plant.

ii. *Removing suckers*: Most anthurium plants naturally produce small suckers at the base of the plant. However, this is highly undesirable for production of flowers. Suckering results in crooked stems and decreased flower size. Moreover, young suckers withdraw energy from parent plant, which does not benefit flower production. It is better to remove the suckers at an early stage. When the suckers become too large, they will cause too much damage to the base of plant when removed. Remove suckers by pulling them out by hand, never use a knife or scissors.

Harvest and packing: The flowers are cut with a sharp knife, leaving about 3 cm of stem on the plant to prevent rotting of stem. After harvest, the basal portion of stalk of flower should be kept in water as soon as possible to prevent drying out. If the flowers are taken to distant markets, a piece of cotton soaked in water is kept at the end of the stem to prevent desiccation and for export purpose, each flower stalk cut end is inserted in small plastic test tube containing water. Soft protective materials can be put in between the spathe and spadix to prevent brushing. The spathe along with the spadix is then inserted into a thin polythene cover of appropriate size. This prevents bruises on the spathe due to pressing of the spadices of adjacent spikes. The open ends of the polythene cover are stapled just outside the spathe to prevent movement of the spathe inside the cover. Corrugated fiber board boxes of 60x30x23cm size are used for packing 90-100 flowers. Harvesting starts 6-8 months after planting. Five-7 flowers/plant/ year for first two years and 10-12 flowers after 2 years of planting can be produced.

Grading:Anthurium flowers are graded based on the spathe breadth which is as follows.

Grade Spathe width

Extra large > 15 cm

Large 13-15 cm

Medium 10-13 cm

Small 8-10 cm

Mini < 8 cm

Plant protection: Mealy bug, scale and spider mites cause damage to anthurium. To control mealy bug and scale, spray melathion 2.0 ml or oxidemeton methyl 1.0 ml or methyl parathion 1.0 ml and for spider mite dicophal 2.5 ml/litre of water. Bacterial blight, anthracnose and foot rot disease are very important. To control bacterial blight, streptocyclin 0.5 g, for anthracnose captan 2.5 g or carbandizen 1.0 g and to control foot rot, drench beds with mancozeb 2.5 g and thiram 2.5 g / litre of water.

Orchid

Orchid belongs to family Orchidaceae and their flower spikes are valued for varied shape and colour of florets and long keeping quality. Demand for orchid flowers in the international market is increasing every year. India is home to nearly 1250 orchid species. Orchids can be divided into two broad groups; the terrestrials and the epiphytes. Terrestrial orchids grow on the ground (e.g. *Cymbidium, Paphiopedilum*). The epiphytes have aerial roots and grow on trees (*Vanda,Dendrobium*). Orchids exhibit two types of growth, monopodial and sympodial. Monopodials have single non-branching stem, which grows upwards. Typical examples are *Phalaenopsis, Vanda,* and *Arachnis.* The sympodial plants have a rhizome, which grows horizontally, producing new growths. Thus a well developed sympodial plant contains a clump of shoots of varying age and size (eg. *Dendrobium, Cymbidium, Oncidium*).

Propagation: Orchids can be propagated by seeds or vegetative means. The vegetative means can be by division, cutting, off shoots (keikis), air layering and by tissue culture for commercial cultivation.

Growing Conditions

Light: Sun light is essential for growth and development, but shade is important for flowering in orchids. Choose a site, which obtains direct sunlight. It is relatively easy to cut down the light, if so required by the use of shading materials. The tolerable light intensities vary with genera. *Arachnis and Vanda* can tolerate very high light intensities. It is lower in the case of *Dendrobium* and very low in *Phalaenopsis*. A level of 50% shade is ideal for most of the orchids.

Temperature: Temperature requirement varies with species and it ranges from 20-30°C during day and 10-25°C during night. Based on the temperature requirement, orchids can be classified as warm (tropical), intermediate (sub-tropical) and cool (temperate) species. Orchids should never be subjected to sudden changes in temperature.

Humidity: Orchids require about 70-80% humidity. A good thumb rule to follow is that when the temperature is high, humidity also should be high. Providing misting units or foggers or even humidifiers will ensure adequate humidity.

Air circulation: Orchids thrive well when the atmosphere is fresh. Air circulation not only keeps the atmosphere clean and fresh but also helps to keep both temperature and humidity at the desired levels. Good ventilation is also necessary to achieve this target.

Media: Varieties of media are used for orchids. Decayed leaves added with humus is ideal media for terrestrial orchids. A potting medium containing equal parts of leaf mould and soil would be desirable. For epiphytic orchids, media used should provide a surface over which the plants can cling to. The medium should be moist but never be soggy. Tree fern fiber, bark, coconut husk, tiles, bricks and charcoal are ideal media for epiphytic orchids.

Planting: Planting in containers and on ground are practiced. All the sympodial and also monopodials, which have limited vertical growth like *Phalaenopsis* can be grown in containers. Miniature orchids also exhibit their best if grown in appropriate type and size of containers. The most widely used containers are clay pots with sufficient holes for drainage and aeration. They are porous and absorb excess water. Plastic containers are light in weight, easy to transport and do not accumulate algae or salts on their sides but they are not porous. Wooden baskets are generally used for *Vanda*. Half or complete portion of coconut husk and wire baskets can also be used as containers.

Irrigation: Mist and foggers are ideal for small units. Micro-sprinkler is ideal for large scale commercial units. Spraying water on the floor and walls helps to maintain high humidity. Water with neutral pH of 6.5 to 7.0 is ideal. Water high in salts is detrimental to the plants. Flushing the pots thoroughly may help to bring down the building up of salts.

Nutrient management: Nutrients, their quantity and frequency of application depend largely on the type of orchid, potting medium, season of the year, growing conditions and stage of development of the plant . Nitrogen is essential for vegetative growth. Phosphorus is needed to initiate flowering process. Potash is required in building up strong roots, which also makes the plants resistant to biotic stresses. Minor nutrients and trace elements are required in very small quantities by orchids. Since most of the cultivated orchids are epiphytic in nature, the media used are highly porous. Hence, foliar feeding is ideal. Here the total quantity required as well as the loss will be low.

Since inorganic fertilizers are applied as foliar spray, the quantity of nutrients used / plant should be relatively low. A fertilizer mixture containing major nutrients and minor nutrients is ideal for general application. The concentration may be adjusted between 0.1 and 0.3% depending upon the situation already explained. A higher dose of nitrogen leads to better vegetative growth, which may be required during early stage. An increase in the proportion of phosphorus will be required to induce flowering when the plants are at maturity. The ratio of NPK suggested is 3:1:1 during pre-flowering stage and 1:2:2 during flowering stages. Since small quantities of fertilizers are recommended for orchids, they are to be applied more frequently. The frequency of application can be twice or thrice a week under tropical conditions. Organic manure is applied once a week.

Harvesting: Depending on varieties, tissue culture plants reach harvesting stage in 1-2 years. Orchid spikes are harvested when more than 50% florets are open. From third year, 5-6 spikes can be harvested/plant and from sixth year 10-12 spikes can be produced. After harvesting, all flowers should be washed with clean water before subjected to grading, pre cooling and storage. Grading is made based on spike length, number of florets/spike and floret size.

Plant Protection

Mealy bug and scale	-	Melathion 2.0 ml or diamethote 1.5 ml
Snail	-	Metaldihyde 5.0% bate or beer (fermented alcohol) or putting salt around beds.
Thrips	-	Oxidemeton methyl 2.0 ml or diamethoate 1.5 ml
Black root	-	Captan 2.0 g
Leaf spot	-	Captan 2.0 g or mancozeb 2.0 g

Seedling Production Under Protection

Production and productivity of crops particularly annuals depend on use of healthy seedlings. Growing plants under a cover protects the crop from damaging rains and scorching radiation, temperature and pests and diseases. Use of protected structures for raising seedlings as a commercial venture is a recent phenomenon especially in the tropics. The most commonly used method to reduce radiation and temperature is by shade net covering and may be with fogging or misting facilities. These are ideal for the production of vegetable seedlings as they are comparatively more economical and feasible. Use of low cost structures involving stone pillar or wooden poles; shade net and 40 mesh nylon net to prevent some vectors which transmit viral disease are in practice.

In India, many young entrepreneurs and progressive farmers are coming forward to take up quality seedling production and supply to the vegetable growers based on the requirement. Seedling production facilities using plug or container seedling production under protected structures viz. shadehouse, nethouse, polytunnel, etc. provide healthy and uniform growth of the seedlings. Since the hybrid seeds are expensive, this method reduces cost by minimizing seed waste. It is a commercial 'venture to produce the seedlings of tomato, capsicum, cauliflower, cabbage, chillies and brinjal. Mostly hybrids are used for raising seedlings in the containers. In the commercial nursery the most commonly used pro-trays are 98 celled and for bigger size seeds like melon and cucurbits 50 celled, the most common growing medium used is sterilized coco peat which helps to prevent diseases especially the seedlings damping off. Coco-peat has high water holding capacity and good texture. Since it is low in mineral nutrients, the media need to be supplemented with nutrient solution.

Procedure for raising seedlings: Use pro-trays having 98 cells for raising seedlings of capsicum, tomato, chilli, cabbage, cauliflower, brinjal and exotic vegetables and 50 celled trays for melons and cucurbits. Plastic pro-trays are disinfected by dipping in thiram or copper fungicide solution prepared by mixing 3 g chemical/ litre of water. Bleaching powder solution can also be used for this purpose. Fill the seedling trays with a sowing medium like coco peat, potting mixture (soil+compost+sand) or vermcompost can also be used with soil in equal proportion. For optimum germination, use well drained sterile medium. Make depressions / holes in the center of each cell in pro tray and sow one seed / cell and cover with the same media. During winter or whenever temperature is low, keep about 10-15 trays one over the other, cover with a plastic sheet and leave for about 3-4 days. This helps in increasing temperature of media and germination will be early. In the protected structures like net house, shade house or greenhouse prepare raised beds of 1 m wide with slope on both the side and leave 0.5 m space for path. Mulch the beds with plastic sheet and arrange the seed sown trays in double row over mulched beds and water daily. Mulching of beds helps in avoiding penetration of seedling roots into the soil, which will damage while lifting trays. In case, seedlings are raised under net house or shade house, rain protection should be made over the beds used for keeping trays. Over the beds use low tunnel structures made by PVC pipe of 3cm diameter and cover with poly sheet. Water the trays thoroughly every day or as needed using a fine sprinkler.

Drench trays with 0.3% 19:19:19 NPK water soluble fertilizer twice at 10 and 20 days after germination for healthy growth of seedlings. In case if there are any insect pests or diseases, follow plant protection as per the requirements. Seedlings will be ready for transplanting in 25-35 days based on the crops.

Capsicum

Capsicum commonly called as sweet pepper or bell pepper is the most popular salad vegetable grown in greenhouse. It is rich in vitamins A, C, and minerals. Capsicum cultivation is gaining popularity in peri-urban production systems because of easy access to urban markets. Capsicum yields in open field cultivation ranges 20-30 t/ha, whereas in greenhouse, the yield range is 100 to 120 t/ha.

Climate: Capsicum is a cool season crop and day temperatures less than 30°C is favourable for growth and yield. Higher temperature results in rapid plant growth and affects fruit set. Lower night temperature (20°C) favours flowering and fruit set. Shading and misting are required during summer to avoid temperature build up in greenhouses. Moderately high RH (50-60%) is preferred, which can be maintained by managing the ventilation.

Varieties: Capsicum hybrids with indeterminate growth habit are suited for greenhouse cultivation, as these hybrids grow to a height of 2 m and above to utilize the vertical space of greenhouse. Commercial green hybrids like Indra, Green Gold and coloured hybrids like Orobelle, Bomby are suitable for greenhouse cultivation with a yield potential of up to 120 t/ha from a crop of six months duration.

Nursery: Seedlings can be raised as described earlier and about 400 g seeds required for one ha area.

Growing medium and beds: Sandy loam soil and well-decomposed farmyard manure mixed in equal proportions are the best suited for capsicum cultivation. In heavier soils mixing sand up to 25% is required to provide proper aeration in the root zone. Loosen and work the soil to fine tilth. Prepare the beds of 100cm width and 15 cm height leaving 50 cm path between the beds. Add 20 kg well decomposed farmyard manure/m^2 of bed for the first crop. Mix thoroughly and level the beds. For subsequent crops, apply 5 kg FYM/m^2 of bed. Apply 50:50:50 N: P_2O_5: K_2O kg/ha to the growing beds before formaldehyde fumigation. Apply neem cake and trichoderma formulation (200 kg +2 kg/ha) just before planting but soon after formaldehyde fumes are exhausted completely.

Sterilization: For disinfecting growing beds, drench the beds with 5% formaldehyde (4 l/m^2) and cover with black polyethylene (400 gauge) sheet.

Close all ventilation spaces. Three to four days after formaldehyde treatment, remove the polyethylene cover. Two days after removing the polyethylene cover, rake the bed repeatedly to remove trapped formaldehyde fumes completely before transplanting. Disinfecting is done once in a year.

Laying of drip line: Place one inline drip lateral at the center of the bed. Inline drip lateral should have an emitting point for every 30-cm interval with a discharge of 2 litres/ hour. Before covering with the polyethylene mulch, check the emitting points for uniform discharge of water.

Mulching: Use 100 micron thick black polyethylene film of 1.2m width to cover the planting bed. Make holes of 5 cm radius on the mulch film as per the recommended spacing (45 cm x 30 cm) then cover the planting beds with mulch by securing the edges of the sheet by burying in the soil.

Transplanting: Water the bed to field capacity before transplanting. Seedlings of 35-40 days old, vigorous and uniform in size are selected for planting. Transplant the seedlings at recommended spacing of 45 x 30 cm at a shallow depth of 2-2.5 cm. Water the beds with rose can or hose pipe with rose head immediately after transplanting and every day until the plants get established. Mist the greenhouse in the seasons of low humidity. Drip irrigation is given daily to replenish 50 % of open pan evaporation.

Fertigation: Fertigate 150:150:150 $N:P_20_5:K_2O$ kg/ha using water-soluble fertilizers. Start fertigation from third week after transplanting. Fertigate twice or thrice a week for 18 weeks, by splitting the dose of fertilizer equally. Micro nutrients can be used as per requirement.

Pruning and training: Prune capsicum plants to four stems. Prune at weekly intervals starting from 15 to 20 days after transplanting. Capsicum plants show dichotomous branching. The tip of the plant splits into two at 5th or 6th node at first and these two branches again split into two giving rise to four branches. Maintain only these four main branches. At every node, the tip splits into two giving rise to one strong branch and one weak branch. Remove the weak branch and retain strong branch. Train the plants along a plastic twine. Each branch should have separate plastic twine to train along.

Harvesting: Harvesting of capsicum fruits starts 60-70 days after transplanting and continues up to 170 to 180 days at 10 days interval. Mature green or 10-15% colour turned capsicum fruits weighing 150 to 250 g are harvested. Keep the harvested fruits in cool place and avoid direct exposure to sunlight. Yield of 100 to 120 t / ha can be expected from one crop of six months duration.

Plant protection: Red spider mite and thrips are the major insect pests, spider mite can be controlled by spraying dicophal 2.5 ml or 3.0 g wetteble sulphur / litre of water and thrips by spray of dimethoate 2.0 ml or oxidemeton methyl 2.0 ml. Powdery mildew is a common disease which can be controlled by spraying wettable sulphur 3.0 g or carbondizim 2.0 g/litre of water.

Tomato

Growing tomatoes in greenhouses can increase the yield and greenhouse grown tomatoes will have superior quality. Fluctuation in the price of tomato is a regular feature. Through greenhouses, production of tomato can be achieved round the year and off season production is more profitable. In addition to the quantum jump in yield and superior quality, substantial reduction in use of pesticide makes it an eco-friendly proposition to grow tomato in naturally ventilated greenhouses/ nethouse/ shadehouse.

Climate: Day temperature of 28°C and night temperature of 18°C is ideal for its growth. Fruit set is affected at temperature higher than 35°C and a relative humidity of more than 90%.

Varieties and hybrids: Tomato hybrids with indeterminate growth habit are suited for greenhouse cultivation. Indeterminate tomato plants grow continually, producing flowers and fruits over a long period of time. Commercial hybrids like SH 7711 are suitable for greenhouse cultivation, with a yield potential up to 180 t/ha from a crop of six months duration.

Nursery: Seedlings can be raised in pro trays as described earlier. About 100 g seeds are required for one ha area.

Growing medium and beds: Sandy loam soil and well decomposed farmyard manure mixed in equal proportions are the best suited for tomato cultivation. In heavier soils mixing with sand up to 25% is required to provide proper aeration in the root zone. Soil has to be brought to fine tilth. Beds of 100 cm width and 15 cm height leaving 50 cm path between the two beds are laid out. FYM 20 kg/m^2 of bed area is added for the first crop. For subsequent crops 5 kg FYM/m^2 has to be applied. Apply N, P_2O_5 and K_2O at the rate of 50:50:50 kg/ha to the growing beds before formaldehyde fumigation. Apply neem cake and trichoderma formulation (200 kg+2 kg/ha) just before planting but soon after formaldehyde fumes are exhausted completely.

Sterilization: For disinfecting the growing beds, 5% formaldehyde (4 litre/ m^2) will be applied and covered with black polyethylene sheet. All ventilation spaces need to be closed after the application of formaldehyde. Three to four days after formaldehyde treatment, polyethylene cover has to be removed. Two days after removing the polyethylene cover, the beds are raked repeatedly to remove trapped formaldehyde fumes completely before transplanting. Sterilization can be done once a year.

Laying of drip line: At the center of the bed, one inline dripper lateral has to be placed. Inline dripper lateral should have an emitting point for every 30 cm interval with a discharge of 2 litre/ h. Before covering with the polyethylene mulch, emitting point has to be checked for uniform discharge of water.

Mulching: Black polyethylene film of 100 micron thickness and 1.2 m width can be used to cover the planting bed. Holes of 5 cm radius are made on the mulch film as per the recommended spacing (60 cm x 45 cm), then cover the planting beds with mulch by securing edges of the sheet by burying in the soil.

Transplanting: Beds have to be watered to field capacity before transplanting. Seedlings of 25 days old, vigorous and of uniform size are selected for planting. Transplant the seedlings on each bed in two rows following a spacing of 60 x 45 cm. at a shallow depth of 2-2.5 cm. Plants are watered with rose can immediately after transplanting and every day until the plants get established. Misting is done in the seasons of low humidity. Drip irrigation is given daily to replenish 50 % of open pan evaporation.

Fertigation: Fertigation is carried out using water soluble fertilizers with a $N:P_2O_5:K_2O$ dose of 250:250:250 kg / ha for six months duration crop from third week after transplanting. Fertigation is carried out twice or thrice a week for 18 weeks. Based on the requirements, micro nutrients can be used.

Pruning and training: The tomato plants are pruned to two stems / plant. Pruning usually starts 20 to 30 days after transplanting. Plants are pruned at weekly intervals. The main stem of tomato plant branches into two after the first flower cluster which are the only two branches (stem) retained and all other branches are removed. Plants are twined along the plastic twine. Separate plastic twine has to be provided to each branch. Remove older leaves periodically.

Harvesting: Harvesting of tomato fruits starts 70 to 80 days after transplanting and continues 170 to 180 days. Harvesting of fruits is done at weekly interval. Fruits should be harvested at colour break stage. A marketable fruit yield of 170 to 180 t / ha can be realized from a crop of 6 months duration.

Plant protection: Red spider mite and thrips are the major insect pests. Spider mite can be controlled by spraying dicophal 2.5 ml or 3.0 g wetteble sulphur / litre of water and trips by spray of dimethoate 2.0 ml or oxidemeton methyl 2.0 ml. Powdery mildew and leaf spot are common diseases of tomato. Powdery mildew can be controlled by spraying wettable sulphur 3.0 g or carbondizim 2.0 g / litre of water and leaf spot by using mancozeb 2.5 g / litre of water.

III. Micro Irrigation and Fertigation

Micro Irrigation

The water requirements of plants are influenced by many factors. The main factors are climate, including temperature, relative humidity, radiation, wind; and vegetative growth, influenced by age and type of plant and leaf shape. A combination of these factors or each one separately, may change the evapotranspiration rate thus changing the water requirements. Irrigation water management is the key input for successful crop production. Efficient irrigation requires uniform application of water at proper time and proper amount. The irrigation system should therefore be designed to ensure satisfactory adoption to the soil and crop conditions so that the desired amount of water may be applied as efficiently as possible. Different irrigation methods are adopted because of variation in soil type, topography, water supply, crop and other management practices. Due to improper methods of irrigation 27 to 42 % of irrigation water is lost by deep percolation beyond the root zone of the crop. This is not only a loss of valuable water but also loss of plant nutrients due to leaching. Therefore selection of proper irrigation method is the most important aspect. The micro irrigation has to be efficient in saving water and also in improving the yield. Micro irrigation means frequent application of required and measured quantity of water directly above or below soil surface, usually as discrete drops, continuous drops, tiny streams or miniature spray through emitters or applicators placed along a water delivery line (laterals).

Types of Micro Irrigation Systems

1. *Drip irrigation system*: It is a standard set of components with the emitting devices which are pre-fitted or inserted during installation along the lateral lines to get continuous flow of water to form the wetting strip on the soil surface or subsurface around the root zone of crops called as drip / line source tubing. Line source tubing may be of built in drippers, inline drippers, integral drippers, porous tube line, pressure compensating drip tube etc.

2. *Bubbler irrigation system*: It is a standard set of system in which the application of water to the soil surface as a small stream or fountain, where the discharge rates from the point source bubbler emitters are greater than for drip or line source emitters because emitter discharge rate normally exceeds the infiltration rate of the soil.

3. *Micro jet irrigation system*: It is a standard set of system in which the application of water in the form of discrete or fan or full circle or half circle or quarter circle or part circle spray on the surface of soil from low height or low angle through air around the crop. Micro jet includes

moving part and having greater discharge rate and greater coverage than dippers and bubblers.

4. *Micro sprinkler system*: It is a standard set of system in which the application of water in the form of discrete or fan or full circle or half circle or quarter circle or part circle spray on the surface of the soil from stake height less than 1 m or low angle through air around the crop. Micro sprinkler incorporates moving part and having greater discharge rate and greater coverage than drippers, micro jet and bubblers.
5. *Sprinkler system*: It is a standard set of system in which the application of water in the form of discrete or fan or full circle or half circle or quarter circle or part circle spray on the surface of soil from riser height 1 m or more and high angle through air around the crop. This system includes moving part and having greater discharge rate and greater coverage than drippers, micro jet, bubbler and micro sprinklers.
6. *Misting or fogger irrigation system*: It is a standard set of system in which the application of water in the form of the full spray to create humidity is done in the air at low level or overhead. Misters or foggers do not incorporate moving part and having relatively low discharge rate and smaller coverage as compared to micro jet and micro sprinkler.

Components of Micro Irrigation System

Distribution Lines: Distribution lines consist of a network of graduated pipe sizes starting with a single, large main line followed by smaller sub main and lateral lines. The buried main and sub main lines are normally permanent and are made up of PVC and HDPE. Main and sub main lines range from 40 mm to 110 mm size depending upon required water flow to the laterals. Both main and sub main lines should have outlets with valves for periodic flushing. Lateral lines are usually made of LDPE and LLDPE and range from 12 to 16 mm diameter.

Control System: Main control for the micro-irrigation system is meant to organize, to measure, filter and to regulate pressure. This includes pump, primary filter, pressure regulator, pressure gauge, and usually chemical/ fertilizer injection equipment.

Filter system: Proper water filtration is essential and is achieved by using either screen, sand or centrifugal filters either individually or in combination.

1. *Screen filter*: This is used where there is problem of clay particles in the water. Screen filters are the most frequently used equipment for removing particles. The method of water entry circulation and exit varies among the manufactures. These filters have back wash arrangements. Corrosion

resistant stainless steel and plastic materials are the more commonly used substances for the screens. Development of an abrupt pressure across the filters means that the screen is beginning to clog. In contrast, if no pressure change occurs for long periods, the screens or seals are broken or mesh size is too large.

2. *Sand filter*: Pressure type, high flow sand filters are more common ones used in the systems. Almost the full depth of the sand is used in the pressurized filters.
3. *Centrifugal filters*: It is used when the source of water is canal or tank and it removes suspended materials with specific gravity greater than water.

Advantages and Disadvantages of Micro Irrigation

It saves water, improves productivity and quality of the produce, reduces labour cost, salt concentrations in the root zone, weed and disease incidence, allows use of water soluble chemicals and fertilizers and saves on fertilizers. Water is applied at frequent intervals in micro-irrigation and hence growth of the crop is uniform. As there is no water/moisture-stress, the crop growth is not affected at all. Water is also controlled and only the required quantity of water is given to each plant based on evapotranspiration requirements.

High investment at the initial stage, frequent clogging of emitters, pilferage of components are a few of its disadvantages.

Fertigation

After realizing the importance of micro irrigation system, the next most essential input to boost productivity and quality is the fertilizer. Fertigation is a simple word coined with union of two words, fertilizer and irrigation, indicating application of fertilizer through an irrigation system. The fertilizers are applied daily or alternate days as per the requirement of crops to the root zone.

Suitable forms of fertilizers for fertigation: The availability of 100 % Water Soluble Solid (WSS) or liquid fertilizer is essential if it is to be applied through fertigation. The fertilizers , traditionally available in the market, are not fully soluble in water and it also contains some non-soluble solids which make them unsuitable for application through micro irrigation systems. For fertigation, the special fertilizers are available in the following forms.

1. *Liquid fertilizers*: These are the solutions which contain one or more plant nutrients in liquids form, but these are not popular in India.

2. *Water soluble solid fertilizers*: These are 100 % water soluble fertilizers, generally containing two or more major nutrients and micro nutrients. These fertilizers completely dissolve in water leaving no precipitate. Therefore, there is no chance of clogging of emitters due to fertigation.

Suitable conditions for fertigation: There are a number of modern techniques suitable for the application of fertilizers, like sprinkler and drip irrigation. The most important criterion for the suitability of an irrigation system for fertigation is accuracy of water application, which largely depends upon proper designing and installation of irrigation system and availability of correct equipment for injecting fertilizers. The success of fertigation not only depends on good irrigation or proper use of equipments, but it also requires adequate information on crop water requirements, soil moisture characteristics, and crop physiology to enable accurate irrigation scheduling and therefore, the maximum water use efficiency. Efficient fertigation requires access to information on soil fertility and an outstanding of nutrient balance in soil.

Advantages of Fertigation

1. The accurate placement of fertilizers in the active root zone with less loss through leaching.
2. The application of nutrients in an available form can maintain the optimum level of nutrients required for crop growth.

   These two points offer potential for much higher fertilizer application efficiencies than with conventional application methods, representing a potential saving of 20-30% in fertilizer use.
3. High frequency fertigation avoids high peaks in salt concentration in the soil solution, thus maintaining a high osmotic potential and aiding nutrient uptake.
4. Use of acidic fertilizers for fertigation on high lime soils can reduce the pH within the zone thus reducing the problems of nutrient fixation.
5. Once information is gathered regarding the peak requirement for nutrient uptake, fertigation can be scheduled to apply the correct proportion of nutrients for optimum yield. Timely application of fertilizer can help to optimize crop yield.
6. The automation of fertilizer application through fertigation can save labour cost and reduce mechanical damage to crop and soil structure by labourers entering the field.

Fertigation methods: There are a number of techniques used to introduce fertilizer into the irrigation system. This may be drip irrigation. micro sprinklers, micro jets or sprinkler irrigation. In the beginning as per the decided schedule, system has to run for 15-30 minutes, then fertigation is done, again followed by irrigation for 15-30 minutes. While fertigating, it should be ascertained that the optimum concentration is maintained taking into consideration the crop reaction. For sensitive crops like tomato, strawberry it should be round about 500 ppm and for other crops maximum is 1000 ppm.

Fertilizer Application Equipments

1. *Ventury*: This system of injection involves very low cost and works by creation of a partial vacuum in the system and suction of the fertilizer solution into the irrigation stream. The vacuum is created by diverting a percentage of water flow from main line and passing it through a constriction which increases velocity of flow thus creating a drop in pressure. When the pressure drops, the fertilizer solution is sucked into the ventury through a suction pipe.

   This is a simple and inexpensive method of fertilizer application, but it has some disadvantages. The pressure loss across a ventury is high, about 1/3 of the pressure existing near inlet of filtration system. Secondly, precise regulation of flow is difficult because the rate of injection is very sensitive to the pressure and rate of flow in the system and it is used for smaller area.

2. *Fertilizer tank*: In this system, a part of the irrigation water is diverted from the main line to flow through a tank containing the fertilizer in a fluid or soluble solid form, before returning to the main line. The pressure in the tank and the main line is the same, but a slight drop in pressure is created between off take and return pipes for the tank by means of a pressure reducing valve. This causes water from the main line to flow through the tank causing dilution and flow of diluted fertilizer in to the irrigation stream.

   Concentration of fertilizer decreases gradually, until it reaches the level of the irrigation water. It is observed that with liquid fertilizers it takes four tank volume displacements to empty the tank. If solid fertilizer is used, at least ten volume displacements are needed to dissolve all the materials. Size of the tank may range from 50 to 100 litres depending on area to be irrigated.

   Fertilizer tank is simple in construction and operation and also low in cost. There is no need of external power supply. However, the varying concentration of nutrients causes bulk of the fertilizer to be applied at the beginning of irrigation cycle. The tank has to be refilled with solution for each irrigation cycle.

3. *Fertilizer injector*: With this method, a pump is used to inject fertilizer solution into main pipe. The solution is normally pumped through the container, and the type of pump used depends on the power source. The pump may be driven by an internal combustion engine or electric motor. Out of them, electric pump is more in use. With these injection pumps, fertilizer may be supplied to the irrigation water at more or less constant rate. Pumping rate and concentration of the solution can be adjusted to attain the desired level of fertilizer application. Changes in water flow, power failure or mechanical failure may cause deviations from the planned concentration.

Nowadays, hydraulic pump, being operated on line pressure is in common use. Such type of injector needs no external power supply. Its linear hydraulic motor is powered by hydraulic pressure of irrigation system. Injector unit should be resistant to chemicals. Since the maximum injection rate of the pump is proportional to the pressure in the main line, the required injection rate is easily adjusted by trotting the injection line using the valve. Injector consists piston and to and fro motion of piston creates suction from container. Piston movement is due to hydraulic pressure of main line. Injection rate may vary from 1-320 litres / hr with corresponding increase in main line pressure.

Liquid and 100% water soluble fertilizers are to be mixed in bucket full of water. Fertilizers and water can be mixed in fertilizer tank. Then this mixture to be mixed for 5 minutes and then this can be used through one of the above methods and equipments.

Fertigation scheduling: While deciding the fertigation schedule, following points need to be taken into consideration.

1. Total nutrient requirements of crops to produce target quality yield.
2. Part of the nutrients supplied in the form of organic manure.
3. Possible savings in the nutrient doses due to higher efficiency of added fertilizer through fertigation.
4. Relative requirement of nutrients at different growth stages of crop.

The requirement of nutrients varies with stage of crop growth. Therefore it is essential to know the nutrient requirements of the crop based on the growth stages. High level of nitrogen is required during vegetative growth stages to facilitate accumulation of dry matter in different plant parts. Its requirement is comparatively low during reproductive stage. The requirement of phosphorus is more during reproductive phase as it induces flowering, fruit set and seed development. Potash requirement is high during crop maturity stage.

Fertilizers Used in Fertigation

1. Straight fertilizers

Urea phosphate (17:44:0), mono ammonium phosphate (12:61:0), mono potassium phosphate (0:52:34), potassium nitrate (13:0:46), sulphate of potash (0:0:50+18%S), calcium nitrate, magnesium nitrate and calcium magnesium nitrate.

2. NPK Formulations

20:10:10+Trace Elements(TE)+Mg, 19:19:19+TE, 16:08:24+TE+Mg, 13:40:13+TE, 20:20:20, 6:12:36, 13 :05:26, etc.

3. Chelated Micronutrients

Fe EDTA (13%), Zn EDTA (15%), Fe EDDH-A (6%), chelated combination of micronutrients, etc.

High Density Planting

Increasing the yield / unit area rather than / plant is the concept of high density planting in perennial horticulture crops. High density planting has a potential to increase yield and reduce cost of production of several horticultural crops. It is already exploited successfully in banana, pineapple, kinnow, orange and to some extent in apple and mango. The productivity for a given land area is related directly to fruit bearing foliage and its photosynthetic capacity. These concepts have converged themselves in the form of high density planting of fruit crops. The continuing decline in the availability of suitable orchard lands, the raising input and labour costs and the mounting demand for fruits have given further impetus to develop technologies to increase productivity / unit area of land, for which the theme of high density planting comes to fore front. High density orchard system is a very intensive form of fruit production with great relevance to food and nutritional security. This differs from traditional way of cultivating fruit crops where very a few plants are accommodated / unit area with a large canopy. These conventional orchards require low establishment cost, involves less labour, but the economic yield can be expected only after 8-10 years and in some cases it is still more. Low and irregular production of fruits is also a problem. The quality is also affected due to meager sunlight penetration into the dense canopies and large sized trees. This is also a problem for orchard management practices. High density orchard on the other hand, accommodates more plants / unit area. It facilitates better utilization of solar radiation and increases bearing surface / unit area.

There are several ways of achieving this objective like selection of dwarf variety, use of dwarfing root stocks, pruning and growth regulators for canopy management. The use of dwarf trees and checking excessive vegetative growth enhance the harvest index and enable early economic yield , thereby increasing productivity and profitability. It is more amenable to modern input use efficient technologies like drip irrigation, fertigation, mulching, mechanical harvesting and horticultural operations like pruning, plant protection that enhances labour use efficiency. Better fruit quality due to optimum exposure to sunlight and easier harvesting of fruits with less injury resulting in better post harvest life are added advantages of high density orchards.

The high density orchard is very much practiced in temperate fruit crops particularly in apple because of use of dwarfing root stocks. It was introduced in apple in Europe during 60's with the development of Malling and Malling-Merton root stocks. The principles of high density orchard, developed over the years for temperate fruit crops, can be attempted for tropical woody perennials, but needs modifications and systematic long term experiments. These modifications are required since it has to be remembered while attempting to restrict size of tropical fruit trees that the evolutionary development of these trees in tropical forests where they originated normally give rise to large trees whose survival depend upon vegetative vigour rather than prolific production of fruits and seeds. Other characteristic of these trees are that they are mostly evergreen, profuse branching and with multiple apical meristems which helped their survival in the original habitat.

High density planting could be maintained throughout by using dwarf varieties and dwarfing root stocks. The normal varieties can also be used in high density planting by following pruning. Pruning can be practiced only in those crops which respond to pruning like guava. The other way of adopting high density planting is to plant at a closer spacing according to the crops. In case of mango, normal spacing adopted is 10m x 10m. In high density planting, 5m x 5m or still less spacing could be followed. Once the plants canopy is covered, alternate plants can be removed in a row. After certain years when plants grow completely, all the plants in the alternate rows can be removed. This helps in getting early yield. This system can be practiced for crops where yield is affected by pruning.

References

Choudhary, M.L.2005, Souvenir of International conference on plasticulture and precision farming. National committee on Plasticulture application in Horticulture, New Delhi, November 17-21, 2005

Iyer, C.P.A. and Kurian, R.M.,2006, High density planting in tropical fruits. Principles and practices. International Book Distributing Co., Lucknow

Jadav, A.S., 2002, Compendium of summer school on Protected cultivation. Organized by Hi-Tech Horticulture and Vegetable Project, College of Agriculture, Pune

More, T.A.,2001, Hi-Tech Horticulture. Mahatma Krishi Viswa Vidhyalaya , Rahuri,MS

Omr Zeidon, 2005, Tomato production under protected cultivation. Mashav, Cinadco, Israel

Prabhakar, B.S., 2002. Compendium of short course on Greenhouse production of flowers and vegetables. Organized by IIHR, Bangalore

Rajeevan, P.K., Valsalakumari, P.K., Geetha, C.K. and Sudhadevi, P.K., 2004, Reflections on Indian Floribusiness.Directorate of Extension. KAU, Thrissur

Singh, D.K. 2004, Hi-tech Horticulture. Agro-tech Publishing Academy, Udaipar

Singh K.P. 2009, Souvenir of National conference on floriculture for livelihood and profitability. Organized by ISOA, at IARI, New Delhi, March 16-19, 2009

The Science of Horticulture : Vol. 1, 2011
© *K.V. Peter (ed.), pp. 65-88*
New India Publishing Agency, New Delhi (India)
E-mail : newindiapublishingagency@gmail.com
Web: www.bookfactoryindia.com

Chapter 3 Science Behind Tracer Studies in Horticulture

KOTUR, S.C.

In the backdrop of a predominantly agrarian economy, India has made tremendous advances in modern agriculture to meet both the food and nutritional security. In recent years, horticulture has assumed great importance mainly to augment nutritional security and secondarily as a highly cost-effective, remunerative and eco-friendly enterprise having vast export potential. To meet the challenges of research and development, tracer techniques have added the much needed fillip as they can be ingeniously applied to resolve issues which are not ordinarily possible using conventional techniques.

The Scope of Tracer Techniques in Horticulture

Tracer techniques have a vast potential of myriad uses. Some important applications are given here.

1) Genetic Improvement of Crop Plants

Success of crop improvement programme depends on availability of large genetic variability which the plant breeder can combine to generate new varieties. In nature, occurrence of natural variability in the form of spontaneous mutations is extremely low (about 10^{-6}), which can be enhanced several fold

(10^{-3}) by using ionizing radiations or chemical mutagens (D'Souza, 2007). Radiation induced genetic variability in crop plants is a valuable resource from which plant breeder can select and combine different desired characteristics to produce better crop varieties. The desirable traits which are bred through induced mutations include higher yield, fruit/vegetable/ flower quality, early maturity, diseases and pests resistance and resistance to abiotic stresses.

Gamma rays have been the most widely used mutagen; more than 90% of the mutant varieties were developed from mutants induced through gamma irradiation (http//www_mud.iaea.org). Irradiating dry seeds, pre-soaked seeds, seedlings, cuttings, young shoots, tissue culture and embryogenic cell suspensions are an accepted practice.

2) Soil-plant-water and Soil Fertility Research

Tracers afford a unique tool to investigate different aspects beyond the realm of conventional techniques of soil-plant-water continuum. Detection of water and mineral contents using neutron probe, NMR and neutron activated techniques are important analytical techniques to study soils. But more important use of tracers in soil and plant nutrition investigations is using both stable and radioisotopes. The criterion of an ideal tracer in such investigations must be physically, chemically and biologically indistinguishable from the tracer and must not disturb the soil-water-plant system (Ramachandran *et al.* 2007). The 'isotope effect' should be practically absent. Radioactivity in such studies should be low enough not to affect either the plants (the experimental material) or the experimenters. Normally, these criteria are easily met in practical situations of soil/crop research.

Mainstay of tracer studies used in soils is the principle of isotope dilution, propounded by Hevesy and Hobbie (1932) and the term first introduced by Rittenberg and Foster (1940). Dilution of the radioactive isotope by its non-radioactive counterpart results in reduction of specific activity, defined as the activity / unit volume or mass in a conserved manner proportional to original specific activity and amount of analyte (Fassett, 1995). When stable isotopes are used, the diluted natural isotopic ratio is related to concentration by the amount and isotopic composition of separated isotope that was added. Quantification of radioactive substances can be similarly done by "inverse dilution" principle where known amounts of inactive materials are added to samples.

The main advantage of the principle of isotope dilution is the luxury of using a non-quantitative isolation procedure in the separation of the analyte. This advantage provides considerable flexibility in design of separations, which can be optimized for simplicity, robustness, speed, cost or some other analytical parameters. The technique is highly accurate because there are minimal sources of potential systematic error.

The principal limitation, however, is availability of a suitable spike or tracers. The half-life and type of radiation emitted by a radio-tracer are very important as also its purity. The half-life must be long enough so that sufficient activity is available during analysis for good counting statistics. Too long half-life can be a problem because of low specific activities and storage and disposal problems. Fortunately, there are good number of radio- and stable-isotopes available to undertakes studies on soil-water-plant system.

Pioneering work carried out at the Isotope Laboratory of Indian Institute of Horticultural Research, Bangalore on both fruit and vegetable crops illustrates how purposefully the tracer techniques can be used to resolve challenges of horticultural crops.

Root Activity Studies in Fruit Crops

Efficient use of fertilizers is a function of both the chemical availability of the nutrient and the volume of soil explored by roots. Information on root activity is useful in standardizing appropriate time and method of fertilizer application, irrigation, planting distance and other cultivation practices (Bojappa and Singh 19730). Root activity studies in important fruit crops were carried out on a Typic Haplustalf having textural horizon (B_t) at 20cm depth. The deeper layers were hard to very hard with dry consistency, blocky structure and slow infiltration rate, all of which increased mechanical resistance to root growth during drier months. Soil injection of carrier-free ^{32}P was used to study root activity distribution in important fruit crops under rain fed conditions. The unique advantage of ^{32}P as a tracer for root activity studies is its relative immobility in soil that makes location of isotope application effective for measurement of root activity and its high mobility within the plant once it is absorbed such that the tracer becomes homogenously distributed within the plant in a couple of weeks after application. Its convenient half-life of ^{32}P of 14.3 days makes it possible for 4 quarterly injections when root activity is required to be monitored during annual growth cycle of perennial plants. Its amenability for Cerenkov counting making the detection convenient is yet another advantage. The results indicated that citrus varieties (*Kagzi* lime, Sweet orange cv. Mosambi, Coorg mandarin and Kinnow mandarin) were shallow rooted (Figure 1). In these soils, the fertilizer applied at shallow depth can be optimally absorbed as also drip irrigation/ fertigation can be efficacious. Both the grape varieties viz., 'Anab-e-Shahi' and 'Thompson Seedless', mango cv. 'Alphonso', guava cv. 'Arka Mridula' and 'Cricket Ball' sapota (Kotur, 2008) are deep-rooted in nature (Figure 2). These deep-rooted crops utilize native nutrients effectively from the entire rooting volume but show limited utilization of nutrients applied at shallow depth and of irrigation water. The spread of active roots in the fruit trees was uniform during North-East monsoon period due to uniformly moist soil water regime. During summer, the active roots moved towards soil surface and away from the trunk towards the periphery of drip-circle. Intensity of root activity (the total radioactivity recovered during the season) was the highest during South-West monsoon due to high volumetric soil moisture and decreased during winter as the rains receeded. The least intensity of root activity was observed during summer due to depletion of soil moisture. Generally, the period of high root activity alternated with that of shoot growth except in guava in which both the periods of high root activity and of shoot activity coincided. In papaya cv. 'Surya', under irrigated conditions, the active roots spread uniformly to a depth of 45cm and radial distance of 100cm from trunk. Both banana and papaya grown under irrigated condition were found to utilize fully the rooting volume both in terms of radial distance (subject to the spacing adopted) and the depth (45cm) fully and uniformly owing to congenial soil physical conditions under irrigation throughout the year. Therefore, these crops can be exhaustive on the native soil nutrients.

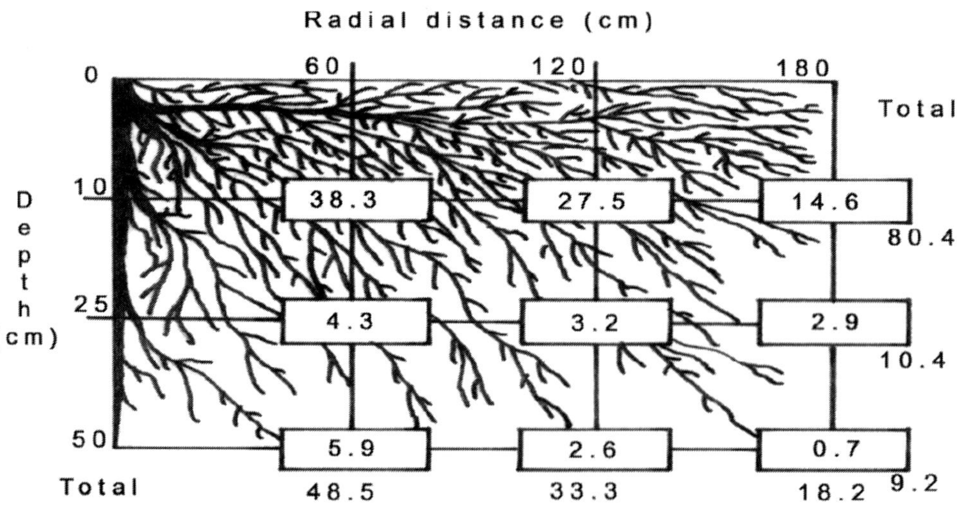

Figure 1 : Surface oriented root activity distribution (Example: 6 year old Coorg mandarin, Iyengar *et al.* 1988)

In the second phase of root activity studies, effect of various production practices on the former was evaluated since the tracer techniques are sensitive for this purpose too. The salient findings are:

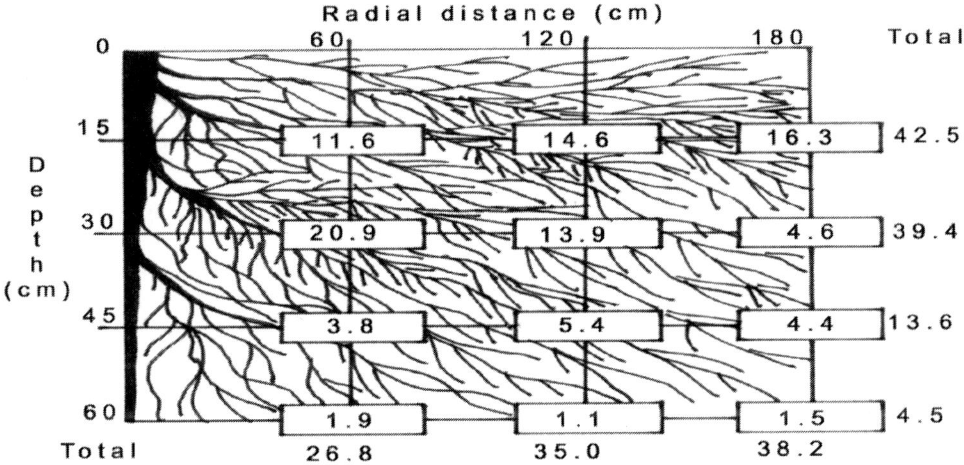

Figure 2 : Deep rooted root activity distribution (Example: 6-year old 'Arka Mridula' guava, Kotur *et al.* 1998)

(i) In 'Ganesh' pomegranate, the kind of planting material showed significant differences in root activity pattern. Air-layers showed predominantly surface oriented root system by confining 98% of active roots up to 50cm depth while seedlings extended roots up to 100cm depth. Conversely, air-layers extended roots to a wider distance of 75cm compared to seedlings (50cm). The seedlings showed better anchorage and were amenable for closer spacing (high planting density) than air-layers (Kotur amd Keshava Murthy, 2004).

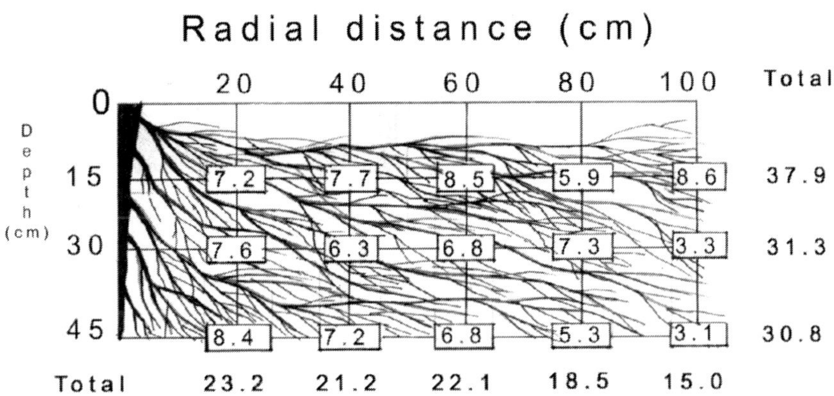

Figure 3 : Uniform root activity distribution (Example: One-year old 'Surya' papaya, Kotur and Keshava Murthy, 2001)

(ii) In 'Arka Mridula' guava trees of 16 years of age, the cultural practices of maintaining the plant bed showed distinct influence on root activity. Intensity of root activity was distinctly higher under polythene mulch followed by under green manure mulch and under non-cultivation. The depth-totals of root activity were substantially higher at 160cm distance under black polythene mulch during winter (50%) and under green manure mulch (41%). In rest of the treatments and seasons, 31-37% of active roots were present at this distance (Kotur, 2007a).

(iii) In 'Arka Mridula' guava, with increasing age, the active roots of guava trees extended sideways predominantly rather than deeper owing to anisotropism of the soil. In 7-year old trees, the bulk of the active roots existed up to 150cm radial distance while in 16-year old trees, the roots extended to distance of 240cm. As a result, to achieve high fertilizer use efficiency, the circular band of superphosphate had to be shifted from 100-160cm in the young trees to a much farther distance of 130-190 cm in older trees (Kotur, 2007b).

(iv) An assessment of root activity distribution of annona as influenced by rootstock-scion interaction showed that in general, seedlings of both *Annona squamosa* and *A. reticulata* showed higher intensity of root activity throughout than when they were grafted with 'Arka Sahan' as the scion owing to a smaller dilution of the isotope in a significantly larger volume of the latter. The root activity distribution of *Annona reticulata* tree closely conformed to that of *A. squamosa*. Grafts of both the species showed a relatively uniform distribution of active roots throughout the year compared to seedlings which promote a better anchorage, drought tolerance and utilization of native soil moisture and nutrients from the rooting volume (Kotur, 2009).

(v) Application of paclobutrazol, a growth retardant, applied to 20-year old 'Alphonso' mango trees to regularize fruit bearing, showed that the active roots tended to bunch close to soil surface and tree trunk compared to 'control' trees. This calls for rationalization of fertilizer placement to such trees to optimize fertilizer use (Kotur, 2006).

High statistical coefficients of variation associated with results of tracer assisted study of root activity distribution studies is a bane. Good comparison of the pattern of root activity distribution from tracer studies with those of the actual using excavation studies was observed (Purohit and Mukherjee 1974). Zapata (1990) listed several sources of error like, spatial variability of the soil across the plantation, plant variability related to genetic origin, sampling factors (type, size, time etc.), injection factors related to unequal probability of contact between root and applied isotope. It is essential to characterize and earmark a fairly uniform land for root activity studies and interpret the results keeping the soil characteristics in view. If the soil depth, shallow water table and physical properties are not a constraint for free growth of the roots, as perhaps in the alluvial soils, the rooting pattern of any plant may be as characteristic and unique as its shoot canopy. A pilot study to standardize the isotope to be applied to each tree to obtain substantial counts during detection of the tracer, the geometry of location of the points of injection for effective evaluation of rooting volume, procedure and frequency of sampling is essential before broaching on detailed studies. Once such protocols are standardized, the root activity distribution of a large number of crops can be studied quickly, efficiently and cost-effectively.

Appropriate Time and Placement of Fertilizers in Fruit Crops

Fertilizers are perhaps, the single costliest item in cost of cultivation of crops and account for 30-40%. Growers use fertilizers in ample measure, notwithstanding the cost, for the simple reason that commensurate responses of yield result from fertilizer application. Indiscriminate use of mineral fertilizers also leads to pollution of soil, water and atmosphere. Fertilizer use aims at supplying the required nutrient to the crop and not to soil. Fertilizer use efficiency is a quantitative measure of actual uptake of fertilizer nutrient by the plant/crop in relation to amount of nutrients added to soil. It is commonly expressed as:

$$\text{Utilization of added fertilizer (\%)} = \frac{\text{Amount of nutrient in the plant derived from the fertilizer}}{\text{Amount of nutrient applied as fertilizer}} \times 100$$

and is the key to achieve the highest possible yield with minimum fertilizer input. In other words, an efficient fertilizer management is the most cost-effective and eco-friendly.

Isotopic techniques are superior to conventional methods as they give direct and quantitative measure of isotope-labelled fertilizer nutrients utilized by the crop quite independent of native nutrients present in the soil. As a result, different fertilizer practices like placement, timing and sources can be evaluated efficiently. The most important parameter determined in this technique is the **f**raction of the nutrient in the plant **d**erived **f**rom the (labelled) **f**ertilizer **fdff**. This fraction is expressed as percentage i.e. fdff, %. Labelling or enriching of fertilizer with the desired isotope is indispensable for this work as the use of labelled soil is not satisfactory (Kotur, 2007b).

In case of ^{15}N studies,

$$\text{Ndff (\%)} = \frac{\text{Percent } ^{15}\text{N atom excess in plant sample}}{\text{Percent } ^{15}\text{N atom excess in labeled fertilizer}} \times 100$$

and in case of ^{32}P studies,

$$\text{Pdff (\%)} = \frac{\text{Specific activity of } ^{32}\text{P in plant sample}}{\text{Specific activity of } ^{32}\text{P in labeled fertilizer}} \times 100$$

are determined as a measure of **fdff**.

Fertilizer use efficiency in relation to application time and placement method were studied in fruit crops raised on red sandy loam (Typic Haplsutalf), weakly acidic in reaction having low cation exchange capacity under semi arid tropic conditions using ^{32}P-labelled superphosphate and ^{15}N-labelled ammonium sulphate. The fertilizers were placed in circular bands around the plant during different seasons and N and P derived from fertilizer (Ndff and Pdff) were determined over a period of time to monitor differences in absorption due to time and placement method. It was found that in citrus crops, there was better absorption of fertilizer during late rainy season when fertilizer was placed in a circular band between 105 and 135cm from trunk (Example: 6-7 years old plants, Figure 4). In grape, there was faster and greater uptake of fertilizer during the vegetative phase than that during reproductive phase. Fertilizer should be applied in two splits but within 15 days after each pruning in a band of 60cm width between radial distances of 60 and 120cm (Keshava Murthy and Iyengar 1992).

In 7-year old guava 'Arka Mridula', to achieve high use efficiency, P fertilizer may be applied between 100 and 220cm during July (Figure 5, Keshava Murthy and Kotur 1998) and in 12-years old mango 'Alphonso' the same may be applied between 60 and 150cm radial distance during July and November for better absorption (Figure 6, Keshava Murthy and Kotur 1999).

In 'Robusta' banana, the recommended P should be applied in 3 equal splits at planting, 8th leaf stage (45-50 days after planting) and at 16th leaf stage (90 days after planting). The recommended N and K should be applied in 4 splits at 8th, 16th and 25th leaf stages and at shooting (Keshava Murthy and Kotur 1998). During early stages (8th and 16th leaf stages) fertilizer may be applied in 25cm circular bands between 25 and 50cm radial distances (Figure 7) whereas during later stages in wider bands of 50cm width between 25 and 75cm radial distances (Figure 8). In 'Surya' papaya, it was the best to apply P fertilizer between 10 and 40cm during pre-bearing early vegetative stage and between 20 and 80cm during flowering and fruit-bearing stages (Figure 9, Kotur and Keshava Murthy 2003). On an overall basis, application of fertilizer during rainy season and placement within drip-circle always resulted in higher absorption of applied fertilizer. By proper placement, P use efficiency of applied fertilizer increased from 0.51 to 16.35% in 'Arka Mridula' guava and 2.90 to 17.70% in 'Robusta' banana. Fertilizer use efficiency of applied N in 'Robusta' banana increased from 6.8 to 60.0%. The dividends of improved fertilizer use efficiency by proper timing and placement based on these studies are deale elsewhere. These results are valid for practical management of other fertilizers as well since nutrient absorption presupposes prevalence of high root activity in the zone of placement at the time of fertilizer application.

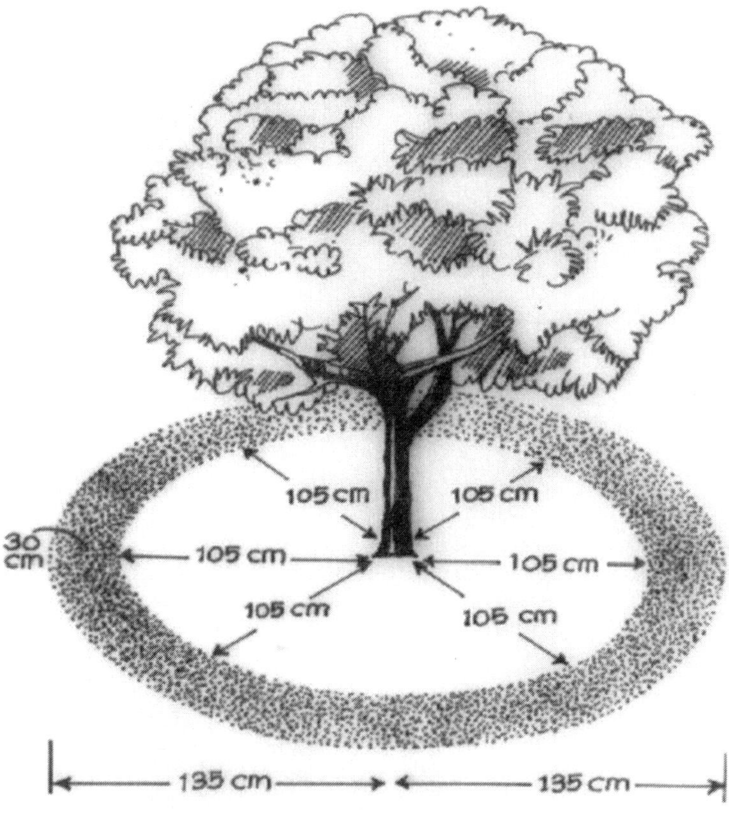

Figure 4: Proper placement of fertilizers in 6-7 years old citrus plants (Iyengar *et al.* 1996).

Figure 5: Proper placement of fertilizers in 7-year old 'Arka Mridula' guava plant.

Figure 6: Appropriate placement of fertilizers in 9-years old 'Alphonso' mango.

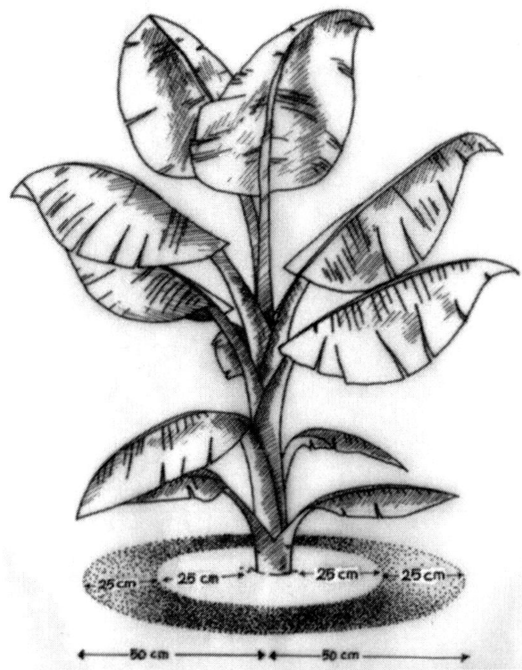

Figure 7: Optimum placement of fertilizers during early vegetative (50-100 days after planting, 8-16 leaf stage), phase.

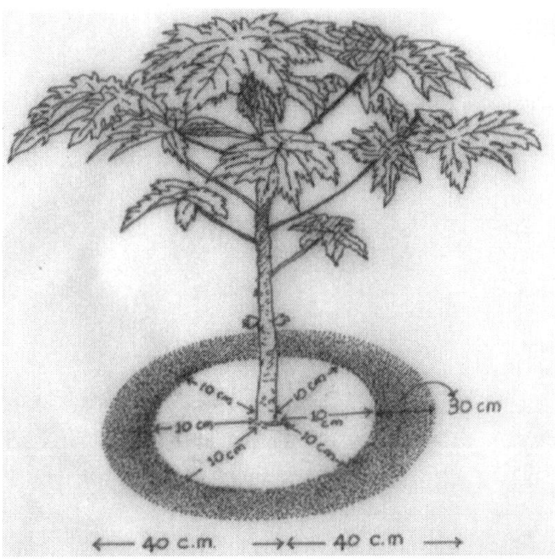

Figure 8: Optimum placement of fertilizers after late vegetative stage (beyond 100 days after planting)

Figure 9: Appropriate placement of fertilizers in 'Surya' papaya during pre-bearing stage (2-6 months after planting).

Fertilizer Management in Vegetable Crops

Efficiency of fertilizer nitrogen use was studied in 6 vegetables namely, tomato, onion, brinjal, okra, chilli and French bean using ^{15}N enriched fertilizer. Utilization of fertilizer nitrogen ranged from 13.9% in chilli to 44.8% in brinjal. Onion showed the highest N use efficiency ratio and physiological efficiency, whereas brinjal showed the highest agronomic efficiency (Iyengar and Shivananda 1992). Nitrogen balance studies in 6 vegetables showed that a large portion of applied N ranging from 31.7% in onion to 80% in French bean remained in the soil, the most of it in the top 30cm of the soil (Iyengar and Shivananda 1991). Loss of N fertilizer from the soil ranged from none in French bean to as high as 50.7% in onion. Recovery of residual N in the soil by the second and third crops following main crop ranged 0.9 to 5.1% and 0.6 to 6.1% respectively. Fertilizer placement studies using radioactive ^{32}P labelled fertilizer in French bean (Iyengar and Shivananda 1988) and okra (Iyengar and Shivananda 1990) showed that deeper placement at 5cm and 10cm depths respectively resulted in the highest utilization in these two crops. Similar studies carried out in brinjal (Iyengar and Shivananda 1990), tomato (Shivananda and Iyengar 1993), 2 cultivars of cabbage (Kotur *et al.* 1995), onion and chilli (Kotur *et al.* 1995) showed that placement of 5cm depth below the plants not only resulted in better utilization of applied P fertilizer but also resulted in substantial saving in the fertilizer P input to the extent of 20% in tomato and 40% in brinjal.

Nitrogen use efficiency in two vegetable cropping sequences showed that the total recovery of applied N was quite low in both the sequences- French bean-tomato-onion (22%) and okra-cabbage-brinjal (18.4%, Kotur and Keshava Murthy 1999). Recovery of fertilizer N applied to it by onion was substantially reduced by considerable recovery of residual fertilizer N left in the soil by previous crop of tomato. Nitrogen management studies in tomato indicated that a saving of fertilizer nitrogen input to the extent of 25% was possible by postponing the basal application after planting and applying the fertilizer in 3 equal splits within 30-35 days after planting (Shivananda *et al.* 1996).

Studies envisaged to develop models of maximizing N and P fertilizer use efficiency (Kotur *et al.* 2006) in (i) high yielding vs. hybrid variety of chilli; (ii) evaluation of vegetable cropping sequences of "okra-tomato-cabbage"; and (iii) evaluation of solo crops vs. crop combinations having enhanced cropping intensity in "capsicum(onion) - water melon(radish) – okra(French bean)" crop combination using isotopic techniques showed that:

- In chilli, the growth habit of the variety was decisive in the manipulation strategy of N management of N fertilizer. It was beneficial to apply full dose of N in 2 equal splits in 'Arka Lohit', a high yielding variety of chilli and increase the number of splits to 3 and applying as basal and 2 top-dresses in 'ECL' hybrid chilli. Any attempt to reduce N dose by deferring the application or increasing number of splits reduced fruit yield. In radish , raised after both the chilli crops, the highest overall recovery of residual fertilizer N was observed under 100% N dose applied to chilli and an overall of only 0.50-1.98% utilization of residual N was seen.
- "Cabbage-tomato-okra" was the most N efficient cropping sequence by showing 39.61% recovery of applied N; (ii) the recovery of N by cabbage was the best when grown in the first season and declined in the later 2 seasons continuously while those of tomato and okra did not show such a trend; (iii) cabbage as the second crop showed the highest recovery of residual fertilizer N of 2.27-4.61% compared to other crops; and (iv) the recovery of N by the third crop was meagre (0.12-0.33%).

- N use efficiency of all the 3 crop combinations was drastically reduced compared to either of solo crops. Recovery of residual N in the second and third crops was low and progressively reduced. In respect of P also, similar trend was seen but the utilization of P by crop combination was intermediate between those of the respective solo crops.

Identification of Nutrient Efficient Varieties of Scion and Rootstocks in Fruit Crops

Inter-specific and varietal differences in kinetic parameters (I_{max}, K_m and C_{min}) leading to different rates of absorption and their response to variations in rhizospheric nutrient concentration on marginal soil situations in respect of P were studied in important fruit crops using ^{32}P as the tracer. Rate of net inflow I_n was calculated by the formula (Neilson and Barber 1978):

$$I_n = \frac{I_{max} (C_s - C_{min})}{K_m + (C_s - C_{min})}$$

where I_{max} denotes maximum rate of nutrient absorption in pmoles/cm/s; K_m, the nutrient concentration at which inflow is ½I_{max} in pmoles/l; C_{min}, the nutrient concentration at which there is no net inflow in pmoles/l; and C_s, the equilibrium concentration of P (at 10µmole/l) in the solution. Although such studies were widely conducted on annuals and short-duration plants, it will be worthwhile to study the phenomenon in perennial fruit trees which conserve nutrients for future use to sustain a cyclic growth pattern (Keshava Murthy and Kotur 2000).

Among polyembryonic mango rootstocks, 'Peach' and 'Prior' may be highly responsive to applied P and likewise the scions budded on them in view of higher values of I_{max} observed . 'Nendran' banana, 'Bangalore Blue' grape and 'Citrumelo' may also be fertilizer responsive. Low values of C_{min} observed in 'Rough Lemon' rootstock of Citrus, 'Vellaikolamban' polyembryonic rootstock of mango and 'Anab-e-Shahi' variety of grape indicate that these rootstocks/varieties of crops may be grown successfully in marginal soils. On the other hand, most Citrus (rootstocks/scion) require a more fertile soil. Phosphorus inflow rate (I_n) which is comparatively high in polyembryonic mango rootstocks particularly in 'Olour', 'Peach', 'Prior' and 'Vellaikolamban' may have better P absorption efficiency and may respond better to its application. Amongst banana varieties, which also have high inflow rates 'Nendran' responds better to fertilization compared to others. Similarly, 'Bangalore Blue' grape responds better to P fertilization. All Citrus rootstocks, which incidentally showed the lowest P inflow rates, show poor response to P-fertilization.

In terms of suitability of a soil to a variety or response of a variety to applied P, the fruit crops and their varieties differ widely. These differences may partly be due to kinetic parameters of absorption or due to root morphology or ontogeny of the crop/variety. The ideal situation for a rootstock or variety would be to have a higher I_{max} and a lower K_m which make it both fertilizer responsive as also suitable to be grown under marginal soil environment. Verification trials in the field are required to confirm influence of these kinetic parameters of P absorption for adoption in future selection or breeding work.

Other Applications

Need based ingenious applications of tracer techniques can be made to address specific experimental requirements. Some instances are mentioned as examples.

Determination of Tree Volume of Fruit Trees using Isotope Dilution Technique

Similar to determining the volume of water in large water bodies using the principle of isotope dilution, ^{32}P was injected into the tree trunk and allowed to attain a complete distribution. Thereafter, its dilution was estimated as a measure of tree volume. The volume of 4-year old 'Alphonso' trees (56.903-68.981 $m^3 \times 10^3$) and biomass (321.70-36.23g × 10^3) showed good agreement with but generally overestimated both by 19% compared to actual values determined physically (43.720-60.693 and 22.88-37.78 respectively). The technique was applied to estimate tree volume of 15-year old 'Alphonso' mango and 12-year old 'Allahabad Safeda' guava trees and held promise in fertilizer use efficiency and nutrient dynamics studies in tree crops in which the total biomass of the tree was an important parameter.

Photograph 1. Enriched cow dung applied to the de-navelled distal end of bunch of 'Robusta' banana

Detection of Unorthodox Movement of Nutrients in Banana

The technology of enhancing size of fingers of banana in the bunch to suit the market demands by de-navelling and post-shooting feeding of N, K and S through the distal stalk-end of rachis was successfully developed (Kotur and Keshava Murthy 2008). The technique involves blending 15 g of (approximately 7.5g of urea) and 7.5 g of sulphate of potash dissolved in 100 ml water in 500 g of fresh cow dung and applying the slurry to the de-navelled stalk-end of 'Robusta' bunch soon after fruit set. About 10-15cm long rachis should be available after the last hand to tie the plastic bag (used milk bag is convenient) with a strong string (Photograph 1). Experimentally it was found that by this treatment, the bunch weight increased by 67% over 'control' in which the male flower was retained till harvest. Using ^{15}N-isotope label the movement of 51% of N into the bunch from ammonium sulphate blended in cow dung was confirmed. In 'Ney Poovan' banana, blending 5 g of ammonium sulphate (or 2.5 g of urea) and 2.5 g of sulphate of potash and applied similarly showed a response of 78% by absorbing 42% of N applied. The quality of banana of both the varieties was not affected in any way by these treatments.

Monitoring of Internal Dynamics of Water in Mango Fruits

Using ^{3}H in tritiated water, the internal movement of water was successfully monitored in investigations on mitigation of spongy tissue malady in 'Alphonso' mango (Kotur, Shivashankar and Ravindra 2008, unpublished).

3. Tracer Studies in Pesticide Biodegradation

Several pesticides are extensively evaluated for their degradation in the environment using labeled compounds. Common labels used for this purpose are ^{14}C, ^{35}S (in sulphur containing pesticides and ^{3}H). The benefits of labeled isotopes in pesticide research (Kale, 2007) are:

i) They help determine the extraction efficiency and efficiency of cleanup procedures in the laboratory.

ii) The path of individual atoms in the biological materials can be traced because of specific labeling sites.

iii) The detection limit is considerably lowered making residue analysis more accurate and complete. As a result, extremely low concentrations can be recognized and determined.

iv) The fate of labeled fertilizers can be followed specifically in presence of background levels of pesticides including the one being studied.

v) Co-chromatographic radio isotope techniques with non-radioactive carriers provide for identification of labeled metabolites or degradation products.

vi) Detection of soil-bound components of residues is possible.

In the present day scenario where massive and extensive doses of pesticides are used in horticultural production and the needs of keeping down the levels of pesticide residues, bioremediation of the contaminated environment and decontamination of fruits and vegetables prior to consumption, tracer studies are of paramount importance.

4. Management of Insect Pests Using Radiation

Tracer techniques are of use to study migration of insects, population dynamics and feeding habits in agricultural entomology. More significant achievement is the use of sterile insect technique (SIT) standardized and popularized by Bhabha Atomic Research Centre, Mumbai (Dongre, 2007). The principle underlying SIT is the genetic effects on insect reproduction in that the reproductive cells of insects are affected rendering them sterile. A continuous release of large number of mass production of sterile insects into the natural population would result in reduction in density of insect population. When assiduously followed, this can render concerned insect extinct. In practice, SIT augments integrated pest management (IPM) significantly. In the realm of horticulture, SIT is demonstrated to control fruit flies, codling moth (*Cydia pomonella*) of apples and pears and potato tuber month (*Phthorimaea opreulella* Zoller).

Possibilities of using tracer techniques in scientific horticultural research are immense in a multi-disciplinary mode. The present treatise provides only a partial glimpse of a vast scenario. The requirement of special design of the isotope-dedicated field and laboratories, trained staff and high cost make it a daunting task in the short run, but its multiple utility by various disciplines, quick and reliable results, when a focused programme is in place, make it a cost-effective and invaluable technique in the long run.

Examples of radio-sensitivity of tissue culture of fruit crops (Suprasanna *et al* 2007)

Species	Mutagen	Explants	LD50 (Gy)
Apple	Gamma rays	Leaves	10 - 20
Banana	Gamma rays	Shoot tips	20 – 30
Grape	Gamma rays	Shoots	20 – 40
Japanese plum	Gamma rays	Shoots	30
Kiwi fruit	Gamma rays	Leaves	40 – 60
Pear	X-rays	Shoots	30
Pear	Gamma rays	Leaves	20 – 30
Prunus	X-rays	Shoots	22 – 29
Strawberry	Gamma rays	Shoots	50

Important isotopes in soil-plant, water and fertility research

Elements	Isotope	Emission	Half-life/or (%) abundance	Measurement*	Use
Hydrogen	^{3}H	Beta	12.3 years	LS	Water movement, metabolism
Carbon	^{14}C	Beta	5720 years	LS	Photosynthesis, organic matter, C balance
Oxygen	^{18}O	Stable	0.204%	MS	Photosynthesis, respiration, hydrology
Nitrogen	^{15}N	Stable	0.366%	MS ES	Fertilizers, BNP, N balance
Phosphorus	^{32}P	Beta	14.3 days	GM Counter, LS	Fertilizers, root distribution, rock phosphates
Potassium	^{40}K ^{41}K	Beta Stable	1.3×10^{9} years 6.77 %	LS MS	Fertilizer, K balance
Calcium	^{45}Ca	Gamma	165 days	LS	Ion uptake, soil exchangeable Ca
Magnesium	^{28}Mg ^{26}Mg	Beta, Gamma stable	21.3 hours 11.29%	LS MS	Movement in plants, environmental pollution
Sulphur	^{35}S	Beta	87 days	LS	Soil availability, S cycle, uptake
Iron	^{55}Fe ^{59}Fe	Beta Gamma Beta	2.6 years 45.6 days	LS GS	Soil erosion, foliar nutrition, soil availability
Manganese	^{54}Mn	Beta	314 days	GS	Foliar nutrition, soil availability
Copper	^{64}Cu ^{65}Cu	Gamma Beta Stable	12.8 hours 30.9%	GS MS	Soil and plant movement
Zinc	^{65}Zn	Gamma, Beta	245 days	GS	Fertilizer, soil and plant movement
Molybdenum	^{99}Mo	Gamma, Beta	66.7 hours	GS	Plant movement
Sodium	^{22}Na	Gamma Beta	2.6 years	GS	Soil tolerance, cell permeability
Chorine	^{36}Cl	Beta	3.08×10^{5} years	GM, LS	Soil tolerance, solute movement in soils
Cobalt	^{58}Co ^{60}Co	Gamma Beta Gamma	71 days 5.3 years	GM GS	Enzymatic studies, vitamin metabolism, plant animal food chain
Selenium	^{75}Se	Gamma	120 days	GS	Animal nutrition, food chain
Chromium	^{51}Cr	Gamma	27.8 days	GS	Soil and plant movement

*GS=Gamma spectrometry; LS=Liquid scintillation; GM=Geiger Muller; MS=Mass spectrometry; ES=Emission spectrometry

Zones of high root activity of some important fruit crops

Crop/ Variety	Age (years)	Radial distance cm	Radial distance %	Soil depth cm	Soil depth %
		Surface oriented root activity distribution			
Citrus/ Coorg mandarin	6	120	80-95	0-15	78-88
Citrus/ Sweet orange	7	120	65-81	0-30	70-90
Citrus/Kagzi lime	4	120	75-80	0-25	94-97
		Deep rooted root activity distribution			
Mango/ Alphonso	8	100	75-80	0-40	69-75
Guava/ Arka Mridula	6	100	55-85	0-30	57-60
Sapota/ Cricket Ball	6	160	43-55	40-60	33-73
		Uniform root activity distribution			
Papaya/ Surya	1	100	100	0-45	100
Banana/ Robusta	7	0-25	85	0-20	64
	months	25-50	15	20-40	36
Banana/ Ney Poovan	7	0-25	76	0-20	82
	months	25-50	14	20-40	18

Scope for improvement of fertilizer use efficiency due to appropriate time and placement in fruit crops

Crop (Variety)	Age (Years)	Range of fertilizer use efficiency observed Least	Highest
		Phosphorus derived from fertilizer (Pdff, %)	
Sweet orange (Mosambi)	9	0.51	2.36
Mandarin (Coorg mandarin)	9	1.08	4.51
Grape (Thompson Seedless)	10	2.35	3.21
Mango (Alphonso)	9	1.32	4.85
Guava (Arka Mridula)	7	2.90	17.70
Papaya (Coorg Honey Dew)	6*	0.72	6.72
Papaya (Surya)	6*	0.83	10.27
Papaya (Surya)	1	5.63	12.96
		Nitrogen derived from fertilizer (Ndff, %)	
Banana (Robusta)	1	6.8	60.00

*Age in months

Kinetic parameters of phosphorus absorption

Crop variety/rootstock	Kinetic paramerters			
	I_{max} (pmoles/ cm/s)	K_m (µmole/l)	C_{min} (pmoles /cm/s)	I_n (pmoles /cm/s)
Citrus (pot grown in quartz sand)				
Rough lemon rootstock	0.057	8.46	0.65	0.030
Rangpur lime rootstock	0.147	37.70	1.94	0.026
'Cleopatra mandarin' rootstock	0.215	32.38	2.55	0.040
'Trifoliate orange' rootstock	0.281	184.90	2.98	0.010
'Troyer citrange' rootstock	0.226	34.60	2.60	0.040
'Citrumelo' rootstock	0.333	138.72	4.33	0.013
'Coorg manadarin' scion	0.369	37.44	1.27	0.014
Citrus (Field grown)				
Sweet orange scion on 'Rangpur lime' rootstock	0.171	26.87	2.45	0.038
'Coorg mandarin' scion on Rough lemon rootstock	0.123	27.22	1.32	0.030
Grape (Field grown)				
'St. George' rootstock	0.047	5.32	0.62	0.030
Thompson seedless scion	0.071	6.03	0.63	0.043
Anab-e-Shahi scion	0.075	5.60	0.44	0.047
Bangalore blue scion	0.125	5.46	0.48	0.080
Mango rootstocks (pot grown in quartz sand)				
'Kurukkan'	2.86	16.88	0.26	1.05
'Moovandan'	3.54	27.74	0.75	0.89
'Nekkare'	2.91	10.65	0.38	1.38
'Olour'	7.15	7.07	0.25	4.14
'Peach'	10.78	8.29	0.39	5.46
'Vellaikolamban'	8.98	10.02	0.18	4.44
Banana varieties (Field grown)				
'Dwarf Cavendish'	1.39	86.80	-	0.144
'Robusta'	6.95	115.00	-	3.021
'Nendran'	13.78	35.62	-	0.688
'Monthan'	3.88	46.61	-	0.881
'Elakki/ Ney Poovan'	5.61	53.69	-	0.881

Nitrogen utilization (%) in vegetable cropping sequences

Cropping sequence	Treated crop	Second crop	Third crop	Total
	French bean – tomato - onion			
F - T – O*	22.20	4.48	1.57	27.11
T - O - F	30.31	11.14	0.71	42.13
O - F - T	14.19	0.72	0.54	15.44
Mean				28.37
	Okra - cabbage - brinjal			
O - B - C	9.46	2.03	1.74	13.24
C - B - O	24.87	2.48	0.56	22.99
B - O - C	20.20	0.62	1091	22.77
Mean				21.82

*F=French bean; O=onion/okra; T=tomato; C-cabbage; B=brinjal

References

Bojappa K M and Singh R N. 1973. Studies on root activity and soil feeding zones of mango (*Mangifera indica* L.) using ^{32}P. *Newsletter, Indian Society for Nuclear Techniques in Agriculture and Biology* 2: 112-3.

Dongre T K. 2007. Sterile insect technique (SIT) for the management of insect pests of economic importance. *Isotope Application in Agriculture,* Indian Association of Nuclear Chemists and Allied Scientists (IANCAS) bulletin, VI(4): 354-62.

D'Souza S F. 2007. Radiation technology for the genetic improvement of crop plants at BARC: An overview. *Isotope Application in Agriculture,* Indian Association of Nuclear Chemists and Allied Scientists (IANCAS) bulletin, VI(4): 285-92.

Fassett J D. 1995. Isotope and nuclear and analytical techniques in biological systems: A critical study. Part X: Elemental isotope dilution analysis with radioactive and stable isotopes. *Pure & Applied Chemistry* 67: 1943-9.

Iyengar B R V and Shivananda T N. 1988. Utilization of fertilizer phosphorus by French bean (*Phaseolus vulglaris* L). as influenced by depth of placement and time of application. *Journal of Nucelear Agriculture and Biology.* 17: 156-61.

Iyengar B R V and Shivananda T N. 1990. Effect of placement and utilization of fertilizer P by okra (*Abelmoschus esculentus* Moench.) *Journal of Nuclear Agriculture and Biology.* 17: 148-51.

Iyengar B R V and Shivananda T N. 1991. Nitrogen balance and residual availability in vegetable cropping sequences. *Journal of Nuclear Agriculture and Biology.* 20: 134-9.

Iyengar B R V and Shivananda T N. 1992. Fertilizer nitrogen uptake and utilization efficiency in vegetables. *Journal of Indian Society of Soil Science* 40:469-73.

Iyengar B R V, Keshava Murthy S V and Shivananda T.N. 1988. Seasonal variations in root activity in mandarin (*Citrus reticulata* Blanco). *Journal of Nuclear Agriculture and Biology.* 17: 67-71.

Iyengar B R V, Keshava Murthy S V and Shivananda T N. 1996. Increasing fertilizer use efficiency in citrus. *Indian Horticulture* 41(4): 21-3.

Kale S P. 2007. Applications of radioisotopes in pesticide biodegradation studies. *Isotope Application in Agriculture,* Indian Association of Nuclear Chemists and Allied Scientists (IANCAS) bulletin, VI(4): 347-53.

Keshava Murthy S V and Iyengar B R V. 1992. Effect of time and method of placement on the absorption of fertilizer P by Thompson Seedless grape. Proceedings of the International Symposium on 'Recent Advances in Viticulture and Oenology', Hyderabad, February 14-15.

Keshava Murthy S V and Kotur S C. 1998a. Phosphorus absorption in guava (*Psidium guajava)* as influenced by time of application and placement method in a Typic Haplustalf. *Journal of Nuclear Agriculture and Biology.* 27: 117-21.

Keshava Murthy S V and Kotur S C. 1998b. Effect of ammonium sulphate and superphosphate on comparative efficiency of N and P absorption and N utilization in 'Robusta' banana *(Musa paradisiaca)* using labeled fertilizers. *Indian Journal of Agricultural Sciences,* 68: 765-8.

Keshava Murthy S V and Kotur S C. 1999. Appropriate time and placement of single superphosphate in 'Alphonso' mango (*Mangifera indica* L.). *Journal of Nuclear Agriculture and Biology.* 28: 107-10.

Keshava Murthy S V and Kotur S C. 2000. Phosphorous absorption in citrus (*Citrus* spp.), grape (*Vitis vinifera*), mango (*Mangifera indica*) and banana (*Musa paradisiaca*). *Indian Journal of Agricultural Sciences,* 70: 186-8.

Kotur, S.C. 2006. Effect of Paclobutrazol on root activity of mango *(Mangifera indica* L.). *Indian Journal of Agricultural Sciences* 76*:* 143-4.

Kotur S C. 2007a. Changes in the spatial root activity distribution and placement of superphosphate for maximum absorption as influenced by age of trees cv. Arka Mridula guava (*Psidium guajava* L.) *Acta Hort.* 735: 413-7.

Kotur S C. 2007b. Evaluation of orchard management practices of guava (*Psidium guajava* L.) for root activity distribution and appropriate placement of fertilizer using trace technique. *Acta Horticulturae.* 735: 419-25.

Kotur S C. 2008. Spatial and temporal root activity distribution of sapota (*Manilkara achras*) raised on an Alfislol. *Indian Journal of Agricultural Sciences* 79*:* 971-3.

Kotur S C. 2009. Spatial and temporal distribution of root activity of *Annona squamosa* (Sugar Apple) and *Annona reticulata* (Bullock's Heart) seedlings and their graft with 'Arka Sahan' scion determined using isotopic technique. *Indian Journal of Agricultural Sciences* 79*:* 422-5.

Kotur S C and Keshava Murthy S V. 2001. Spatial and temporal root activity distribution of papaya (*Carica papaya*) determined by isotopic technique. *Indian Journal of Agricultural Sciences* 71: 571-5.

Kotur S C and Keshava Murthy S V. 2003. Standarization of method of fertilizer placement in "Surya" papaya *(Carica papaya)* using ^{32}P-labelled superphosphate. *Indian Journal of Agricultural Sciences* 73: 269-71.

Kotur S C and Keshava Murthy S V. 2004. Influence of kind of planting material and planting density on partitioning of dry matter and of nutrients and root activity distribution pattern in "Ganesh" pomegranate. *Indian Journal of Agricultural Sciences.* 74: 2002-3.

Kotur S C and Keshava Murthy S V. 1999. Fertilizer management in vegetable crops. *Indian Horticulture* 43(1): 23-5.

Kotur S C, Iyengar B R V and Kacker N K. 1995. Effect of depth of placement and levels of fertilizer phosphorus on its utilization in two genotypes of cabbage *(Brassica oleracea* var *capitata). Journal of Nuclear Agriculture and Biology.* 24: 49-52.

Kotur S C, Keshava Murthy S V and Iyengar B R V. 1993. Spatial distribution of active roots in 'Arka Mridula' guava (*Psidium guajava)* as influenced by season, soil-moisture and growth. *Indian Journal of Agricultural Sciences* 68: 744-6.

Kotur S C, Iyengar B R V, Shivananda T N and Kacker N K. 1995. Utilization of fertilizer phosphorus by onion and chilli as influenced by depth of placement. *Indian Journal of Horticulture* 52: 206-9.

Kotur S C, Anjaneyulu K, Ramesh P R, Ramachandran V and Murthy N B K. 2006. Development of models of efficient fertilizer use in cropping systems of vegetables using isotopic techniques. Final Report. Board of Research in Nuclear Sciences, DAE, Mumbai.

Nielson N E and Barber S A. 1978. Differences among genotypes of corn in the kinetics of phosphorus uptake. *Agronomy Journal* 70: 695-8.

Purohit A G and Mukherjee S K. 1974. Characterizing root activity by radiotracer technique. *Indian Journal of Agricultural Sciences* 44: 575-8.

Ramachandran V, Srivastava M and D'Souza S F. 2007. Role of isotopes in soil-plant, water and fertility research. *Isotope Application in Agriculture,* Indian Association of Nuclear Chemists and Allied Scientists (IANCAS) bulletin, VI(4): 341-6.

Shivananda T N. and Iyengar B R V. 1993. Effect of placement on the utilization of fertilizer P by brinjal (*Solanum melongena* L.) and tomato (*Lycopersicon esculentum* Mill). *Journal of Nucelear Agriculture and Biology.* 22: 94-9.

Shivananda T N and Iyengar B R V and Kotur S C. 1996. Nitrogen management in tomato (*Lycopersicon esculentum* L.) using ^{15}N-enrich fertilizer. *Indian Journal of Agricultural Sciences* 66: 151-4.

Suprasanna P, Ganapathi T A and Kulkarni V M. 2007. Radiation induced mutagenesis of plant cell and tissue culture. *Isotope Application in Agriculture,* Indian Association of Nuclear Chemists and Allied Scientists (IANCAS) bulletin, VI(4): 330-40.

Zapata F.1990. Isotope techniques in soil fertility and plant nutrition studies. pp. 61-129 In *Use of Nuclear Techniques in Studies of Soil-Plant Relatioships.* International Atomic Energy Agency, Vienna.

The Science of Horticulture : Vol. 1, 2011
© K.V. Peter (ed.), pp. 89-111
New India Publishing Agency, New Delhi (India)
E-mail : newindiapublishingagency@gmail.com
Web: www.bookfactoryindia.com

Chapter

4 Principles of Plant Germplasm Conservation

DOIJODE, S.D.

Horticultural crops are ideal and valuable assets of nature to mankind. They are popular among masses since time immemorial but their importance was realized only in the recent past due to their high nutritive value, good taste and pleasing quality. They provide essential nutrients and minerals for proper functioning of human body and aesthetic surroundings for peaceful living. India is an important centre of diversity and several horticultural crops originated here. The horticultural crops constitute 13.08% of total crop area and contribute about 28% to GDP. The sub-continent extends from 8° 04′ - 37° 06′ North latitude and 68° 07′- 97° 25′ East longitude and it has tropical, sub-tropical and temperate climates which favour successful growing of a large number of horticultural crops. India has a wide variety of climate and soil on which a large number of horticultural crops like fruits, vegetables, potato and other tropical tuber crops, ornamentals, medicinal and aromatic plants, plantation crops, spices, cashew and cocoa are grown. Of late, major emphasis was laid on development of high yielding varieties and high production technologies and their adoption in areas of assured irrigation paved the way towards food security. However, gradually it became clear that horticultural crops for which the Indian topography and agro climates are well suited are ideal way of achieving sustainability of small holdings, increasing employment, improving environment, providing an enormous export potential and above all achieving nutritional security. As a result, due emphasis was given on diversification of horticultural crops only during the last one decade.

Germplasm provides base for sustained plant improvement and it comprises of genetic variability within and among species. Genetic resources of horticultural crops are distributed in the agro-ecosystem of the Indian sub-continent and controlled by the National Agricultural System. Indian Council of Agricultural Research (ICAR) is the nodal agency under the Ministry of Agriculture for collection, inventorization, conservation and documentation of horticultural genetic resources. Under ICAR, National Bureau of Plant Genetic Resources is the nodal agency for exploration, collection, evaluation, characterization, conservation and documentation of plant genetic resources. The Division of Horticulture under the ICAR system is the primary user and active custodian of horticultural plant genetic resources, under its Institutions like IIHR, CIAH, CISH, CITH, CPCRI, CTCRI, IISR, IIVR and National Research Centres for banana, cashew, citrus, grapes, onion and garlic, medicinal and aromatic crops, orchids and seed spices. The Indian gene centre is rich in genetic diversity of several horticultural species, both indigenous as well as introduced from exotic sources.

Genetic Diversity and Conservation

A wide genetic diversity exists in nature for different species of horticultural crops. It helps the plant to survive under heterogeneous environment. Also it provides raw material to plant breeders in evolving new varieties with objectives of improved yield, quality and resistance to various biotic and abiotic stresses. The increase in crop production is largely due to gene contribution from existing gene pool. Genetic uniformity makes a crop vulnerable to epidemics of insect pests and diseases, as it happened in past in several agricultural crops. Conservation of genetic resources is vital not only for crop improvement but also for the very survival of the species in time and space. Species may reproduce sexually or asexually or combination of the two. If sexual, it may be self or cross-pollinated and the level of which may vary with species and environment. Depending on mode of reproduction, the genetic composition of the population and individual plant differs and which in turn has a bearing on the germplasm conservation. Most perennial species undergoing domestication are still close to their natural evolutionary origins. Further, populations of many fruit species still exist in little changed native stands while a few selections were clonally propagated for cultivation. An important insight from early genetic-architecture studies was that large amounts and proportions of genetic variation are typically maintained within the most local populations of plants with respect to a wide range of both qualitative and quantitative traits. The need for germplasm conservation of horticultural crops is widely realized due to increasingly economic importance, narrow genetic base, gaps in existing collections and prevalence of genetic erosion. The genes, species, eco-systems and traditional knowledge and wisdom are being lost on increasingly accelerated pace. Habitat destruction, alien species invasion and genetic homogeneity in farming systems are the major threats to biodiversity (Swaminathan, 1999). There are two approaches to conservation of plant genetic resources, *ex situ* and *in situ*. *Ex situ* conservation approaches generally comprise of the following methods- seed storage, field gene bank, *in vitro* storage, DNA storage and botanical gardens. Conservation of plant diversity using reserve/protected areas, on farm conservation and home gardens are *in situ* conservation approaches.

The biodiversity is recognized *viz.,* ecosystems, species and genetic diversity (Anon., 1993). Higher diversity is found more among cross pollinated species than self pollinated species. Annuals, short lived perennials and endemic species normally have a lower genetic diversity than long lived perennials. Dicots generally have a lower genetic diversity than monocots. The genetic diversity occurred in cultivated species mainly due to Mendalian variation, inter-specific hybridization and polyploidy. Mendalian diversity arises basically from gene mutations and recombination resulting from hybridization. The inter-specific hybridization depends on the crossing of distinct taxonomic species with the preservation of important types from segregation products. The large number of gene differences are likely to occur in inter-specific hybrids leading to bewildering complexity in the segregating generations. Another source of variation arises through reduplication of chromosome sets. The normal plant has the diploid number of chromosomes ie. two complete sets of chromosomes, one derived from each of its parents. Misadventures in the reproductive processes may give rise to haploids with one set of chromosomes, triploids with three, tetraploids with four sets or even higher levels of polyploidy. The outstanding feature of sexual reproduction in cross-pollinated crops is the creation of an effectively infinite array of genetic diversity, even within populations and between siblings. In contrast, asexually propagated plants will only vary within and between lines by *de novo* mutations. These mutations will tend to accumulate within an asexual line and the only way, they can be lost is through natural selection. If they are recessive and slightly disadvantageous, they will tend to persist. In large random mating population with no selection, mutation or migration, the gene and genotype frequencies are constant from generation to generation.

A wide genetic base helps in development of new varieties for sustainable and high crop production. Genetic diversity is being used as source of genes in the crop improvement for production of high yielding varieties, hybrids and to effect ecologically sustainable economic and social development. The increase in crop productivity is largely due to gene contribution from existing gene pool. Breeder improves crop productivity through genetic introduction of traits. Gene identification and access to a diverse germplasm pool are crucial for the successful incorporation of novel traits in commercial horticultural crops. Most of breeding programmes focused mainly on aesthetic qualities causing thereby advertent erosion and narrowing of genetic diversity of many economically important horticultural crops. This narrowed genetic base increased the potential for epidemic diseases and reduced our ability for crop improvement. Unlike most food and other crops, the germplasm of horticultural plants are neither collected nor maintained suitably. Genetic uniformity makes a crop vulnerable to epidemics of pests and diseases as it occurred in past in several agricultural crops. The plant genetic resources have gained global attention in the recent past by agricultural and environmental institutions. The conservation and use of genetic resources attracted considerable attentions. India is one of the major centres of diversity in world and based on agro-climatic conditions and vegetation, it is divided into 8 regions namely Western Himalayas, Eastern Himalayas, Northeast regions, Gangetic plains, Indus plains, Western Ghats, Eastern Ghats and Islands regions. Rich diversity occurs in these regions for several horticultural plants and their wild relatives. Further continuous introduction of new crops from different parts of world helped in enriching the total genetic wealth. They show wider genetic variations for growth habit, flowering characteristics and resistance to certain insect pests, diseases and stress conditions, which have greater potential in breeding programme.

Over exploitation and replacement by hybrids and improved cultivars as well as exotics caused a greater harm to native genetic diversity. Three main processes *viz.*, genetic erosion, genetic vulnerability and genetic wipeout cause loss of genetic diversity of cultivated crop species. These are not mutually exclusive but are in fact interlocked by the demand of increasing population and the rising demands. Preservation of genetic diversity in plants requires effective application of storage technologies. The priority for conservation should be to those plants, having valuable heritable characteristics. Different storage technologies are presently used for existing collections of various horticultural crops. Conservation programmes are being implemented in India at genotypes, species and ecosystem levels. *In-situ* conservation of wild flora through protection of habitats and ecosystems is being implemented by the Ministry of Environment and Forests,Govt. of India. *Ex-situ* conservation of genetic variability of cultivated plants and their wild relatives are being followed in several horticultural crops based institutes, SAU's and NBPGR. Botanical gardens are being managed by Botanical Survey of India and other organizations to help in *ex-situ* conservation of endangered, threatened and rare plant species.

Principles of Gene Pool Sampling

The sampling strategy largely depends on the crop species, self or cross pollinated or vegetatively propagated and the extent of gene exchange between populations. Most species exhibit extensive geographical variation and superimposed on this is the variation within populations. Ecological habitat and/or factors are a major determinant of genetic diversity and agro-ecotypes are the most clearly distinguished in old cultivars. It is also realized that geographical variation patterns include such characteristics as disease resistance, morphological features and other conspicuous differences as well as variation in quantitative characters relevant to breeders. Though varieties or strains may look alike, they would differ greatly in useful attributes especially in physiological characters. Sampling methods may be used to ensure the collection of representatives within population variation as well as that associated with geographical patterns of variation. Plant populations that originated and continued to evolve under natural conditions without any interference from mankind are called wild forms or *taxa*. Further wild progenitors constitute important gene pools, especially for adaptation to stress conditions and resistance to pests and diseases, which could be shared by cultivated forms. Genetic differences between populations, particularly those of long lived perennials are usually the result of a long term evolution. Alleles which are common in a particular population can be very rare, even totally absent in another population. As in sampling of allelic diversity, in maintenance of populations, alleles which are locally common are critical since they represent the variation which makes populations adapted to their environment. In field gene banks, accessions collected from different environments are located for maintenance in one environment. Sampling of vegetatively propagated crops is more difficult than that of field crops. Random sampling of vegetatively propagated population in the field may not be effective because many populations originate from a single or a few clone. As it is impossible to conserve all possible alleles, it is also an impossible task to conserve all the genotypes of clonal species. Therefore, selective sampling to obtain distinct morphotypes from different collection sites would be more appropriate. Whenever possible, the collection of vegetative propagules should be supplemented with collection of seeds.

Germplasm Collection: The exploration and collection activities need intensification with the emphasis based on a) the priority crops of national importance b) under explored/un-explored areas c) wild relatives of crop plants d) endangered economic species e) genes for biotic and abiotic stresses and other specific traits in the hot spot areas of diversity f) under-utilized/ lesser known crops of tribal importance.

The priority crops and specific areas for collection:

Crops	Areas to be explored
Fruits	
Mango	Manipur, Uttar Pradesh, Himachal Pradesh, Bihar, Orissa, Maharashtra, Karnataka, Goa
Annona spp	Deccan plateau, Madhya Pradesh, Andhra Pradesh, Karnataka, Eastern Uttar Pradesh
Citrus spp	Sikkim, Assam, Meghalaya, Manipur, Mizoram, for hills of Himalaya, North Bengal
Musa spp	North east regions, West Bengal, Orissa, Maharastra, Tamil Nadu, and Kerala
Jackfruit	Bihar, Uttar Pradesh, Madhya Pradesh, Orissa, Tamil Nadu, Kerala, Goa, Andaman and Nicobar
Pomegranate	Jammu and Kashmir, Himachal Pradesh and Uttar Pradesh hills
Ber	Central and Eastern Uttar Pradesh, Maharastra, Goa, Karnataka
Amla	Uttar Pradesh, Madhya Pradesh
Pyrus and Prunus	Temperature regions of Jammu and Kashmir, Himachal Pradesh, Uttar Pradesh, Sikkim and North Bengal
Vegetables	
Cucumber	Indo-gangetic plains, sub-Himalayas tract, Western ghats, Eastern peninsular tract
Melon	Jammu and Kashmir, Himachal Pradesh, Uttar Pradesh, Rajasthan, Karnataka, Andhra Pradesh
Pointed gourds	Bihar, Bengal plains and Assam valley
Ivy gourd	Eastern HP, West Bengal, North Eastern region
Luffa, Lagenaria	Indo-gangetic plains, Terai region, North eastern plains
Bittergourd	Uttar Pradesh, Bihar, Kerala, Karanataka
Cucurbita, Benincasa, Cho-cho	North Eastern regions
Sweet potato	Bihar, Orissa, Uttar Pradesh and West Bengal
Colocasia	Kerala, Andhra Pradesh, Orissa, Bihar, Uttar Pradesh, West Bengal and North Eastern regions, Nilgiri and Annamalai hills

Need for Introduction from Exotic sources:

Name of crops	Sources
Fruits	
Mango	Australia, Philippines, Brazil, USA
Banana	Jamaica, France, Papua Guinea, Trinidad, Philippines, Cameroon
Citrus	Japan, China, Indonesia, Malaysia, USA, Spain, Brazil
Grapes	Russia, Afghanistan, France
Pomegranate	UAE, Iran, Afghanistan, Algeria, Cyprus, Iraq, Sudan
Ber	Africa, Argentina, Brazil, China, Mexico, Vietnam, USA
Date palm	Canary island, Sri Lanka, Malaysia, Philippines, Senegal, Taiwan, Algeria, Iraq, Saudi Arabia, Sudan, Tunisia
Annona sp	Brazil, Mexico, USA, Australia
Fig	Algeria, USA, Egypt, Saudi Arabia, Tunisia
Vegetable crops	
Phaseolus	Belgium, Colombia, USA, Germany
Vigna	Belgium, Philippines, Taiwan, Nigeria, USA
Potato	Peru
Sweet potato	USA, Taiwan, Japan
Allium	UK, USA, Hungary, Japan, USA
Capsicum	Costa Rica, Netherlands
Cabbage	Netherlands, UK, Germany
Radish	UK, Japan, Spain
Egg plant	Netherlands, USA
Okra	USA
Cucurbits	Philippines, USA, Spain
Tomato	Costa Rica, Germany, USA, Philippines
Beetroot	Germany, Sweden

Characterization : Characterization of elite materials based on morpho-taxonomic, biochemical, multimedia and molecular methods are helpful for the registration of important elite materials. Plant Genetic Resources programmes are expected to promote and facilitate the use of germplasm through the maintenance of sufficient healthy, readily accessible and adequately characterized and evaluated material as well as through proper documentation of relevant information. Besides being useful for analyzing genetic diversity, molecular techniques may play a major role in utilizing this diversity for plant improvement programmes in future. RFLPs are used for fingerprinting to generate genetic maps and to enable the identification of specific genotypes and horticultural traits. Gene mapping using RFLPs etc will promote better use of available diversity. The ease with which a genetic map can be developed and applied to a target crop depends on the genetic complexity of the species and the extent of DNA polymorphism present in the species. It is obvious that the application of modern molecular techniques will help germplasm curators enormously to increase significantly the knowledge base of individual accessions, if an entire crop species and of its gene pool are taken note of.

Documentation : The rising number and extent of collections as well as the amount and variety of the data accompanying the samples, documentation becomes an increasingly important requirement for the internal management and rationalization of collections in the conserving institutions. The same is true for collecting, storage and regeneration of material as well as the exchange of materials between institutions. Documentation of passport, characterization and evaluation data are necessary conditions for successful use of the collections by other interested group.

Genetic Basis of Germplasm Conservation

The main objective of genetic resource conservation is the maintenance of broad based genetic diversity of a species. It is necessary to reduce changes due to contamination through mutation, genetic drift etc. during process of regeneration. In case of asexual reproduction, it leads to perpetuation of the same genotype with great precision. It can be a great advantage since an indefinitely large number of genetically identical individuals can be obtained irrespective of the degree of heterozygosity of the genotype. However, when we go for budding or grafting, the influence of the rootstock should be considered as rootstocks are often raised from heterozygous seeds. The experience of plant breeding indicates that horticultural varieties of asexually propagated species are highly heterozygous and segregate widely upon sexual reproduction. Azygotic seeds like that of a pure line can be multiplied by growing only a few plants, whereas a heterozygous population would need to be multiplied from a much larger population and much care should be taken to ensure maintenance of genetic integrity. Seeds from one or a few plants are unlikely to contain a representative sample of the genes of the population and therefore cannot reproduce its structure. Thus seeds from many plants of a cross pollinated species should be collected in order to sample adequately its genes and retain its population structure. A sufficiently large effective population size must be preserved and whenever the population is regenerated, a sufficiently large number of plants must be grown to avoid the loss of genetic variation. The environmental conditions of the multiplication site should be near as possible to those under which the accession evolved or was cultivated for long period. The population under conservation may lose variability owing to natural selection, which is likely to lead to deterioration in the economic qualities of the crop. If germplasm is maintained correctly, the genetic constitution of the population at the time of its collection should be preserved. It is presumed that the original population was efficiently sampled and that the sample contains representative allelic frequencies and adaptive complexes, the maintenance of genetic integrity depends on how closely population genetic theory, biology, ecology and agricultural properties of the collections, the intended uses and the practices of multiplication are taken into account.

Native Horticultural Plants

Plant genetic resources are the most valuable and essential basic raw materials of crop improvement programme. A wider genetic base, thus, assumes priority in development of new crop plants for sustainable crop production. They are our heritage, which need conservation for posterity. Over 15000 species of higher plants occur in India, of which about 33% are endemic. These 4900 endemic species belong to 141 genera and 47 families (Nayar, 1980) and mainly distributed in the Himalayas, Southern Peninsular region and Northeastern region. Agro climatically and floristically, India is divided into 8 regions, 1) Western Himalayan, 2) Eastern Himalayan, 3) North-eastern, 4) Gangetic plains, 5) Indus plains, 6) Western Ghats, 7) Eastern Ghats and 8) Island region. Mankind has used only about 5000 plant species world wide to meet food and other needs. Today only about 150 plant species are important and meeting the food needs of mankind worldwide. Amongst horticultural crops, the widely cultivated crops comprise of banana, orange, apple, pear, mango, melon, tomato, cabbage, onion, squash, potato, sweet potato and beetroot.

India is one of the twelve centres/regions of diversity of crop plants in the World (Zeven and de Wet, 1982). Rich diversity occurs in several crop plants and their wild relatives, about 166 species of agri-horticultural crop plants and about 320 species of wild relatives (Arora and Nayar, 1984). In addition, the genetic diversity is enriched by continuous introduction of new crops and their cultivation from other parts of Globe. Introduced Crops are apple, pear, grapes, cherry, peach, apricot, potato, onion, cauliflower, cabbage, carrot and tomato. India is primary centre of diversity of crops like mango, jackfruit, banana, acid lime, phalsa, wood apple, jamun, bael, aonla, kokam, egg plant, cucurbits (Smooth gourd, ridge gourd, pointed gourds, cucumber, melon), rhododendron, orchids, and jasmines and secondary centre of diversity for crops like cow pea, cluster bean, okra, tomato, pumpkin, chayote, chilli and amaranths. The Indian gene centre assumes great importance for conservation and utilization of plant genetic resources. The process of indiscriminate exploitation of natural resources results in degradation, depletion or even destruction of natural environment. The genetic diversity is a source of gene(s) in crop improvement to produce high yielding varieties, hybrids etc for sustainable increase in food production, for food security and poverty-alleviation.

Need for Germplasm Conservation

Crop vulnerability and genetic erosion of horticultural plants are major concerns for plant breeders and growers. In the past 40-50 years, plant breeders have used a few wild species as a germplasm source but instead have relied on genetic variation within segregating cultivars (Ewart, 1981). Domesticated materials like breeding lines, clones and cultivars always remain an important germplasm source, but may not contain the genes necessary for pest resistance or improvement of other traits. It is likely that the genetic diversity found in wild or unimproved germplasm plays an increasingly integral role in breeding programme. Horticultural crops are susceptible to several insect pests and disease causing microorganisms. Further growers are constantly searching for environmentally safe and cost effective methods of pest control which are not damaging to the finished product or the customer. The several horticultural crops are cultivated in different parts of India for local consumption and for export purposes. Overexploitation and large scale replacement of present cultivars with hybrids and improved cultivars as well as greater introduction of exotics caused serious threat to native genetic wealth. The process of indiscriminate exploitation of natural resources results in degradation, depletion or even destruction of natural environment. Thus germplasm conservation is inevitable for crop security to meet any eventuality in future. Germplasm conservation benefits both horticultural industry and consumers by preserving unique genetic materials essential for present and future crops, resistant to pests and diseases, requiring a fewer economic inputs and promoting consumer-products appeal through expansion of crop diversity. The processes such as genetic erosion, genetic vulnerability and genetic wipeout are largely responsible for loss of genetic diversity of cultivated crops. These are not mutually exclusive but are in fact interlocked by the demand of increasing population and the rising demands. Genetic homogeneity enhances genetic vulnerability to various biotic and abiotic stresses. An ideal conservation programme demands combination of methods and technologies for effective maintenance of genetic diversity. It should be capable of retaining genetic stability, practicable and less expensive. The priority for conservation should be to those plants having vulnerable heritable characteristics and native ones.

It is always to have collection in more than two or three places for security and for ready utilization in near by places.

Germplasm Conservation

Several horticultural crops grown commercially are herbaceous or woody annuals and/or perennials. The genetic resources of horticultural crops have not received as much attention as that of food crops. These crops comprise of numerous *taxa* and different groups utilize its genetic resources. There are two main approaches for germplasm conservation *viz., in situ* and *ex situ* means. However, an ideal one should be the simple to follow, able to provide adequate viable materials, capable of maintaining genetic stability and economical to use. Conservation programmes are followed at genotype, species and ecosystem levels. *In situ* conservation largely confines to conservation of wild flora through protection of habitats and ecosystems while *ex situ* conservation is for cultivated plants and their wild relatives in different horticultural species.

Conservation Methods

In situ conservation: The *in situ* conservation approach to conservation, though the most desirable , has only recently received increased attention (Hodgkin *et al* 1993). It is well recognized that farmers are conserving high levels of diversity as part of traditional management practices. The conservation of eco system and natural habitats and the maintenance and recovery of viable populations of species in their natural surroundings and in the case of domesticated or cultivated species, in the surroundings where they have developed their distinctive properties are defined as *in situ* conservation (Anon., 1992). It has two approaches:

1. *Genetic reserve conservation* – The location, management and monitoring of genetic diversity in natural wild population within defined areas designated for active long term conservation.
2. *On farm conservation* – The sustainable management of genetic diversity of locally evolved traditional crop varieties with associated wild and weedy species or forms by farmers within traditional agricultural, horticultural or agri-silvicultural cultivation systems-. *In situ* conservation maintains both gene pool of target species and its eco system. It allows continuity of evolutionary processes so that the population continually adapts to the changing environment. It is beneficial especially in tropical perennial crops and recalcitrant seeded species.

In situ conservation provides the species or populations with an opportunity to evolve under natural conditions. One of the main reasons given for choosing *in situ* conservation over *ex situ* is the need to maintain the evolutionary potential of species and populations.

Home gardens: In most urban homes, one can find large collections of horticultural plant species, which not only beautify the environment but also conserve the genetic wealth. Although, the system is vulnerable to changes in management practices, it continues to play a significant role in genetic diversity conservation as well as development. Home gardens are similar to on farm conservation; however, the scale is smaller. In most rural situations, home gardens tend to contain more of a wide spectrum of species like vegetables, fruits, medicinal plants and spices than on farm plots. Home garden as a single unit, has a very little value in terms of conservation, but a community of them in a given area may contribute significantly to the conservation and direct use of genetic diversity. Most of such diversity could be somewhat unique/rare, as the people tend to grow unique materials in their gardens and also underutilized species or even un-domesticated species. However, the system is vulnerable to changes in management practices. In rural areas, the home garden will continue to play a role in genetic diversity conservation as well as development.

Ex situ conservation: *Ex situ* conservation conserves the biological diversity outside their natural habitats. Since the germplasm is grown outside its natural habitats, evolutionary process ceases increasing the risk of genetic drift. This system comprises various conservation methods like field gene bank, seed storage, botanical gardens, DNA storage, *in vitro* and cryopreservation.

Field gene bank: This is the most popular method for perennial germplasm conservation like field and ornamental woody species. Several horticultural plant seeds do not store well and their heterozygous nature invites to conserve as a living plant. Hence these are being conserved in field gene bank. They are prone to being damaged by natural calamities, infection, neglect or abuse. This is the most popular method for fruit germplasm conservation, which is readily usable by breeder and other researchers. In case of seed and *in vitro* conservation the genetic material cannot be used in the present form. Field gene bank will play a major role in conservation and utilization of fruit genetic resources. *Ex situ* conservation of perennial species requires more space and is relatively expensive to maintain. In view of advantages like easy and ready access to conserved material for use are being adopted. Normally three plants/ accession are conserved in field gene bank.

Seed conservation: Seed conservation is widely practiced for preservation of genetic diversity and it is popular amongst conservationist due to easy handling, economical and free from somaclonal variations. Seeds are readily equipped to survive for a period that varies in different species. As a living entity, it retains viability to a certain period and eventually deteriorates and dies. The process of seed deterioration is irreversible and cannot be totally eliminated. Further, loss of seed viability during storage can neither be completely stopped nor eliminated. The process of seed deterioration is minimal under proper storage conditions, thereby it lives for a longer period. Seeds of various horticultural plants are able to survive much longer under proper conditions (Doijode, 2001). Seed storage conservation method is an efficient and reproducible technique. The seeds are dried to lower moisture contents and stored at a low temperature. This method is almost universally applied to the orthodox seed species. It is well recognized that seeds should be maintained under conditions in which the life processes in seeds are minimized so that they can be stored for a longer period with little loss in genetic diversity, genetic integrity and viability. The limits are set for loss of viability as indicator of genetic shift in stored seed material and for regeneration in field. Due to the extended life span of seeds stored under optimum conditions, it will not be necessary to regenerate seeds at frequent intervals. Many horticultural plants produce seeds that are used for propagation. It contains valuable genetic material that controls various morphological, physiological and biochemical phenomena in plants thus serve as an important tool for the germplasm conservation. Seed quality represents genetic, physiological and physical characteristics of seeds. High quality seeds are genetically and physically pure, vigourous and free from insect pests and diseases causing organisms. Two important parameters namely seed viability and vigour determine the seed quality in different plants. It also ensures good seedling establishment and adequate plant population in field. High seed quality is desirable for crop improvement and production and also for seed germplasm conservation. Improper handling of seeds and storage under unfavourable conditions affect the seed quality. Pre-factors like stage of harvest, nutrition, irrigation and plant health also affect the seed quality and storability of seed lots to be conserved. Short term seed storage usually is conducted under ambient conditions, while medium term storage from 1 to 4°C and long term conservation from –10 to –20°C. Lowering the storage temperature to –20°C greatly enhances the longevity and is economic to achieve as well as use for effective conservation of large number of horticultural crops for a longer period. Some orthodox seeds are desiccation tolerant but are sensitive to low temperatures as seen in most of temperate fruit seeds. Several tropical fruit seeds exhibit recalcitrant nature, where life span ranges from a few weeks to several months (Roberts, 1973). They are sensitive to desiccation, show chilling injury on exposure to very low temperate and invade microorganisms readily at higher seed moisture during storage. Mango, jamun and jackfruit are the examples. The viability is retained under imbibed conditions at low temperature with assured supply of oxygen. Storability ranges from weeks to months under these conditions and are too short for adequate germplasm preservation (Roberts and King, 1981). Therefore, these species mainly are stored as vegetative plants.

Seed Preparations for Germplasm Conservation

Seed is a fertilized ovule botanically; dormant entity represents and holds a miniature plant, which emerges into a seedling under favourable conditions. It contains genetic material, that controls various morphological, physiological, and biochemical phenomena in plants. The genetic potential of genotype determines the plant characters, quality and productivity. Seed quality represents genetic, physiological and physical characteristics of seeds. It determines the crop performance and length of storage. The following aspects are considered while in storage.

Fruit Maturity: Seed quality is largely influenced by time of harvest of crop. Seed maturity normally coincides with that of fruit maturity. Matured seeds are selected for storage, which are heavier, lustrous with high germination and high seedling vigour. The maturity level affects seed germination and vigour.It determines seed longevity also(Eguchi and Yamada, 1958). Seed viability is higher at mature stage and it reduced on early or late harvest of crop. In certain horticultural crops like cucurbits, seeds continue to mature in fruit even after harvest. Further, storage of fruit for 4 days improved seed quality in bitter gourd (Krishnaswamy, 1991). Marrow seeds showed certain amount of dormancy, which was eliminated on or after ripening of seeds for 1-2 months (Ingold, 1960). Removal of seed coat also overcomes dormancy shortly after mass maturity (Demir and Ellis, 1993). Araujo *et al* (1982) reported that pumpkin seeds extracted from fruits 15 to 65 days after anthesis and stored for 7 weeks, seed germination and seedling vigour increased with increasing fruit age and storage length. Fruit turns to lighter in colour on maturity in different cucurbits. Fruits are left on vine for a comparable period till vine becomes dry. In cucurbits, seed coat completes growth earlier to embryo and that plays an important role in inhibiting germination of immature seeds. Decoating of seed increases germination (%) in immature seeds (Nerson, 1991). Fresh muskmelon seeds show dormancy and an after-ripening period is required at all stages of seed development. Seed germination increases in fully matured seeds after one year of storage (Welbaum and Bradford, 1991). In cucumber ,seed quality is poor in seeds obtained from immature fruits even after subjecting to post harvest ripening. Maximum seed quality is attained at the end of the seed filling phase in bean. The slow desiccation occurring during natural maturation is beneficial for development of high quality seeds. Seeds of early harvested pods (60 days after sowing) exhibited low germination (25%) and seeds harvested at later dates (90 days) showed higher germination (99%) (Pimentel and Miranda, 1982). Pods became brown or yellow on maturity in different legume vegetables and harvested in time to avoid shattering of seeds. The ripe pods are collected and threshed and the seeds are separated, cleaned, graded, dried and used for storage.

Fruits' position on mother plant: Fruits form in cluster or continuous on plant, and its number restricts on allowing fruits to ripe. Earlier formed fruits normally contain bold and relatively large seeds due to better source-sink relationship and adequate nutrients. Subsequently formed fruits were of smaller size with poor seed quality. Bold seeds with higher specific gravity produce healthy and vigourous seedlings and out yielded seeds of lower specific gravity (LeRon and Wyatt, 1977). Similarly higher seed germination and more vigourous seedlings were obtained from larger and heavier seeds (Doijode, 1984). A direct positive correlation existed between seed size and seedling size in French bean (Ries, 1971). In muskmelon, position of fruits on vine influences the seed quality. There was reduction in fruit weight, seeds / fruit, seed weight and viability with increasing fruit numbers on plant (Incalcaterra and Caruso, 1994). Seedling emergence is delayed and uneven by the presence of hard seeds as well as by slow imbibitions of moisture in certain seeds.

Plant Health: Seeds collected from healthy plants preserve well for a longer period. Plants are commonly infested with different insects and pathogens; some of them carry microorganisms thereby affecting the seed quality, productivity and storability. Infested seeds show discolouration, shriveling, low germination, low vigour and decaying. The activity of pathogen depends on level of injury to seed, availability of food materials, seed moisture, storage temperature and aeration that occur in storage. Seeds extracted from infected rotten fruits have inferior seed quality. The fungal pathogens are mainly responsible for seed deterioration and reduction of germination potential in stored seeds. Seed quality is improved either by providing unfavourable conditions to pathogens during storage or treatment with suitable pesticides.

Fruit Storage: Fruit storage is a low cost tool for seed preservation and it provides natural packaging for maintenance of seed quality and it is useful in certain horticultural crops for short term storage of seeds especially during plant/seed collection. In cucurbits especially gourds, seeds are left in fruit for a shorter period in order to improve and preserve seed quality. Matured fruits of marrow and watermelon were stored for 48 months and seed showed high viability and vigour (Nerson and Paris, 1988). Such fruits are easily handled and transported to distant places without affecting seed quality.

Seed Storage: Seeds are readily equipped to survive for fairly long period until the time and place are right for germination. As a living entity, they cannot retain their viability indefinitely and eventually deteriorate and die. Seeds of many horticultural species are able to survive much longer under proper conditions. Favourable conditions are created to minimize the deterioration, thereby seeds remain alive for longer period. The purpose of seed storage is to curtail the loss of viability and vigour of seeds during storage. Neither loss of viability during storage can be completely stopped nor the viability of seeds during storage can be enhanced. Seed deterioration is an irreversible process and cannot be stopped.

Seeds are preserved for a few days to years depending on need in different horticultural crops. Farmers desire to maintain high seed quality till next growing season while breeders aim at maintaining genetically viable and stable seed materials for very long period, avoiding repeated growing in field, which is a cumbersome and labor intensive process. The main objective of seed storage is to retain high seed quality. Seeds start deteriorating immediately after separation from mother plant. In an ideal storage method, process of deterioration is lowered so that seeds remain viable for relatively longer period. Seeds are broadly classified into orthodox and recalcitrant based on drying and storage behavior. Those seeds, retaining viability on drying and able to store at low temperature, are called orthodox seeds while those seeds whose viability is affected on drying and unable to preserve at sub-zero temperatures are called as recalcitrant. Seeds of many vegetable crops belong to orthodox group except seeds of *Sechium edule*, which exhibits recalcitrant storage behavior. Orthodox seeds are relatively easier to handle and capable of preserving high seeds quality in terms of viability and vigor for longer period. Quality of seeds is affected by changes in factors like seed moisture, storage temperature, relative humidity, and oxygen content during storage. Several insects and fungal pathogens affect the seed quality in storage by damaging embryo and causing discoloration and decaying of seed materials. Such seeds are unfit for sowing. The damage is more under greater insects or microbial population, higher seed moisture and higher storage temperatures. Insects feed on different kinds of seeds and damage the growing point. It is common in seeds of pod vegetables that are affected by pulse beetle during storage. Its activity is rapid at high seed moisture and temperature and effectively controlled by use of fumigants like ethylene dibromide or malathion. Microorganisms are present on surface or inside seed tissues. In onion, reduction of seed viability and vigor during storage are attributed to production of fungal metabolites. Seed treatment with fungicides effectively controles the infestation and helps in retaining high seed quality.

Sub-zero temperature storage: Handling of seed quantity is comparatively small in vegetables than field crops. They are packed either in moisture semi-pervious or impervious containers during storage. Commonly used containers are cloth, jute, paper and polyethylene bags and metal cans or laminated pouches. These are selected based on duration, quantity and storage conditions. Seed quality was preserved well under sealed conditions, thus well dried seeds with 5-7% moisture were packed in thick polyethylene bags or in laminated aluminium foil pouches and stored at low (5°C) or sub-zero (-20°C) temperatures for various period (Bass, 1980). Seed storage in sealed containers maintains seed quality by protecting seeds against extraneous environment like high humidity, insects and pathogens. Large quantity of seeds is normally stored under open conditions using dehumidifiers. Low temperature storage was effective in maintaining high seed quality by reducing the rate of respiration. The temperature in cold storage was selected based on the duration of storage material. Normally temperatures of 0 to 10°C are used for short to medium term storage and -20°C for long-term conservation. Seed treatment with chemicals or pesticides is avoided during seed germplasm conservation and also it is not required at very low temperature as conditions are not favourable for microbial activity.

Seed storability in different vegetables under ambient and low temperature storage

Name of crops	Storage period (yr) Ambient conditions	-20°C
Tomato	3	20
Brinjal	2	20
Chilli	1	20
Okra	2	25
Radish	1	25
Carrot	1	20
Onion	1	25
Amaranth	1	20
French bean	2	25
Garden peas	1	33
Broad bean	1	3
Hyacinth bean	2	5
Cluster bean	1	25
Cowpea	1	5
Muskmelon	2	20

Contd...

Watermelon	2	25
Cucumber	2	20
Pumpkin	2	25
Winter squash	2	20
Bottle gourd	2	15
Ash gourd	2	20
Ridge gourd	2	20

The retention of seed viability was greater at –20°C than at 5°C storage. Seeds stored in moisture impervious containers like laminated aluminium foil pouches preserve high seed quality both in terms of viability and vigour. Further, stored seeds did not show any morphological variations in seedlings and they were healthy and vigourous. Seeds of cowpea and hyacinth bean are severely damaged by pulse beetle under ambient condition. Insect activity is rapid in high moisture seeds at high temperature. Further occurrence of heavy rains during harvest increases the seed mycoflora and decreases germination %. Seed treatment with chemicals or pesticides is avoided during seed germplasm conservation and also it is not required at very low temperature as conditions are not favourable for microbial activity. Onion seeds are poor under storage in ambient conditions and are viable for 12 months. Thereafter seed quality declined rapidly. Seed quality was preserved for 25 years at sub-zero temperature (-20°C) storage. Similarly seeds of radish, pepper, okra, French bean, cluster bean, pumpkin, and watermelon were stored for 25 years and tomato, eggplant, amaranth, longmelon, bottle gourd and muskmelon for 15 years successfully without affecting the quality.

Modified atmosphere storage of seeds: Modified atmosphere storage of seeds is effective for short-term preservation of germplasm and beneficial particularly in absence of cold storage facility. Seeds are packed in laminated aluminium foil pouches in presence of partial vacuum or replacing air with nitrogen or carbon dioxide to reduce metabolism in seeds. The advantages of this method are limited to a few crops only and seed quality was further affected by presence of high seed moisture and storage under warmer conditions (Tao, 1992). High seed quality was preserved in seeds of garden peas, cluster bean, cowpea, pepper, bitter gourd, cucumber and French bean in carbon dioxide atmosphere while muskmelon, watermelon, okra, and onion seeds in nitrogen atmosphere. Seeds require oxygen for different physiological and biochemical processes. The rate of metabolism increases at higher level of oxygen content and results in greater chemical reactions, breakdown of food material and insects and fungal activity thereby promoting seed deterioration. The deleterious effects of oxygen are prominently expressed in presence of high level of seed moisture. Onion seeds stored in partial vacuum exhibited twice the % of germination than in air after 18 months of storage in ambient conditions. In peas, high oxygen content in environment affected seed quality and reduced storability of seeds. Cabbage seeds also maintain high seed quality under partial vacuum, while lettuce seeds did not perform well under vacuum conditions, which depict that response varies in different crops. As a result, watermelon seeds remain viable for 7 years and muskmelon seeds for 4 years under ambient conditions

Invigouration of stored seeds: Seed viability and vigour are reduced gradually during storage. The vigour can be revived to certain extent in low vigourous seeds. Mid-storage treatment of seeds with growth substances or hydration followed by drying or seed priming with chemicals improve seed quality (Savino *et al* 1979). It helps in usages of expensive, low vigourous seed material in crop improvement programme.

Moist storage: Seeds of *Sechium edule* show recalcitrant nature of storage. Seeds are viable for a very short period (2 weeks), loss viability rapidly on drying, and do not withstand low temperature for storage. The only alternative method available is the moist storage of seed. Here un-dried seeds are treated with fungicides to protect against pathogen and packed preferably in thin and perforated polyethylene bags and stored at 15°C (Doijode, 1995). It helps in preserving seed quality to certain extent and extends seed storability by some more weeks.

Storage of asexual propagules: Asexual propagules like cuttage, bulbs, corms etc are stored for a short period by placing them in media like saw dust at low temperatures (5°C).

Botanical gardens: Botanical gardens preserve aesthetes' beauty and genetic variability to limited extents. These gardens play a complementary role of conserving limited amount of living germplasm. It also maintains wild species of both commercial and non-commercial plants. A few of tropical fruits and ornamental plants are being preserved in this garden. The objective of most of the gardens are 1) maintain essential ecological processes and life support systems 2) preserve genetic diversity and 3) ensure sustainable utilization of species and eco system. The botanical gardens may play a limited role in the context of conservation and propagation and probably a greater role in public awareness and education. It displays a great number of different exotic species.

In-vitro conservation: Tissue culture offers great potential for the conservation of germplasm of vegetatively propagated materials and species with recalcitrant seeds. There are two major technical problems. First, the genetic instability of the materials conserved as tissue culture due to somaclonal variations at the time of regeneration of the tissue into seedlings. Second, the length of storage as tissue is limited. For the purpose of conservation of plant genetic resources, the growth of cultures is kept to minimum, if not completely arrested. This is essential to avoid frequent transfer to fresh media, which would require a high level of inputs and would make *in vitro* conservation expensive. There are several ways to achieve slow growth like 1) Use of immature zygotic embryos 2) modification of culture medium by adding osmotic or hormonal inhibitors or other growth retardants 3) reduction of storage temperatures for tropical species 4) mineral oil overlay 5) reduced oxygen tension 6) defoliation of shoots. *In-vitro* plants are useful for preservation of active collection. Disease free plants can be maintained with less cost. Germplasm can be exchanged easily. *In vitro* conservation also includes cells, tissues, buds or stem buds in culture. *In vitro* storage of plants is conducted either under normal growth conditions (20-25°C under lights, reduced temperature (-3 to 12°C with or without light) or on growth restrictive nutrient medium at ambient or reduced temperature. It preserves for a few weeks or months before sub culture is needed. Certain difficulties are faced in storage through *in vitro* application like culture establishment, micro propagation, rooting and acclimatization

Cryopreservation: In this, tissue or seed is conserved at ultra low (-196°C) temperature in liquid nitrogen, which arrests all metabolic activities resulting in storage of materials indefinitely. This method is particularly recommended for rare or endangered and recalcitrant seed species. Theoretically, cryopreservation *i.e.* using liquid nitrogen either by immersion or in vapour phase as a storage medium, is ideal for long term storage since it virtually suspends all the metabolic activities in any living tissue, be it seed, cell suspensions, callus, cultured tissue, pollen or shoot tip. Cryopreservation is the most successful for cell cultures since differentiated cultures may be damaged by cryopreservation due to structural injury. One promising method varying such effects is "Vitrification". This method involves immersion of the culture in a very high concentration of cryoprotectants solutions, followed by rapid cooling. Another promising method for the conservation of species that are clonally propagated or species with recalcitrant seed is the possibility of producing the synthetic seeds and conserving them as true seeds. This involves encapsulation of shoot tips and somatic embryos in semi-solid material, which serves as an artificial seed coat and endosperm producing 'beads'. Beads may also contain nutrients and pesticides. Somatic embryos and shoot tips are active and lack desiccation tolerance. Cryopreservation of meristem tips or buds is useful for clones of base collections since it provides a cheap and space saving method for long term storage. Matured buds are subjected to preservation in liquid nitrogen and thawed; the excised buds survive and can be grafted. Germplasm of apple, pear, *Prunus* spp, blue berry, gooseberries and currents are successfully preserved with this technique. The treated buds are grafted to suitable rootstocks to generate a shoot.

Pollen storage: Pollen is mainly stored for hybridization purposes for plants, which do not flower at the same time. It requires less space and can be stored at low temperatures. Its main limitation is that the whole genotype of a plant is not conserved.

DNA storage: Storage of DNA is, in principle, simple to carry out and widely applicable. The storage of DNA seems to be relatively easy and cheap. DNA libraries store total genomic information of germplasm. However, strategies and procedures have to be developed on how to use the materials stored in the form of DNA. Therefore role and value of this method for plant genetic resources conservation are not completely clear yet.

Conclusion

A single method is not useful for effective conservation of gene pool, however it has to be complemented with other conservation strategies. Various methods of conservation, along with seed and *in situ* conservation can be combined into a complementary conservation strategy for any given gene pool. Conserving gene pool should employ a combination of methods from nature reserve to gene banks as no single method can conserve all the diversities. The appropriate balance between different methods employed depends on factors like biological characteristics of the gene pool, infrastructure and human resources, number of accessions in a given collection and their geographic sites and the intended uses of conserved germplasm.

References

Anonymous, 1992. Convention on biological diversity. Environment Law and Institutions programme Activity Centre UNEP.

Anonymous, 1993. Diversity for development, the strategy of the International Plant Genetic Resources Institute IPGRI, Rome.

Araujo, E.F., Mantovani, E.C. and Silva, R.F.D. 1982. Influence of fruit age and storage period on squash seed quality. *Revista Brasileira de Sementes*, 4:77-87.

Arora, R.K and Nayar E.R. 1984. Wild relatives of crop plants in India NBPGR Sci Mongr No. 9 p 90

Bass, L.N. 1980. Seed viability during long term storage. *Hort Rev.* 2:117-141.

Chandel, K.P.S. and Agrawal A. 1996. *In vitro* conservation and cryopreservation of fruit germplasm. (In) Genetic Resources of Tropical Fruits (Eds Gupta P N, Mathura Rai and Chandel K P S) NBPGR. New Delhi pp 145-152.

Demir, I. and Ellis, R.H. 1993. Changes in potential seed longevity and seedling growth during seed development and maturation in marrow. *Seed Sci. Res.*, 3:247257.

Doijode, S.D. 1984. Effect of seed size on the longevity of seeds in French bean. *Singapore J. Pri. Indust.*, 12:62-69.

Doijode, S.D. 1995. Short-term conservation of mango seeds. *FAO/IPGRI Plant Genetic Resources Newsletter*, 104:24-25.

Doijode, S.D. 2001. Seed Storage of Horticultural Crops. Food Product Press, New York USA.

Eguchi, T. and Yamada, H. 1958. Studies on the effect of maturity on longevity in vegetable seeds. *Natl. Inst. Agr. Sci. Bull. Ser. E Hort.* 7:145-165.

Ewart, L. 1981. Utilization of flower germplasm. *HortScience* 16:135-138.

Hodgkin T.H., Ramanath Rao V and Riley K.W. 1993. Current issues in conserving crop land races. On Farm Conservation Workshop Bogor, Indonesia.

Incalcaterra, G. and Caruso, P. 1994. Seed quality of winter melon (*Cucumis melo* var. *indorus*) as influenced by the position of fruits on the mother plant. *Acta Hort.* 362:113-116.

Ingold, M. 1960. Contribution a lethde de la germination des Semences d' *Allium cepa* L, et *Cucurbita pepo* L. *Proc. Inter. Seed Test. Assoc.*, 25:787-799.

Krishnaswamy, V. 1991. Post-harvest seed maturation in bitter gourd. *J. Appl. Seed Prod.* 9:41-43.

LeRon, M. R. and Wyatt, J.E. 1977. Specific gravity of snap bean seeds: Influence on quality attributes. *HortSci.*12:233

Nayar M.P. 1980. Endemism and patterns of distribution of endemic genera. *J. Econ Tax Bot* 1 99-110.

Nerson, H. 1991. Fruit age and seed extraction procedures affect germinability of cucurbit seeds. *Seed Sci. Technol.*, 19:185-195.

Nerson, H. and Paris, H.S. 1988. Effects of fruit age, fermentation and storage on germination of cucurbit seeds. *Scientia Hort.* 35:15-26.

Pimentel, M. DeL. and Miranda, P. 1982. Physiological maturation and conservation of bean (*Phaseolus vulgaris*) seeds. *In Anais I reuniao nacional de pesquisa de feijao.* 7: 333-335.

Ries, S.K. 1971. The relationship of size and protein content of bean seeds with growth and yield. *J. Am.Soc.Hort.Sci.* 96:557-560

Roberts E. H. 1973. Predicting the storage life of seeds. *Seed Sci. Technol.* 1:499-514.

Roberts E. H. and King M.W. 1986. Storage of recalcitrant seeds. In crop Genetic Resources- The conservation of different material (Ed Withers L.A. and Williams J.T.) IBPGR Paris.

Savino, G., Haigh, P.M. and Leo, P.De. 1979. Effects of presoaking upon seed vigour and viability during storage. *Seed Sci. Technol.* 7:57-64.

Swaminathan M.S. 1999. Government – industry- civil society partenship in integrated gene management. *Indian J. Plant Genetic Resources* 12:277-289.

Tao, K.L. 1992. Should vacuum packing be used for seed storage in gene banks? *Pl Genetic Resources Newsletter* No. 88/89:27-30

Welbaum, G. and Bradford, K.J. 1991. Water relations of seed development and germination in muskmelon (*Cucumis melo*), VII. Influence of after-ripening and ageing on germination response to temperature and water potential. *J. Exp. Bot.* 42:1137-1145.

Zeven A.C. and Wet J.M.T. 1982. Dictionary of cultivated plants and their regions of diversity. Wageningen 259.

The Science of Horticulture : Vol. 1, 2011
© K.V. Peter (ed.), pp. 113-143
New India Publishing Agency, New Delhi (India)
E-mail : newindiapublishingagency@gmail.com
Web: www.bookfactoryindia.com

Chapter 5 Advances in Research on Male Sterility

DHALL, R.K.

Vegetables are the most extensively utilized plants for exploitation of heterosis for development of hybrids. As the importance of heterosis breeding increased, male sterility proved an asset particularly in crops like onion and carrot which produce many but small sized flowers making hand emasculation tedious. Large scale hybrid seed production sometimes remains handicapped because of high labour cost, unavailability of trained labour at crucial times i.e. at blooming, bad weather conditions like continuous rains etc. The phenomenon of male sterility has always been of long term interest for researchers of various disciplines of applied, strategic and basic sciences. It is of special interest for plant breeders to produce more efficient and economic hybrid seed. Discovery of certain male sterile mutants which eliminate more laborious operations of emasculation combined with various marker genes further facilitate identification of undesirable types even at seedling stage has widened the very basis of hybrid seed production. This technique has also reduced the total cost of hybrid seed production. Onion crop provides one of the rare examples of very early recognition of male sterility (Jones and Emsweller 1936), its inheritance and use in hybrid seed production (Jones and Clarke 1943). Since then male sterility is reported in fairly large number of crops including vegetables. These male sterile plants were either isolated in natural populations or were artificially induced through mutagenesis (Kaul 1988). In recent past, male sterility systems are also developed through engineering (Williams *et al.* 1997) and protoplast fusion (Pelletier *et al.* 1995). Presently crops like muskmelon and chilli are very successful examples of utilization of male sterility system in India (Kalloo *et al.* 1998).

Several compiled reports on the availability, characterization, mechanism and utilization aspects of male sterility in vegetable crops are available (Kalloo 1988, Kalloo and Bergh 1993), but critical review on vegetable crops is not available. An attempt is made to illustrate different aspects of male sterility and critically review the reports on male sterility mechanism in vegetable crops. Special emphasis is being laid on the review of the reports on identification, characterization and utilization aspects of male sterility in tomato, chilli, cole vegetables, carrot, onion and muskmelon.

Classification of Male Sterility

Kaul (1988) classified male sterility in to two major groups *viz.,* genetic (spontaneous or induced) and non-genetic (induced) male sterility. On phenotypic basis, genetic male sterility is classified in three classes i.e. sporogenous, structural and functional. Similarly, non-genetic male sterility is classified as chemical, physiological and ecological male sterility. Further, on genotypic basis genetic male sterility was grouped as genic, cytoplasmic and gene-cytoplasmic male sterility (Kaul 1988). In this article only genetic (inherited) male sterility is described.

1. Genic Male Sterility (GMS) or Nuclear Male Sterility

GMS is reported in about 175 plant species (Kaul, 1988) including important vegetable crops. As the name suggests, this type of male sterility is controlled by the gene(s) from the nuclear compartment. Most of the naturally occurring or induced male sterile mutants are recessive in nature with a few exceptions in cole vegetables (e.g. cabbage and broccoli) and genetically transformed male sterile lines (Kaul 1988, Williams *et al.* 1997). Occurrence of predominantly recessive male sterility clearly indicates that GMS is the result of mutation in any gene(s) either controlling microporogenesis (pollen development process), stamen development or micro-gametogenesis (male gamete development process) and accordingly the phenotype of GMS mutants may vary.

All the transgenic male sterile lines developed till date are GMS, since they are developed through transformation of male sterility causing gene construct(s) inside the nuclear genome. Certain mutants, which although produce functional pollen, fail to self fertilize, either due to non-dehiscence of pollen or their special flower morphology. These mutants are often termed as functionally sterile, for example genotypes with exerted stigma in tomato (Georgiev 1991), brinjal and several other vegetables (Kaul 1988). Discussion of such functional sterile mutants is beyond the scope of this article, however,a few of them are mentioned under the respective crop section. Genetic male sterility is subdivided in 2 broad groups: (i) Environment sensitive genetic male sterility (EGMS) (ii) Environment insensitive genetic male sterility (GMS). In addition, genetic engineering has produced a novel type of dominant genetic male sterility referred to as transgenic male sterility.

EGMS : In EGMS, ms gene expression occurs within specified range of temperature and photoperiod regimes. Certain GMS lines are conditional mutants, meaning thereby, in particular environment male sterile mutant plants turn into male fertile. After determination of critical environment (usually temperature or photoperiod) for sterility and fertility expression, such EGMS is further divided into two groups i.e. (i) Temperature sensitive genetic male sterility (TGMS) (ii) Photoperiod sensitive genetic male sterility (PGMS). EGMS lines are reported in several vegetable crops. A majority of these, however, were previously identified as normal genic male sterile lines. Initially EGMS lines were of very less practical value, as they were unstable. But now they represent the most efficient system for hybrid seed production. From practical viewpoint, it is necessary to identify critical temperature or photoperiod for the fertility/sterility expression in temperature and photoperiod sensitive genetic male sterility, respectively.

Environmental sensitive male sterile mutants

Vegetables	Mutants	References
Cabbage	TGMS, PGMS	Rundfeldt 1961
Brussels sprout	TGMS	Nieuwhof 1968
Broccoli	TGMS	Dickson 1970
Pepper	TGMS, TCMS*	Daskalov 1972, Shifriss 1997*
Carrot	TGMS	Kaul, 1988
Tomato	TGMS	Sawhney 1983

* TCMS- Thermo sensitive cytoplasmic male sterility

Transgenic Genetic Male Sterility : A gene introduced into the genome of an organism by recombinant DNA technology or genetic engineering is called *transgene*. Many transgenes produce genetic male sterility, which is dominant to fertility. Consequently, it is essential to develop effective fertility restoration system if these are to be utilized for hybrid seed production. An effective restoration system is available in at least one case called *Barnase/Barstar* system.

The *Barnase* gene of *Bacillus amyloliquefaciens* encodes an RNAase. When *Barnase* gene is driven by TA29 promoter, it is expressed only in tapetum cells causing their degeneration. Transgenic tobacco and *Brassica napus* plants expressing *Barnase* were completely male sterile. Another gene, *Barstar,* from the same bacterium encodes a protein, which is a highly specific inhibitor of *Barnase* RNAase. Therefore, transgenic plants expressing both *Barstar* and *Barnase* are fully fertile.

The *Barnase* gene is tagged with *bar* gene, which specifies resistance to be herbicide phosphinothricin. This male sterile line is maintained by crossing it with a male fertile line. The progeny so obtained contains 1 male sterile: 1 male fertile plant; the latter is easily eliminated at seedling stage by a phosphinothricin spray. The male sterile plants are crossed with the *Barstar* line to obtain male fertile hybrid progeny (Fig 1). This system of male sterility is yet to be commercially used.

BARNASE-BAR/- X
Male sterile, phosphinothricin resistant ↓ Normal fertile,phosphinothricin sensitive

50% BARNASE-BAR/- : 50% -/-
Male sterile, herbicide resistant : Male sterile, herbicide sensitive

↓ Phosphinothricin spray eliminates male fertile plants

BARNASE-BAR/- X
Male sterile ↓ BARSTAR/BARSTAR male fertile restorer

50%- BARNASEBAR / BARSTAR : 50%/ -BARSTAR
Male fertile (BARSTAR product inhibits BARNASE RNAase) : Male fertile

Figure 1 : Male sterility due to the bacterial gene *barnase* and its restoration by another bacterial gene *barstar,* Gene *barnase* produces RNAase, which kills the cells expressing this gene, while *barstar* encodes a specific inhibitor of this RNAase. Herbicide resistance gene bar is linked with *barnase* to eliminate the male fertile plants from the male sterile line.

Utilization of GMS: First step in the utilization of GMS system is seed multiplication of male sterile line, which is not possible through selfing or sibing, since pollen is either not viable or not produced. Commonly utilized monogenic recessive GMS (*msms*) is maintained by back crossing it with heterozygous isogenic line (*Msms*) for male sterility. Therefore, in hybrid seed production field, 50% male fertile segregants (*Msms*) need to be identified and removed before they shed pollen (Figure 2). Hence the stage of identification of male sterile/fertile plants in hybrid seed production field is very important. In general, male sterile plants are morphologically not distinguishable from the sister fertile plants, except in a few cases, where male sterile flower size is smaller than that of fertile flowers e.g. tomato (Sawhney 1997) and chilli (Hundal and Khurana 1993). In certain GMS lines, *ms* gene is tightly linked with the recessive phenotypic marker gene. Such marker genes especially those that express at seedling stage are good proposition for the identification of sterile/fertile plants at seedling stage. Thus labour and time required for identification and removal of fertile plants in hybrid seed production field can be avoided. Removal of fertile plants before transplanting also facilitates proper production management especially the maintenance of female plant (male sterility) population per unit area increased productivity of F_1 seeds.

For hybrid seed production through EGMS line, seeds of EGMS line can be multiplied in an environment where it expresses male fertility trait while hybrid seeds can be produced in other environment, where it expresses male sterility. Male (pollen) fertility in hybrid crop is not affected, as male parent contributes normal (wild) allele of the environmental sensitive mutant gene.

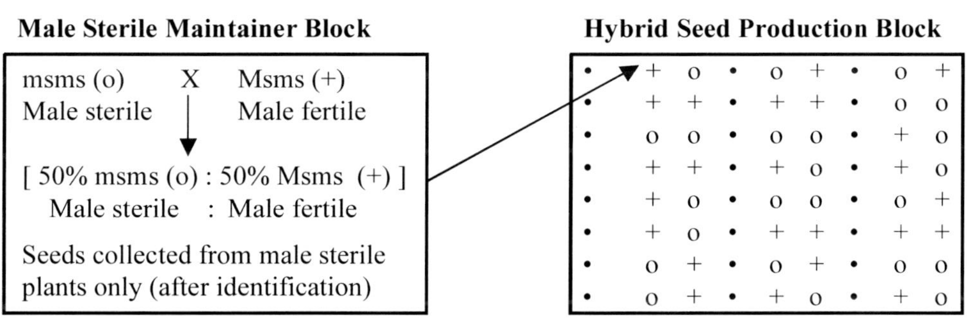

Male parent (•)
Male fertile plants (+) in the female rows are removed after identification and seed harvested from male sterile (o) plants are hybrid seeds.

Figure 2 : General scheme of hybrid seed production utilizing monogenic recessive GMS

Limitation of GMS: Because of more tedious maintenance process and non-availability of suitable marker genes among the vegetable crops, GMS is utilized commercially only in chilli and muskmelon. Moreover, a number of crops e.g. tomato, brinjal, pea etc. which are highly self pollinated, free out crossing is prohibitive thus leading to poor seed and/or fruit set. This method could gain popularity in practice only if suitable insect pollinators or other means are found to raise the % of cross pollination.

Linkages of ms gene with marker gene in vegetables

Vegetable	Marker gene for	Reference
Broccoli	Bright gene hypocotyle	Sampson 1966
Tomato	Potato leaf shape and green stem colour	Kaul 1988
	Parthenocarpic fruit	Soressi and Salamini 1975
	Enzyme marker	Tanksley *et al.* 1994
Watermelon	Delayed green seedling	Zhang *et al.* 1996

2. Cytoplasmic Male Sterility (CMS)

This type of male sterility is determined by cytoplasm. Since the cytoplasm of a zygote comes primarily from egg cell, the progeny of such male sterile plants would always be male sterile. Cytoplasmic male sterility can be transferred easily to a given strain by using that strain as a pollinator (recurrent parent) in the successive generations of backcross programme. After 6-7 backcrosses, the nuclear genotype of male sterile line would be almost identical to that of the recurrent pollinator strain.

Utilization of CMS: Cytoplasmic male sterility can be maintained by crossing a male sterile line (A line) with the pollinator strain (maintainer line) used as the recurrent parent in the backcross programme since the nuclear genotype of the pollinator is identical with that of the new male sterile line. Such a male fertile line is known as maintainer line or B line as it is used to maintain the male sterile line. The basic method of utilization of a cytoplasmically male sterile parent for hybrid seed production is illustrated in Fig-3. Cytoplasmic male sterility can be utilized for producing hybrid seeds in those vegetables where vegetative part is of economic value e.g. onion, carrot, radish, cole crops etc. But in those vegetables crops where seed is the economical part like tomato, chilli, melon etc, it is of no use because the hybrid progeny will be male sterile.

Limitations of CMS: This technique has certain limitations to which are enumerated hereunder:

1. Cytoplasmic male sterility (CMS) is sensitive to environmental factors e.g. a line may be completely male sterile in one set of environment and may have partial fertility in another set of environment. This phenomenon may lead to mixture of selfed seed in an otherwise hybrid seed.

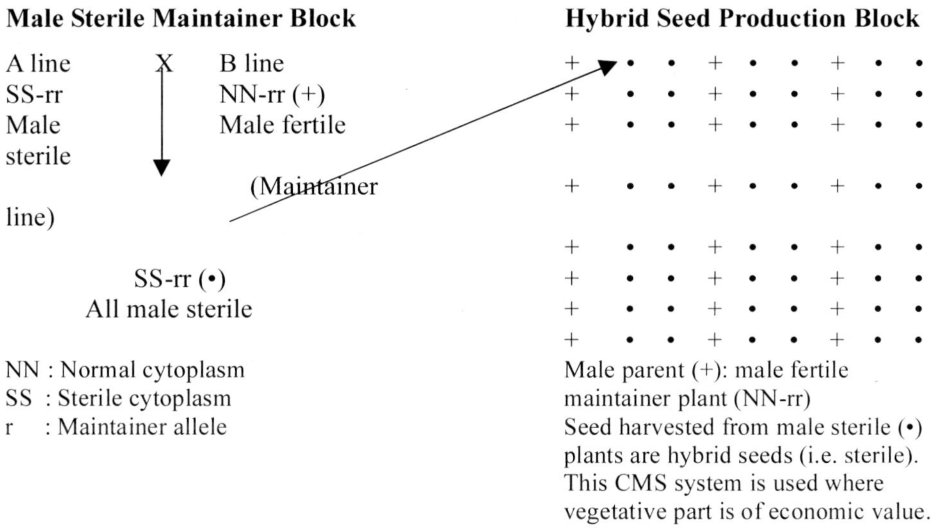

Figure 3: General scheme of hybrid seed production utilizing CMS

2. Certain diseases or disorders are associated with particular type of cytoplasm which leads to genetic vulnerability e.g. T-cytoplasm is associated with Southern Corn leaf blight in corn.
3. Continuous incorporation of small amount of male parent cytoplasm through each backcross during maintenance may lead to partial or complete break down of male sterility.

3. Cytoplasmic – Genetic Male Sterility (CGMS)

There is case of cytoplasmic male sterility where a nuclear gene for restoring fertility in the male sterile line is known. The fertility restorer gene, RR is dominant and is found in certain strains of the species or may be transferred from a related species. Thus gene restores male fertility in the male sterile line, hence it is known as restorer gene. The sterility factor is determined by interaction of nuclear genes and cytoplasm but none of them singly can control sterility. This type of male sterility is reported in carrot, onion, sugarbeet, chilli, capsicum and *Brassica napus*.

New male sterile line may be developed following the same procedure as cytoplasmic system, but the nuclear genotype of the pollinator strain in such a transfer must be NN-rr otherwise the fertility would be restored. The development of new restorer strain is somewhat indirect. First a restorer strain (NN-rr) is crossed with a cytoplasmic male sterile line A (SS-rr). The resulting male fertile plants (SS-rr) are used as a female parent in repeated backcrosses with the maintainer or B line (NN-rr) used as the recurrent parent into which the transfer of restorer gene(s) is desired. In each generation, male sterile plants are discarded and male fertile plants are used as females for backcrossing to the strain B (maintainer line). This acts as a selection device for the restorer gene RR during the backcross programme. At the end of backcross programme, a restorer line isogenic to strain B would be recovered.

Utilization of CGMS: Cytoplasmic-genetic male sterility can be maintained by crossing a cytoplasmic male sterile line (SS-rr) or A line with the pollinator strain (NN-rr) used as recurrent parent in the backcross programme since the nuclear genotype of the pollinator is identical with that of new male sterile line (Fig. 4). Such a male fertile line (NN-rr) is known as maintainer line or B line as it is used to maintain the male sterile line.

For hybrid seed production, 2 to 3 rows of line A (SS-rr) are alternated with 1 row of line C, which is generally expected to NN-RR. The seed is harvested from line A for use as the commercial hybrid seed. The line C may have genotypes NN-rr, NN-Rr, SS-Rr, SS-RR, but the hybrid developed by using any of the first 3 genotypes will be sterile and can be used where seed is not a commercial product but where seed is important, pollen parent should have genetic constitution "NN-RR" i.e. have restorer genes. The main advantage of CGMS system over the genic male sterility is that we can get 100% male sterile plants for direct use as female.

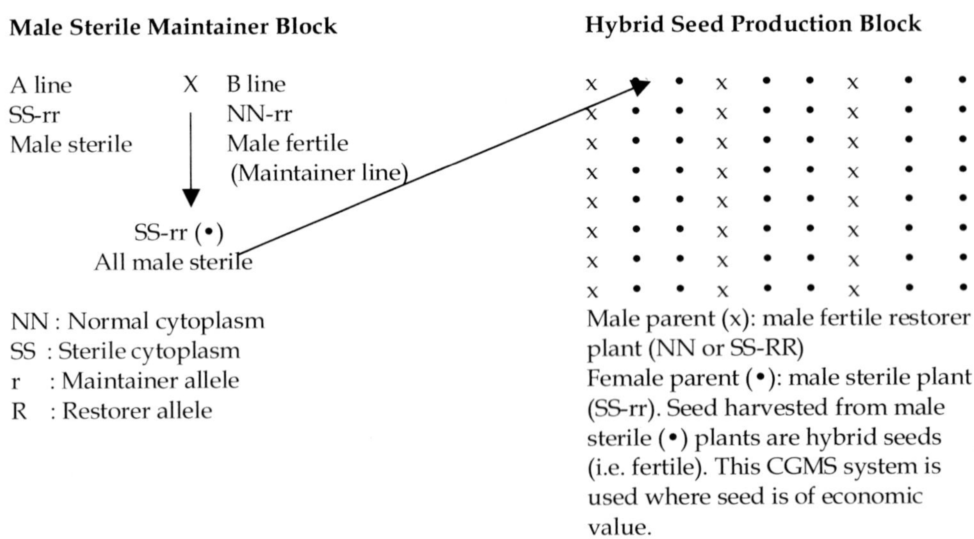

Figure 4: General scheme of hybrid seed production utilizing CGMS

Limitation of CGMS : CGMS system though is the most commonly utilized male sterility, its utilization is restricted in specific species, because of certain limitations. These are non-availability of CGMS in many crops and their wild relatives; need of fertility restorer allele in fruit producing vegetables; undesirable pleiotropic effect of sterile cytoplasm on horticultural qualities; breakdown of male sterility in particular environment; highly unstable sterile cytoplasm in several cases; poor cross pollination ability of flowers of plants with sterile cytoplasm due to altered morphology and technical complexity involved in seed production and maintenance of parental lines.

Male Sterility in Selected Vegetables

1. Tomato

GMS : More than 55 male sterile (*ms*) alleles causing sporogenous, structural and functional sterility are reported (Kaul, 1988). Chromosomal location of some of these genes is also known . The list of artificially induced and spontaneously isolated male sterile mutants is increasing constantly. There are four types of male sterility in tomato. Each one is governed by a single recessive gene. Stamenless type produces misshapen fruit in the F_1 hybrid generation where as positional sterility is not stable. Pollen abortive type and functional sterilities are often used in F_1 hybrids production.

Chromosomal location of some *ms* genes in tomato (Kaul 1988)

Chromosome	ms genes
1	*ms*-6, *ms*-32, *ms*
2	*ms*-2, *ms*-5, *ms*-10, *ms*-15, *ms*-26, *ms*-35, *ms*, *ps*
3	*ms*-9
4	*sl*
6	*ms*-16, *ms*-32, *cl-2*
8	*ms*-8, *ms*-17, *vms*
10	*ms*-31
11	*ms*-3, *ms*-7, *ms*-12, *ms*-14, *ms*-42, *ms*, *ap*

Description of different male sterile mutants in tomato

Mutant	Description	Inheritance	Governing genes
Pollen sterile	Pollen abortive	Monogenic recessive (Except MS-48), monogenic dominant)	*ms* series (- 49 independent genes)
Stamenless	Stamens absent	Monogenic recessive	*sl-1*, *sl-2*
Positional sterility	Stigma exerted	Monogenic recessive	*ps*
Functional sterility	Anthers do not dehisce	Monogenic recessive	*ps-2*

Two stamenless mutants, viz., stamenless-1(*sl-1*) and stamenless-2 (*sl-2*) produce flowers with carpel like structures instead of stamens (stamenless) when grown at higher temperature(28°C day/23°C night), while those grown at low temperature (18°C day/15°C night), produce flowers with abnormal stamens and often with viable pollen. More importantly, at intermediate temperature (23°C day/18°C night), *sl-2* plants produced flowers with half stamens and half carpals. The expression of *sl-2* allele is significantly modified at 5°C change in temperature (Sawhney 1997). In vegetative and floral parts (except pistil) of *sl-2* mutant, abscisic acid (ABA) content was more than the normal, especially in stamens. The increase in ABA content in stamens coincided with first sign of abnormality in anther. At low temperature, when fertility was restored, there was a drop in ABA levels in leaves and stamens. Male sterility in *sl-2* is a manifestation of hormonal imbalance (high ABA) and low temperature regulation of male sterility is mediator through reduction in ABA content (Santokh and Sawhney 1998). Similarly, in *ms-15* and *ms-33* mutants, low temperature (<30°C) is associated with fertility restoration (Sawhney 1997). The variable male sterile (*vms*) plants express male sterility at a temperature of 30°C or above (Rick and Boynton 1967).

Pollen abortive type is maintained through back crossing (ms ms x Ms ms), while functional male sterility is maintained by manual selfing. The efficiency of hybrid seed production by involving pollen abortive type male sterile lines are further improved by incorporating morphological markers e.g., anthocyanin less gene '*aa*'. The gene is lined with *ms-10*[35] and male sterile plants are identified from the mixed population of sterile and fertile plants in nursery itself (Philouze 1974). Consequently only male sterile plants are transplanted in hybrid seed production block economizing area and labour. The efficiency of this method depends upon the power of the linkage between the two factors viz. male sterility and anthocyanin less. The male sterility genes can be exploited advantageously only if the style is accessible to cross pollination. Otherwise disturbing anther cone to expose stigma for pollination will neutralize advantage of male sterility. According to Georgiev (1991) utilization of male sterility in combination with exerted stigma and with anthocyanin less gene, faces some inherent problems. These include appearance of 2-3% reciprocal recombinants, i.e. absence of anthocyanin in fertile plants, low seed yield due to poor receptivity of stigma and loss of exerted style in 20-60% of the flowers depending upon the genotype and the environmental factors. The exerted style character is more stable in small fruited genotypes than in large fruited ones. The efficiency of male sterility (pollen abortive types) genes can further be improved if the homogenous stand of the seed parent is available. This should be made possible by regenerating male sterile plants from *in vitro* cultured explants. In such a case, there is no need of genetic markers. Phenotypic fertility of genetically sterile plants can also be restored by application of growth regulators. Progeny of the self seed thus produced will be all male sterile. Partial success with application of gibberellic acids A_3+ or A_4+A_7 was reported by Yordanov *et al.* (1971).

The functional male sterility governed by *ps*-2 had indisputable advantage over other types. These include maintenance of male sterility under homozygous state by self pollination, possibility for simultaneous emasculation and pollination of blossomed flowers when the stigma is the most receptive and high yield of hybrid seed with low cost. The main disadvantage of this sterility is that up to 5% selfing is possible which is not tolerated in hybrid varieties known for their extreme uniformity. This was overcome to great extent by the development of lines combining functional sterility (*ps-2*) with short style (*shst*), which decreased chance selfing to less than 0.2% and enabled emasculation to be done very easily by non-skilled workers. The cost of hybrid seed by using the *ps-2* female lines decreased by approximately 50% as a result of easiness in emasculation and pollination and by '*ms* 10^{35}*aa*' approximately 20% as a result of elimination of emasculation of flower (Georgiev 1991). Atanassova and Georviev (2002) used potential sterility (*ps-2*) as sterile seed parent in tomato hybrid seed production. Comparative study on time required for the emasculation of floral buds and flowers at anthesis made it clear that emasculation during anthesis was easier and almost two times more rapid than emasculation applied on the floral buds. Moreover introduction of *ps-2* sterile parent in breeding programme also increased seed yield by 1.5-2.0 times and hybridity % of the seeds (100).

CMS: Through protoplast fusion of *Lycopersicon esculentum* with *Solanum acaule* and *S. tubersoum*, cytoplasmic male sterile cybrid plants were isolated with different flower morphology than tomato. Restriction analysis of mt-DNA of these sterile plants revealed that these plants originated due to recombination between the parental mt-genome (Melchers *et al.* 1992). Recently, sterile cytoplasm from *L. peruvianum* was transferred into the L. *pennellii*. Subsequently, CMS *pennellii* was successfully crossed with *esculentum*. This hybrid provides basis for the development of CMS system in tomato derived from sterile cytoplasm of *peruvianum* (Petrova *et al.* 1999). However, the practical utility of these CMS would depend upon the identification of restorer gene in tomato. Incorporation of parthenocarpy trait in parental line would be the other alternative for desired amount of fruit set without restorer gene in the F_1 derived from CMS.

2. Chilli

GMS: The genetic male sterility is an important pollination control mechanism which is exploited commercially for hybrid seed production in chilli. More than a dozen monogenic recessive male sterile lines were identified either in natural population or induced through mutagenesis (Shifriss and Frankel 1969, Pochard 1970, Daskalov 1971, Shifriss and Rylski 1972, Shifriss 1973, Hirose and Fujime 1980, Deshpande *et al.* 1983, Milikova and Daskalov 1984, Prakash *et al.* 1987, Meshram *et al.* 1992, Patel *et al.* 1998). But, very little information is available on the allelism of these male sterile alleles. The induced male sterile in France (*mc*-509; Pochard 1970) renamed *ms*-10 was allelic to *msk* allele isolated spontaneously. The *ms-2* line identified by Shifriss and Rylski (1972) was non-allelic to *ms*-1 isolated by Shifriss and Frankel (1969).

Although there are reports on tight linkage of *ms* gene with phenotypic marker traits (Murty and Lakshmi 1979, Meshram and Narkhede 1982, Pathak *et al.* 1983) but till date none of them was exploited for early identification of male sterile plants. To establish linkage between *ms* gene and seedling marker, several strategies were proposed by the chilli breeders to overcome problems associated with maintenance of GMS line. Shifriss and Pilowsky (1993) increased the ratio of *ms* plants in a line. They crossed two isogenic lines that have different male sterility genes. Their intention of this digenic cross was to produce male sterile plants containing both *ms-1* and *ms-2* genes. This plant was then crossed to a fertile plant that was heterozygous for both genes, i.e. *ms-1ms-1 ms-2ms-2 x Ms-1ms-1 Ms-2ms-2*. The progeny of this cross segregated in a ratio of three male sterile and one male fertile plant. Only one quarter of the plants have to be removed from the hybrid seed production field. The problem was that both parents have to be maintained asexually with proper protection from viruses (Shifriss 1997). Based on the principle of XYZ system of male sterility in wheat (Driscoll 1972), Shifriss (1983) tried to develop XYZ system using *Capsicum annuum* (*msms*) and *C. chinense* (*Ms Ms*), but desirable 12 II *msms* (ann.) + 1 II *Ms Ms* (chi.), maintainer line (Y) could not be obtained. We presume that now such attempts or proposal may not be attractive in the light of fact that (i) chromosome decay in such system would be a handicap and (ii) advancement of genetic engineering and development of EGMS system offer more attractive and easy methods to overcome maintenance problem.

The Punjab Agricultural University (PAU) developed ms-12, which carried genetic male sterility (GMS) controlled by recessive gene (msms). The plants having recessive gene in homozygous state (msms) are male sterile whereas those in heterozygous (Msms) and homozygous dominant (MsMs) state are male fertile. The progeny of male sterile plant pollinated by male fertile plant segregates in 1:1 ratio for male sterility and male fertility. The male sterile line (ms-12) is developed by transferring sterility gene from France (ms-509) into the cultivar 'Punjab Lal' through back crossing (Singh and Kaur 1986). By using this male sterile line (ms-12), PAU released two chilli hybrids viz. Chili Hybrid-1 (CH-1) and Chilli Hybrid-3 (CH-3). This male sterile parent (ms-12) was later on utilized by Kalloo *et al.* (1998) to develop commercial male sterility based chilli hybrid. Patel *et al.* (1998) also reported genic male sterile line $ACMS_2$, having mongenic recessive gene ($acms_2$ $acms_2$) but no hybrid was developed by using this male sterility. Male sterile line *ms-3* introduced from Hungary is maintained at Asian Vegetable Research and Development Centre (AVRDC), Taiwan. This line is being utilized for hybrid seed production in Hungary (Berke 1999).

CGMS: Cytoplasmic-Genic Male Sterility (CGMS) in chilli was first reported by Peterson (1958) in an introduction of *C. annuum* from India (PI-164835), however, it is not exploited commercially, because of an instability under fluctuating conditions, particularly temperatures and a low rate of natural cross pollination in cultivated peppers. Novak *et al.* (1971) working with the Peterson's source of male sterility and broader genetic material found digenic inheritance; they also noted variable degree of sterility from complete male sterility to partial fertility. In Peterson cytoplasm, pollen fertility was restored under 25°C and 17°C day and night temperature, respectively (Shifriss 1997). The instability of sterility expression is attributed to the interaction between temperature and sterility modifier genes. It is observed that at low temperature, meiotic breakdown is either completely stopped or delayed, resulting in pollen fertility (Shifriss 1997). Like EGMS system, in this CGMS system, seed increase of male sterile line can be done in cool season and hybrid seed can be produced during late summer when expressivity of sterility is absolute (Shifriss 1997).

Shifriss and Frankel (1971) found S-type cytoplasm following inter specific crosses in *C. annuum*, probably identical to that found by Peterson (1958). Since these previous studies indicated lack of stability of cytoplasmic genic male sterility, the character was not utilized in hybrid seed production. Shifriss and Guri (1979) developed several cytoplasmic genic male sterile (partial fertile) cultivars in pepper and concluded that cultivar with male sterility stability, like that of "Bikura" can serve as "A" line using natural cross pollination in hybrid seed production. Woong Yu (1985) suggested to combine both genic and CMS components of sterility to obtain double cross hybrids. Such a programme can exploit the yielding heterosis well known among hot pepper accessions. Daskalov and Mihailov (1988) successfully tested an improved method for hybrid seed production, based on using a cytoplasmic male sterile line pocessing a lethal gene with action that can be easily inhibited by a female sterile pollenizer. The lethal gene ensures 100% purity of the F_1 crop. The female sterile pollenizer produces a permanent abundant flowering with excess of pollen grains that leads to increased hybrid seed production without additional labour expenses. A hybrid 'Jingla No 2' of chilli was developed by crossing CMS line 181 A with the restorer line 98199 (Geng *et al.* 2005). Shen and Shi (2005) also developed a hot pepper F_1 hybrid 'Nongda 082' by crossing the female parent S200243A (male sterile line) with the male parent S200244C (restorer line). Yang *et al* (2007) developed a new hybrid 'Chuanjiaozidantou' by crossing a cytoplasmic male sterile line E16A with the restorer E04C, which is early maturing and highly resistant to TMV and CMV.

Presently, CGMS system is not utilized at commercial scale mainly due to instability of Paterson type sterile cytoplasm. Several seed companies in South Korea started utilizing CGMS system (Shifriss. 1997). Work is also under progress at AVRDC to develop stable CGMS system in chilli.

3. Bell pepper (Capsicum)

GMS: In bell pepper, genic male sterile (GMS) lines are used to a limited extent for hybrid seed production. GMS lines are maintained by hand pollination using heterozygous fertile plants as males. Shifriss and Frankel (1969) found spontaneous genetic male sterile plants in the cultivar 'All Big'. Male sterility was determined by a single recessive gene. Due to this sterility, parthenocarpic fruits were developed in unfertilized flowers throughout the season. Shifriss and Rylsky (1972) discovered a second gene encoding genetic male sterility in the cultivar 'California Wonder'. It too, was inherited as a single recessive gene. The authors suggested that the sterility discovered in 1969 may be described as *ms-1* and second as *ms-2*. The use of genetic male sterility is limited in hybrid seed production due to inefficiency in rouging out male fertile plants, since 50% of the plants are sterile & 50% are fertile. Ideally you would use a molecular marker to identify male sterility in the early growth stages so that fertile plants can be rouged at the seedling stage. Shifriss and Pilowsky (1993) increased the ratio of ms plants in a line. They crossed two isogenic lines that have different male sterility genes. Their intention of this digenic cross was to produce male sterile plants containing both *ms-1* and *ms-2* genes. This plant was then crossed to a fertile plant that was heterozygous for both genes, i.e. *ms1ms1 ms-2ms-2 x Ms-1ms-1 Ms-2ms-2*. The progeny of this cross segregated in a ratio of three male sterile and one male fertile plant. Only one quarter of the plants were removed from the hybrid seed production field. The problem was that both parents have to be maintained asexually with proper protection from viruses (Shifriss 1997). Poulos (1994) reported that genic male sterility was the first controlled pollination strategy adopted by seed industry in the production of F_1 hybrid seed after a long history of hand pollination and emasculation. Although much hybrid seed is still produced by hand emasculation and pollination, sterile lines are being used to reduce the labour cost and to improve hybrid purity. A number of GMS genes are reported and spontaneous mutants occur frequently. The discovery of molecular and physical seedling markers that are linked to GMS alleles is pending.

CGMS: CGMS is another system of sterility by which hybrids are produced in capsicum. Peterson (1958) described a CGMS system in peppers but it was found unstable, and fertile pollen developed under cool conditions. Several studies (Novak *et al.* 1971, Shifriss and Frankel 1971, Shifriss and Guri 1979) indicated using Petesons CMS material that additive factors affect pollen sterility and stability. In general, seed set on pollinated male sterile plants was only half that of normally pollinated plants and hand pollination has to be performed although emasculation is avoided. A few red pepper hybrids were developed by using CMS and exhibited heterosis for early and total yield (Diku and Anakecko 1975). CGMS was tested by Daskalov and Mihailov (1988) and Yoo (1990) for commercial hybrid seed production in bell pepper. They reported that there were some problems in the stability of sterility and its restoration. Daskalov and Mihailov (1988) proposed a scheme using a CMS seed parent containing a lethal gene and a female sterile pollen parent to improve the efficiency of hybrid seed production without hand pollination. Chemical gametocides may complement the use of male sterility of bell pepper hybridization but limited success was reported (Salgare 1989). Novel sterility systems are likely to be adopted to reduce the labor cost of producing hand pollinated hybrids and to assure hybrid purity.

4. Cabbage

There are four types of male sterile lines in cabbage, two types of male sterility - cytoplasmic male sterility and dominant genetic male sterility - have recently become economically significant, and are used by plant breeders in some countries. The main advantage to use male sterile line lies in two aspects: the low labor cost and high purity hybrid.

GMS: The first recessive genic male sterility was reported by Cole (1959) and Sampson (1966). No marker genes linked to male sterility were found, so it was difficult to use this kind of male sterility.

The Second type is dominant nuclear male sterility reported by Fang *et al.* (1997) in cabbage from a natural population. The sterility was controlled by one dominant gene, and is environmental sensitive. The male sterile lines could become partial fertile at relative lower or higher temperature situations. In the first step, seeds of male sterile line with homozygous dominant gene (*Ms Ms*) are proposed to be multiplied by clones, then further increased by crossing it (*Ms Ms*) with male fertile sister line (*ms ms*) to obtain larger quantity of 100% male sterile seeds (*Ms ms*). In second step, seeds of male sterile lines can be used for hybrid seed production. Utilization of dominant male sterility by above scheme will be feasible only in those vegetables, where vegetative parts are of economic importance (Kalloo *et al*, 1998). Several superior hybrids from male sterile lines were developed and used for commercial seed production.

CMS: The third type, Ogura male sterility, was transferred from radish cytoplasm. The main problem of such type is chlorosis at low temperatures i.e below 12°C and suppressed nectar gland development. In recent years, scientists in France and US used the protoplast fusion to improve male sterile lines with radish cytoplasm. More recently, through protoplast fusion, Ogura cytoplasm from broccoli was transferred into cabbage (Sigareva and Earle 1997). In addition, sterile 'Anand' cytoplasm from *B. rapa* (originally derived from wild species *B. tournefortii*) was transferred in *B. oleracea* through protoplast fusion (Cardi and Earle 1997). Several other sterile cytoplasm sources were identified in other *Brassica* spp., having potential to be transferred in cole vegetables (Hinata and Konno 1976, Quiros 1987, Crisp and Tapsell 1993). While the chlorosis problem at seedling stage was solved through protoplast fusion, the improved materials showed reduced nectar production, inability to attract insects and low seed setting by open pollination (Pearson 1972, Dickson 1975). The CMS cybrid plants developed by Pelletier and his associates (Pelletier *et al.* 1995) were of normal flower morphology, with good nectar production and found highly stable. These promising cybrids, contained genomes resembling more to the *B. oleracea* type than the Ogura. Restriction fragment profiles of mt-DNA revealed that some part of Ogura genome was systematically deleted. These CMS lines are being successfully utilized by seed companies in France to develop hybrids in cabbage and cauliflower (Pelletier *et al.* 1995). Jian and Ding (2005) utilized CMS line (introduced from abroad) and were backcrossed with cabbage inbred line to transfer the CMS gene and developed sixteen cabbage CMS lines which are later on utilized for production of six perfect hybrids in cabbage with uniform and good characters. Jian *et al* (2007) developed a new hybrid 'Qiugan No.1' by crossing CMS line (CMS021) with inbred line (97025-B). A new dominant male sterile line (79-399-3) of cabbage was developed from the cytoplasmic male sterile line (R3625) and this male sterile line has stable sterility, normal growth and development and good combining ability (Fang *et al* 2004) and five new cabbage hybrids were developed by utilizing this new dominant male sterile line.

CGMS: The forth type of male sterility controlled by nigra nuclear-cytoplasm interaction was reported by Pearson (1972). The defect of this type of material is that the petals cannot open fully and the nectar gland is less developed, and doesn't attract insect pollinators.

By now, the most widely used male sterile lines in the hybrid seed production are improved radish CMS and dominant GMS.

5. Cauliflower

GMS : Male sterility in cole crops is mainly a recessive character. A single *ms* gene mutated from male fertile *Ms* gene was reported in cauliflower by Nieuwhof (1961), Borchers (1966), Nieuwhof (1968) and Ahluwalia *et al.* (1977) and was designated as *ms-4* and ms-C. Van der Meer (1985) reported male sterility under the control of duplicate dominant genes with cumulative effect. Ruffio-Chable *et al* (1993) and Crisp and Tapsell (1993) also reported that dominant genes control the male sterility in cauliflower. This has some possible practical value in hybrid seed production programs, because of inadequate and unreliable nature of self in-compatibility system in some of the cauliflowers. This sterility can be responsive to temperature and humidity.

CMS : Cytoplasmic male sterility is not apparently found in cauliflower or other cole crops but was introduced from several other sources. Cytoplasmic male sterility was reported in an identified cultivar of Japanese radish by Ogura (1968) and was introduced by transferring to *Brassica oleracea* genomes through repeated back cross with broccoli (Bannerot *et al.* 1974 and McCollum 1981). Later Dickson (1975) and Hoser-Krauze (1987) transferred it from broccoli to cauliflower. The Ogura type cytoplasmic male sterility was transferred into heat tolerant Indian cauliflower from kale and broccoli through repeated backcrosses, four lines, MS-91, MS-51, MS-11 and MS-110 from the former and five lines, MS-01, MS-04, MS-05, MS-09 and MS-10 from the later were developed, which are now being used in heterosis studies by Sharma (2003).

Both genic as well as cytoplasmic male sterility are associated with physiological problems. Some forms of genetic male sterility are temperature sensitive and result in self-pollination when used for F_1 seed production (Nieuwhof 1968). Pearson type of cytoplasmic male sterility functional notaries are not developed making them unsuitable for commercial hybrid seed production (Pelletier *et al.* 1983). Both the Ogura and McCollum types of cytoplasmic male sterile plants or their hybrids when grown at low temperature less than 12°C, show chlorosis and loss of vigor at their early stage of growth (Dickson 1985, Hoser Krauze 1989). High regeneration capacity from cultured mesophyll cells of a cauliflower line having Ogura system was reported by Jourdan *et al.* (1985). This was useful step in the possible production of cytoplasmic mutants, transgenics or recombinants of superior male sterile genotypes. Non-chlorotic male sterile lines, using cybrids followed by protoplast fusion between sterile and normal genotypes were also developed in cauliflower and other Cole crops. Male sterile cybrids with normal photosynthesis and improved nectar secretion were obtained through chloroplasts exchange and mitochondrial recombination. The Ogura CMS system was first improved in *B. napus* in 1983, and then in *B. oleracea* in 1989 (Delourme and Budar 1999).

6. Broccoli

GMS : This was isolated from calabrese and purple heading broccoli within broccoli group (Borchers 1971, Cole 1959). Dunemann and Grunewaldt (1987) induced male sterility in broccoli by *in vitro* mutagenesis. Genic male sterility is conditioned by recessive male sterility (ms) genes, of which several designated as *ms-1, ms-2, ms-4, ms-5* and *ms-6* were identified (Dickson 1970), which are non-allelic. Among several non allelic *ms* genes in broccoli (var. italica), *ms-6* was sensitive to low temperature. After 30 days exposure to low temperature (i.e. 10°C-11°C (continuous) for 15°C days and 7°C night), the sterile plants (*ms-6 ms-6*) became fertile. It was observed that even 1°C increase in temperature (i.e. 12°C) was sufficient to revert male fertile plants into male sterile plants and vice versa (Dickson 1970). Thus, *ms-6 ms-6* broccoli line as such is not suitable for hybrid breeding, because the range of critical temperature is very less (1°C). Borchers (1966) recognized male steriles by their small flowers, short stamens and shriveled anthers. Their pollen grains were characteristically small, abortive, non-staining and often clump together. Borchers (1971) studied hybrid broccoli seed production utilizing the *ms-6* gene for male sterility and observed that the use of temperature sensitive male sterility such as that conferred by *ms-6* gene could enable hybrid seed to be produced at relatively high temperature with a few sibs, while Dickson (1970) reported that low temperature could be used to maintain homozygous male sterile lines by self-pollination.

CMS : In *Brassica oleracea* L. first CMS system was developed by Pearson (1972) through inter-specific hybridization between *B. nigra* (wild mustard) having sterile cytoplasm and *B. oleracea* var. *italica* (broccoli). Artificially produced tetraploid amphidiploid plants were backcrossed with pollen from broccoli, to transfer broccoli nuclear genome inside the sterile cytoplasm derived from the *B. nigra*. Backcrosses were also made between the amphidiploid and cabbage (var. *capitata*) cultivar Green Globe and from these materials Pearson established two CMS systems, viz., petaloid and vestigial, anther male sterility. When *B. nigra* is used as sources of CMS, the derived broccoli lines are considerably altered in flavour and the characteristic broccoli flavour being replaced by a sharper, maternally inherited mustard flavour (Pearson, 1972). These problems slowed the pace of utilization of CMS system in production of commercial F_1 hybrid broccoli.

Petaloidy CMS has a wide range of expression as in its extreme; its all stamens are converted into narrow incurved petals (Pearson 1972). Pistils are enlarged and malformed, and nectarines are generally absent. Pollinators are less attracted to these flowers. Female fertility varies in broccoli petaloid male sterile but can be improved by selection. Dickson (1975) utilized cabbage male sterile stocks to develop CMS broccoli lines that set seed freely, had the largest nectarines, although these were still reduced when compared with those of the male fertile maintainers.

CMS from *Raphanus sativus* (Ogura 1968) is a vestigial anther type with normal flowers and fully seed fertility. Nectary development may not be normal but there seems to be scope for selection. The *Brassica oleracea* CMS lines based on radish cytoplasm develop chlorosis under low temperatures (below 12°C) (Bannerot *et al.* 1977). The intensity of chlorosis depends upon the recurrent pollen parent used (McCollum, 1981). In recent years, scientists in France and US used the protoplast fusion to improve male sterile lines with radish cytoplasm. This technique could also be used to create unique and possibly heterotic materials by combining different cytoplasm to form cybrids as reported by Pelletier *et al.* (1983) in which the male sterile cytoplasm from *R. sativus* was combined with the chloroplasts of *B. napus*. Recombination between mitochondria is useful in separating CMS from 'defective nectary', as both characters are mitochondrially determined. A lack of suitable restorers prevents use of this source of CMS in rape (*B. napus)* but not in vegetable brassica's where restoration is unnecessary. Conventional sexual hybrids contain only maternally derived cytoplasmic factors but somatic fusion combines both parental cytoplasms. Subsequent cytopasmic segregation results in elimination of one or the other cytoplasm (Aviv *et al.* 1980) but evidence from restriction analysis of mitochondrial DNA suggests that recombination occurs between parental mitochondrial DNAs prior to cytoplasmic segregation (Belliard *et al.* 1979). Thus, protoplast fusion can give rise to unique nuclear – cytoplasmic combinations, not available using conventional methods (Sharp *et al.* 1984). Pearson (1972) also stated that in vestigial anther male sterility, the flowers are normal in appearance but are smaller, with reduced anthers but apparently functional nectaries, and are therefore more attractive to pollinating insects. Dickson (1975) reported that homozygous vestigial anther lines were, however, not recovered after six generations of back crossing and testing in broccoli. Lin-Bi Ying *et al.* (1977) developed CMS line 5A of broccoli after a single crossing generation and four generations of backcrossing using male sterile cauliflower line i.e MSA as CMS source. This line is genetically stable and has excellent horticultural characteristics. The CMS cybrid plants developed by Pelletier and his associates (Pelletier *et al.* 1995) were of normal flower morphology, with good nectar production and found highly stable. These promising cybrids, contained genomes resembling more to the *B. oleracea* type than the Ogura. Restriction fragment profiles of mt-DNA revealed that some part of Ogura genome has been systematically deleted. These CMS lines are being successfully utilized by seed companies in France to develop hybrids in cabbage and cauliflower (Pelletier *et al.* 1995). Sharma and Vinod (2002) developed CMS lines in broccoli suitable for growing at a temperature range of 20-27°C using male sterile backcross progeny (Kale EC 173419 x G-SB^2 x 005559^2) as a source after three backcrosses. These showed normal or near normal floral characteristics with good seed setting ability. Honeybees were frequent visitors as these had functional nectaries. Zhu *et al.* (2001) successfully developed cytoplasmically male sterile line of broccoli with radish cytoplasm, BC7-19, by transferring male sterility of radish (*Raphanus sativus* L. Ogura) to rape (*Brassica napus*) and further develop a broccoli hybrid 'Huqing-1'.

7. Carrot

GMS : This is one of the few species in which male sterility was documented very early i.e. in 1885 (Kaul 1988). Welch and Grimball (1947) reported isolation of genic male sterile plant within the commercial variety 'Tender sweet'. Its anthers were shrivelled, brown and non-exerted and female fertility was normal. Several other genic male sterile mutants were described (Kaul 1988). However, none of them were utilized for commercial seed production due to the availability of more efficient CMS system in carrot.

CMS: The hybrid development in carrot was facilitated by the cytoplasmic male sterility (CMS) which is of two types as described by Riggs (1987) viz. (i) brown anther type, in which the anthers degenerate and shrivel before anthesis, based on S-cytoplasm and at least two recessive genes with complementary action and (ii) petaloid, in which five additional petals replace anthers, based on S-cytoplasm and at least two dominant genes with complementary action. In commercial hybrid development, petaloid steriles are employed more widely and then the brown anther type (Riggs, 1987). If genetic and environmentally stable brown anther steriles were available, then they would be preferred over petaloids because of their higher seed yielding potential. Suh *et al.* (1999) reported that American seed companies mainly used petaloid type male sterility with Imperator group varieties whereas European companies used brown anther type male sterility mainly with Nantes group but some companies used petaloid type also in varieties that were different from the Nantes group.

The hybrids in carrot are normally three way crosses, (A x B) x C, because the hybrid vigor in a single cross F_1 female seed parent normally results in much greater seed production than that of inbred male sterile parent. Single cross hybrids, A x B are on an average more uniform than three way crosses. Moreover, they do not require an extra year to produce F_1 seed parent stock. So the single crosses can be used, if their productivity is adequate. Reduced uniformity of three way crosses can be overcome if backcrosses are utilized as seed parents in hybrids as a result the final product attains the form {(A x B) x B} x C. Although compared to three way crosses, it consumes an additional year of seed parent production, but it permits utilization of less similar seed parent inbreds, A x B, than what is required.

Cytoplasmic inheritance of petaloidy, was first described by Thompson, in carrot lines initiated from a male sterile wild carrot plant found near Orleans, Massachusetts (USA) in 1953. It was named the Cornell cytoplasm and used to produce the majority of hybrid carrots in the United States (Goldman 1996). Thompson (1962) developed a complex model for inheritance of pollen sterility from the study of wild carrot (petaloid type) along with the brown anther material from Welch and three additional brown anther sources from Gabelman and concluded that there are at least two and probably three duplicate dominant maintainer genes and an epistatic restorer operating in cytoplasm of these material. But genetics of fertility restoration is complicated. This is because, the nuclear gene responsible for male sterility has not been well characterized (Peterson and Simon 1986) and the structural variants of mt-DNA are numerous among the species (Scheike *et al.* 1992, Ranfort *et al.* 1995). However, an efficient RAPD based technique and STSs (sequence-tagged sites) marker were identified to characterize mt-DNA (Nakajima *et al.* 1997, Nakajima *et al.* 1999), which may be useful in the characterization of variants of sterile cytoplasm. Recently, three novel alloplasmic CMS types were isolated, all are with altered flower morphology (Linke *et al.* 1999, Nothnagel *et al.* 2000). Chen *et al.* (2008) developed a new carrot hybrid 'Jinhong No 5' by technology of male sterile three lines concerting.

Taki Seed Company of Japan developed the first F_1 hybrid variety in 1982 using CMS (Pelletier *et al.* 1995). In USA, vast majorities of hybrids are produced from one cytoplasm i.e. Cornell cytoplasm. Considering the risk of disease vulnerability of hybrid variety due to monopolistic use of Cornell sterile cytoplasm, USDA released a petaloid type new CMS line derived from sterile cytoplasm of Wisconsim (Morelock *et al.* 1996).

8. Onion

CGMS : First male sterile plant was reported within the progenies of an onion cultivar Italian Red (Jones and Emsweller 1936), which was cytoplasmically inherited and male sterility was under the control of single recessive nuclear restorer locus (Jones and Clarke 1943). Male sterile plants occur in natural populations, but proportion of male sterile plants varies with the site and season. Plants of some A line populations were completely male sterile at 14°C, but predominantly fertile at 23°C (Van Der Meer and Van Bennekom 1969). Berninger (1965) identified a new sterile cytoplasm (T-cytoplasm) in French cultivar 'Jaune Paille des Vertus'. Fertility restoration in T-cytoplasm is controlled by three independent genes (Scheweisguth 1973).

In the most S-cytoplasm male sterile lines, anther morphology is normal, but at anthesis these are green, small and indehiscent. Contrastingly, in T-cytoplasm lines, anthers morphology is disrupted (Kaul, 1988). Recently, Sato (1998), identified a specific RAPD marker, in mt-DNA of S-cytoplasm. Such markers are very useful in the characterization of newly identified sterile cytoplasm in individual plants during diversification process of sterile cytoplasm. S-cytoplasm is widely utilized for production of hybrid seeds despite the problem that the expression of male sterility in some A lines is temperature sensitive. Male sterility was isolated from indigenous germplasm by several workers in India-Patil *et al.* (1973) in cv. Niphad 2-4-1, Pathak *et al.* (1980) in cv. Nasik White Globe (IIHR 20)~. Further studies indicated strong cytoplasmic factor responsible for male sterility in cv. Bombay White Globe (Pathak *et al.* 1987). This male sterility was transferred to several breeding lines by backcross breeding method.

Transfer of cytoplasms from related species into cultivated populations may produce new sources of CGMS (Havey, 1999). In an attempt to diversify the cytoplasm conditioning male sterility, the cytoplasm of *Allium galanthum* was backcrossed for seven generations to bulb-onion populations. The flowers of galanthum-cytoplasmic populations possess upwardly curved perianth and filaments with no anthers, making identification of male sterile plants easier than for either S-or T-cytoplasmic male sterile onion plants. Mean seed yield / bulb of the galanthum-cytoplasmic populations was measured in cages using blue bottle flies (*Calliphora eythrocephala* Meig.) as pollinators and was not significantly different from one of the two S-cytoplasmic male sterile F_1 lines, a T-cytoplasmic male sterile inbred line, or N-cytoplasmic male fertile lines. Male sterile lines possessing either the S galanthum-cytoplasm were each crossed with populations known to be homozygous dominant and recessive at the nuclear locus conditioning male fertility restoration of S-cytoplasm and progenies were scored for male fertility restoration. Nuclear restorers of male fertility for S-cytoplasm did not condition male fertility for the galanthum-cytoplasmic populations. It is intended that these galanthum-cytoplasmic onion populations used as an alternative male-sterile cytoplasm for the diversification of hybrid onion seed production.

World wide more than 50% onion varieties currently cultivated are F_1 hybrids derived from S-cytoplasm (Pelletier *et al.* 1995). In India, public sector bred commercial hybrid has not been recommended so far, however, in a few areas, farmers are cultivating male sterile based hybrids developed by seed companies.

9. Muskmelon

GMS : First recessive male sterile lines were reported by Bohn and Whitaker 1949, since then at least four additional male sterile recessive alleles, viz.

ms-2 (Bohn and Principe 1962), *ms-3* (McCreight and Elmstrom 1984), *ms-4* (Pilrat 1990) and *ms-5* (Lecouviour *et al.* 1990) were identified. The phonotype of *ms-4* plant is different from the *ms-1, ms-2, ms-3 and ms-5* (McCreight *et al.* 1993). Nandpuri *et al.* (1982) were the pioneers for its commercial utilization in India and developed male sterile line (*ms-1*) which was successfully utilized to develop first commercial hybrid (Punjab Hybrid-1). Though, this system eliminates emasculation, timely identification and rouging of male fertile plants make it cumbersome and dexterous. In Punjab, at many places even farmers are producing hybrid seeds utilizing ms-1 male sterile line (Kalloo *et al.* 1998). Despite the complex maintenance process and additional labour requirement to remove fertile segregants in hybrid seed production field, production of male sterile based hybrid seeds is more economical than the seeds produced by manual emasculation.

Conclusion

The research on male sterility in vegetables is a never ending process due to rapid advancement of molecular techniques and their implementation. Substantial progress has been made in understanding the mechanism of male sterility in selected vegetable crops. At global level, CMS and CGMS are the most widely utilized in most of the vegetables. Unfortunately in India we do not have any released hybrid vegetables developed on CMS and CGMS, even in vegetables like onion, where more than 50% cultivars grown world wide are F_1 hybrids derived from S-cytoplasm. The only successful examples are in case of chilli and muskmelon, where GMS is being utilized to develop F_1 hybrid seed commercially (Hundal and Khurana 1993, Nandpuri *et al.* 1982). In the recent past, chilli CMS lines were introduced at IIVR from AVRDC, which are utilized directly or indirectly to produce CMS based hybrids. Similarly in tomato, work on GMS lines is in progress at PAU. At many other institutes and state agricultural universities, male sterility in vegetables is being also utilized to develop experimental hybrids. But the development of commercial hybrids on the pattern of chilli and muskmelon should be the immediate aim of the vegetable breeders. In fruit bearing vegetables like tomato, brinjal, chilli, muskmelon etc., identification and utilization of functional male sterility are more attractive. This is because unlike in sporogenous genic male sterility (50:50 ratio of male sterile and male fertile plants), 100% male sterile progenies can be obtained after forced selfing of male sterile plants, since functional pollen can be extracted from the non-dehising anther. The attempts should be made to collect GMS lines of important vegetables at one place, so that these can be evaluated against sensitivity to temperature or photoperiod. This helps in the identification of EGMS lines, which are like functional sterility, believed to be having more potential for the production of commercial hybrid seeds. EGMS lines with a wide range of critical values will be practically useful because selfing risk of revertant male sterile plants due to slight environmental fluctuation can be avoided in seed production field. In this context, photoperiod sensitive lines are more attractive than the temperature sensitive lines, as photoperiod variation is steadier than the temperature. In India, research on transgenic male sterility system was initiated in selected vegetables but our first priority should be utilization of existing and established but unexploited male sterility systems especially in chilli, onion, tomato, muskmelon etc. This will not only promote adoption of hybrid vegetable technology by economizing the cost of hybrid seeds but also provide basic material and scope for the development of more efficient male sterility system in respective vegetable crops.

References

Ahluwalia KS, Swarup V and Chatterjee SS. 1997. Inheritance of qualitative characters in Indian cauliflower. *Vegetable Science* 4: 67-80.

Atanassova BB and Georgiev H. 2002. Using genic male sterility in improving hybrid seed production in tomato (*Lycopersicon esculentum*). *Acta Horticulture* 579: 185-188.

Aviv D, Fluhr R, Edelman M and Galun E. 1980. Progeny analysis of the interspecific somatic hybrids : *Nicotiana tabacum* (CMS) + *Nicotiana sylvestris* with respect to nuclear and chloroplast markers. *Theoretical and Applied Genetics* 56: 145-50.

Bannerot H, Boulidard L and Chupeau Y. 1977. Unexpected difficulties met with the radish cytoplasm in *Brassica oleracea. Eucarpia Cruciferae Newsletter* 2: 16-18.

Belliard G, Vedel F and Pelletier G. 1979. Mitochondrial recombination of cytoplasmic hybrids of *Nictiana tabacum* by protoplast fusion. *Nature* 281: 401-403.

Berke T. 1999. Personal communication.

Bernninger E. 1965. Contribution a letude de la sterilite male de loifnon (*Allium cepa*). *Annals of Amelior Plant* 15: 183-199.

Bohn GW and Principe GA. 1962. A second male-sterility gene in the muskmelon. *Journal of Heredity* 55: 211-215.

Bohn GW and Whitaker TW. 1949. A gene for male sterility in the muskmelon (*cucumis melo* L.). *Proceeding of American Society of Horticultural Sciences* 53: 309-314.

Borchers EA. 1966. Characteristics of male sterile mutant in cauliflower (*Brassica oleracea* L.) *Proceeding of American Society of Horticultural Sciences* 88: 406-410.

Borchers EA. 1971. Hybrid broccoli seed production utilizing the ms_6 gene for male sterility. *Journal of American Society of Horticulture Science* 96: 542-546

Cardi T and Earle ED. 1997. Production of new CMS *Brassica* oleracea by transfer of 'Anand' cytoplasm from *B. rapa* through protoplast fusion. *Theoretical and Applied Genetics* 94: 204-212.

Chen YM, Wang Y and Zhang YP. 2008. A new carrot F_1 hybrid – 'Jinhong No. 5'. *China Vegetables* 4: 40-41.

Cole K. 1959. Inheritance of male sterility in green sprouting broccoli (*B. oleracea* var. italica). *Journal of American Society of Horticulture Science* 95(1): 13-14.

Crisp P and Tapsell CR. 1993. Cauliflower. In : *Kalloo G and Bergh BO (eds.) Genetic Improvement of Vegetable Crops*. Pergamon Press, U.K., pp. 157-177.

Daskalov S. 1972. Male sterile pepper ((*C. annuum* L.) mutants and their utilization in heterosis breeding. *Proceeding of Eucarpia Meeting on Capsicum* 7: 202-210.

Daskalov S and Mihailov L. 1988. A new method of hybrid seed production based on cytoplasmic male sterility combined with a lethal gene and a female sterility pollenizer in *Capsicum annuum* L. *Theoretical and Applied Genetics* 76: 530-532.

Delourme R and Budar F. 1999. *Male sterility*. In Gomez Campo (ed), Biology of *Brassica Coenospecies*. Elsevier Science. pp 185-216.

Deshpande AA, Pathak CS and Singh DP. 1983. Types of male sterility in chilli pepper, *Capsicum* spp. *Capsicum Newsletter* 2: 104-105.

Dickson MH. 1970. A temperature sensitive male sterile gene in broccoli. *Brassica oleracea* L. var. italica. *Journal of American Society of Horticulture Science* 95: 13-14.

Dickson MH. 1975. G1117A, G1102A and G1106-A cytosterile broccoli inbreds. *Horticulture Science* 10: 535-537.

Dickson MH. 1985. Male sterile persistent white curd cauliflower NY7642A and its maintainer NY7642B. *Horticulture Science* 20: 397-399.

Diku SP and Analkecko VS. 1975. Heterotic hybrids of red pepper bred using male sterility. In : *Testy Doks Knuf Selektsiyaigenet Ovosch. Kultur*, Vol. 3, Suinea, Kishiner, Moldavian, SSR, 71.

Driscoll CJ. 1972. XYZ system of producing hybrid wheat. *Crop Science* 12: 516-517.

Dunemann F and Grunewaldt J. 1987. Male sterile broccoli (Brassica oleracea var. italica) induced by in viro mutagenesis. *Eucarpia Cruciferae Newsletter* 12: 50-52.

Fang Z, Sun P, Liu, Yang L, Wang X, Hou A and Bian C. 1997. A male sterile line with dominant gene (*Ms*) in cabbage (*Brassica oleracea* var. capitata) and its utilization for hybrid seed production. *Euphytica* 97: 265-268.

Fang ZY, Liu YM, Yang CM, Wang XW, Zhuang M, Zhang YY and Sun PT. 2004. Breeding and Seed production technique of dominant genic male sterile (DGMS) line and cytoplasmic male sterile (CMS) line in cabbage. *Scientia Agriculture Sincia* 37(5): 717-723.

Geng SS, Chen B and Zhang XF. 2005. A new hot pepper F_1 hybrid "Jingla No. 2". *China Vegetables* 10/11: 41-42.

Georgiev H. 1991. Heterosis in tomato breeding. In : *Genetic improvement of tomato* by G. Kaloo (Ed) springer – verlag, Berlin, Heidelberg, pp 83-98.

Goldman H. 1996. A list of germplasm release from university of Wisconsin carrot breeding programme 1964-94. *Horticulture Science* 31: 882-883.

Havey MJ. 1999. Seed yield, floral morphology and lack of male fertility restoration of male sterile onion (*Allium cepa*) populations possessing the cytoplasm of *Allium galanthum*. *Journal of American Society of Horticulture Science* 124: 626-629.

Hinata K and Konno N. 1976. Cytoplasmic male sterile strain of yukina (*Brassica campestris* L.) produced by nucleus substitution. *Japanese Journal of Breeding* 26 (2) : 127.

Hirose T and Fujime Y. 1980. Studies on the hybrid seed production by using a new male sterile line in pepper (*Capsicum annuum* L.) I. Morphology and the mode of inheritance. *Journal of Japanese Society of Horticulture Science* 48: 453-458.

Hoser-Krauze J. 1989. Comparison of suitability of sources of cytoplasmic male sterility (CMS) and self-incompatibility for breeding F_1 hybrids of cauliflower. *Biuletyn Warzyniczy* supplement I: 29.

Hundal JS and Khurana DS. 1993. 'CH-1'-A new hybrid of chilli. *Progressive Farming* 29: 11-13.

Jian YC and Ding YH. 2005. Breeding and utilization of *Brassica oleracea* cytoplasmic male sterile lines. *China vegetables* 6: 4-6.

Jian YC, Ding YH and Qu GQ. 2007. A new cabbage F_1 hybrid – 'Quigan No-1'. *China Vegetables* 12: 29-31.

Jones HA and Clarke AE. 1943. Inheritance of male sterility in onion and the production of hybrid seed. *Proceeding of American Society of Horticulture Science* 43:189-194.

Jones HA and Emsweller SL. 1936. A male sterile onion. *Proceeding of American Society of Horticulture Science* 34: 582-585.

Jourdan PS, Earle ED and Mutscher MA. 1985. Efficient plant regeneration from mesophyll protoplast of fertile and CMS cauliflower (*Brassica oleracea* cv. *botrytis*). *Eucarpia Cruciferae Newletter* 10: 94.

Kalloo G. 1988. *Vegetable Breeding Vols. I, II and III*. Panima Educational Agency, New Delhi.

Kalloo G and Bergh BO. 1993. *Genetic Improvement of Vegetable Crops*. Pergamon Press, UK.

Kalloo G, Banerjee MK, Kumar Sanjeet and Parkash C. 1998. Hybrid vegetable technology in India – an overview. In : *Souvenir on National Symposium Emerging Scenario in Vegetable Research and Development*, PDVR, Varanasi, pp. 42-52.

Kaul MLH. 1988. Male Sterility in Higher Plants. Monographs on *Theoretical and Applied Genetics* 10, Springer-Verlag, Berlin.

Lecouviour M, Pitrat M and Risser GA. 1990. Fifth gene for male sterility in *Cucumis melo*. *Cucurbits Genetic Cooperation Rep.* 13: 34-37.

Lin-Bi-Ying, Wei-Wenlin and Gao-shan. 1997. The breeding of a cytoplasmic male sterile line of broccoli. *Journal of Fujian Academy Science* 12(1): 24-26.

Linke B; Nothnagel T and Borner T. 1999. Morphological characterization of modifier flower morphology of three novel alloplasmic male sterile carrot sources. *Plant Breeding* 188: 543-548.

McCollum GD. 1981. Induction of an alloplasmic male sterile *Brassica oleracea* by substituting cytoplasm from 'Early Scarlet Globe' radish (*Raphanus sativus*). *Euphytica* 30: 855.

McCreight JD and Elmstrom GW. 1984. A third male sterile gene in muskmelon. *Horticulture Science* 19: 268-270.

McCreight JD, Nerson H and Grumet R. 1993. Melon. In : *Kalloo G and Bergh BO (eds.). Genetic Improvement of Vegetable Crops.* Pergamon Press, U.K. pp 287-294.

Melchers G, Mohri Y, Watanabe K, Wakabayasshi S and Harada K. 1992. One step generation of cytoplasmic male sterility by fusion of mitochondrial inactivated protoplasts with nuclear inactivated *Solanum* protoplasts. *Proceeding National Academy of Sciences (USA)* 89: 6832-6836.

Meshram LD and Narkhede MN. 1982. Natural male sterile mutant in hot chilli (*C. annuum* L). *Euphytica* 31: 1003-1005.

Meshram LD, Chaudhari RV, Kudade BK and Marawar MW. 1992. Functional male sterility in hot chilli (*C. annum* L.) *Eucarpia VIIIth Meeting on Genetics and Breeding on Capsicum and Eggplant*, Rome, Italy, pp 61-65.

Milikova L and Daskalov S. 1984. Ljulin-a hybrid cultivar of pepper based on induced male sterility. *Mutation Breeding Newsletter* 24: 9-12.

Morelock TE, Simon PW and Peterson CE. 1996. Wisconsin Wild : another petaloid male-sterile cytoplasm for carrot. *Horticulture Science* 31: 887-888.

Murty NSR and Lakshmi N. 1979. Male sterile mutant in *Capsicum annuum* L. *Current Science* 48: 312-315.

Nakajima Y, Yamamato T, and Oeda K. 1997. Genetic variation of mitochondrial and nuclear genomes in carrots revealed by random amplified polymorphic DNA (RAPD). *Euphytica* 95: 259-267.

Nakajima Y, Yamamato T, Muranaka T and Oeda K. 1999. Genetic variation of petaloid male sterile cytoplasm of carrots revealed by sequence tagged sites (STSs). *Theoretical and Applied Genetics* 99: 837-843.

Nandpuri KS, Singh S and Lal T. 1982. 'Punjab Hybrid' – a variety of muskmelon. Progressive Farming 18(6): 3-4.

Nieuwhof M. 1961. Male sterility in some cole crops. *Euphytica* 10: 351-356.

Nieuwhof M. 1968. Effect of temperature on the expression of male sterility in Brussels sprout (*Brassicca oleracea* var. gemmifera). *Euphytica* 17: 265-273.

Nothnagel T, Straka P and Linke B. 2000. Male sterility in populations of Daucus and the development of alloplasmic male sterile lines in carrot. *Plant Breeding* 119 (2): 145-152.

Novak F, Betalch J and Dubovsky J. 1971. Cytoplasmic male sterility in sweet pepper (*Capsicum annuum* L.) Phenotype and inheritance of male sterile character. *Journal of Plant Breeding* 65: 129-140.

Ogura H. 1968. Studies on the new male sterility in Japanese radish with special reference to the utilization of this sterility towards the practical raising of hybrid seeds. Mem. Fac. Agric. Kagoshima Univ. 6: 39.

Patel JA, Shukla MR, Doshi KM, Patel SA, Patel BR, Patel SB and Patel AO. 1998. Identification and development of male sterile in chilli (*Capsicum annuum* L). *Vegetable Science* 25: 145-148.

Pathak CS, Singh DN and Deshpande AA. 1983. Male and female sterility in chilli pepper, *Capsicum annuum* L. *Capsicum Newsletter* 2:102-103.

Pathak CS, Singh DP and Deshpande AA. 1987. A new type of cytoplasmic male sterility in onion. *First All India Conference of Cytology and Genetics*, Bangalore University, India, 1987, Asstr 46.

Pathak CS, Singh DP and Deshpande AA. 1980. Annual report of Indian Institute of Horticultural Research. 34-36.

Patil JA, Jadhav AS and Rane MS. 1973. Male sterility in Maharashtra onion. *Research Journal of Mahatma Phule Agricultural University* 4: 29-31.

Pearson OH. 1972. Cytoplasmically inherited male sterility characters and flavour components from the species *Brassica napus* (L.). *Journal of American Society of Horticulture Science* 97: 397-402.

Pelletier G, Ferault M, Lancelin D, Boulidard L, Dore C, Bonhomme S, Grelon M and Budar F. 1995. Engineering of cytoplasmic male sterility in vegetables by protoplast fusion. *Acta Horticulture* 392: 11-17.

Pelletier G, Primard C, Vedel F, Chetrit, P, Remy R, Rouselle P and Renard M. 1983. Intergeneric cytoplasmic hybridization in cruciferae by protoplast fusion. *Molecular and General Genetics* 191: 244-250.

Peterson CE and Simon PW. 1986. Carrot breeding. In : *Basset MJ (ed), Breeding Vegetable Crops*, AVI Publishing Co., West Port, Connecticut, pp. 321-356.

Peterson PA. 1958. Cytoplasmically inherited male sterility in *Capsicum*. *American Naturalist* 92: 111-119.

Petrova M, Yulkova Z, Gorinova N, Izhar S, Firon N, Jacquemin JM, Atanassov A and Stoeva P. 1999. Characterization of a cytoplasmic male sterile hybrid between *Lycopersicon peruvianum* Mill. x *Lycopersicon pennellii* Corr. and its crosses with tomato. *Theoretical and Applied Genetics* 98 : 560-566.

Philouze J. 1974. Genes marquehrs lies aux genes de sterilite male *ms10* et ms 32 che 2 1a tomato. *Annals of Amel. Plant* 24 : 77-82.

Pochard E. 1970. Obtaining three new male sterile mutants of pepper (C. *annuum*) through application of mutagenes on monoploid material. Eucarpia, Versailles, France, pp. 93-95.

Poulos MJ. 1994. Pepper breeding (*Capsicum* spp.): Achievements, challenges and possibilities. Plant Breeding Abstracts 64(2): 143-154.

Prakash NS, Lakshmi N and Harini I. 1987. Male sterility and manipulation of yield in *Capsicum*. *Capsicum Newsletter* 6: 35-38.

Quiros CF. 1987. Duplicated isozyme loci and cellulose compartmentation of their products in *Brassica*. *Eucarpia Cruciferae Newsletter* 12: 24-26.

Ranfort J, Saumitou-Laprade P, Cuguen J and Couvet D. 1995. Mitochondrial DNA diversity and male sterility in natural populations of *Daucus carota* ssp. Carota. *Theoretical and Applied Genetics* 91: 150-159.

Rick CM and Boynton JE. 1967. A temperature sensitive male sterile mutant of tomato. *American Journal of Botany* 54: 601-611.

Riggs TJ. 1987. Breeding of F_1 hybrid varieties of vegetables. In : *Hybrid production of selected cereal oil and vegetable crops* (Eds. Feistritzer WP and Kelly AF), FAO, Rome, pp 149-173.

Ruffio-Chable V, Bellis H and Herve Y. 1993. A dominant gene for male sterility in cauliflower (*Brassica oleracea* var. botrytis). Phenotype expression, inheritance and use in F_1 hybrid production. *Euphytica* 67: 9-17.

Rundfeldt H. 1961. Untersuchungen zur zuchtung des kopfkohs (*B. Oleracea* var. *Capitata*). *Z Pflanzenzucht* 44: 30-62.

Salgare SA. 1989. Induction of male sterility by dalapon in *Capsicum frutescens* L. *Advances in Plant Science* 22: 527-529.

Sampson DR. 1966. Linkage of genetic male sterility with a seedling marker and its use in producing F_1 hybrid seed of *Brassica oleracea* (cabbage, broccoli, kale etc.). *Canadian Journal of Plant Science* 46: 703-706.

Santokh S and Sawhney VK. 1998. Abscisic acid in a male sterile tomato mutant and its regulation by low temperature. *Journal of Experimental Botany* 49: 199-203.

Sato Y. 1998. PCR amplification of CMS-specific mitochondrial nucleotide sequences to identify cytoplasmic genotypes of onion (*Allium cepa* L). *Theoretical and Applied Genetics* 96: 367-370.

Sawhney VK. 1983. Temperature control of male sterility in a tomato mutant. *Journal of Heredity* 74: 51-54.

Sawhney VK. 1997. Genic male sterility. In : *Shivanna KR and Sawhney VK (eds), Pollen Biotechnology for Crop Production and Improvement*. Cambridge Univ. Press, pp. 183-198.

Scheike R, Gerold E, Brennike M, Mehring-Lemper M and Wricke G. 1992. Unique patterns of mitochondrial genes, transcripts and proteins in different male sterile cytoplasm of *Daucus carota*. *Theoretical and Applied Genetics* 83: 419-427.

Scheweisguth B. 1973. Study of a new type of male sterility in onion. *Genetics* 29: 569-572.

Sharma SR. 2003. Unpublished data.

Sharma SR and Vinod. 2002. Breeding for cytoplasmic male sterility in broccoli (*Brassica oleracea* L. var. *italica* Plenck). *Indian Journal of Genetics* 62(2): 165-166.

Sharp WR, Evans DA and Ammirato PV. 1984. Plant genetic engineering: Designing crops to meet food industry specifications. Food Technology, February 1984, 112-119.

Shen HL and Shi ZQ. 2005. A new hot pepper F_1 hybrid "Nongda 082". *China Vegetables* 10/11: 43-44.

Shifriss C. 1973. Additional spontaneous male sterile mutant in *Capsicum annuum* L. *Euphytica* 22: 527-529.

Shifriss C. 1983. An attempt to maintain genic male sterile (*MSMS)* lines of pepper *Capsicum annuum* L. via XYZ system. *Eucarpia Vth Meeting on Genetics and Breeding on Capsicum and Eggplant*, Plovdiv, Bulgaria, pp, 42-47.

Shifriss C. 1997. Male sterility in pepper (*Capsicum annuum* L). *Euphytica* 93: 83-88.

Shifriss C and Frankel R. 1969. A new male sterile gene in *Capsicum annuum. Journal of American Society of Horticulture Science* 94: 385-387.

Shifriss C and Frankel R. 1971. New sources of cytoplasmic male sterility in cultivated peppers. *Journal of Heredity* 64: 254-256.

Shifriss C and Guri A. 1979. Variation in stability of cytoplasmic genic male sterility in *Capsicum annuum* L. *Journal of American Society of Horticulture Science* 104: 94-96.

Shifriss C and Pilowsky M. 1993. Digenic nature of male sterility in pepper (*Capsicum annuum* L). *Euphytica* 67: 111-112.

Shifriss C and Rylski L. 1972. A male sterile (*ms*2) gene in California Wonder pepper (*Capsicum annuum* L). *Horticulture Science* 7: 36-38.

Sigareva MA and Earle ED. 1997. Direct transfer of a cold tolerant Ogura male sterile cytoplasm into cabbage (*Brassica oleracea* ssp. *Capitata*) via protoplast fusion. *Theoretical and Applied Genetics* 94: 213-220.

Singh J and Kaur S. 1986. Present status of hot pepper breeding of for multiple disease resistance in Punjab. *Proceeding of VI EUCARPIA Meeting on Genetic and Breeding on Capsicum and Eggplant,* held at Zaragoza (Spain), Oct 21-24 (1986): 111-114.

Soressi GP and Salamini F. 1975. A monomendelian gene inducing parthenocarpic fruits. *Rep. Tomato Genet. Coop* 25: 22-23.

Suh YK, Young GH, Cho YH and Poek KY. 1999. Utilization of male sterility type and frequency of male sterile genes in carrots. *Korean Journal of Horticulture Science and Technology* 17(4) : 470-472.

Tanksley SD, Rick CM and Vallejos CE. 1994. Tight linkage between a nuclear male sterile locus and an enzyme market in tomato. *Theoretical and Applied Genetics* 68: 109-113.

Thompson DJ. 1962. Studies on inheritance of male sterility in carrot (*Daucus carota* L.). *Proceeding of American Society of Horticulture Science* 78: 332-338.

Van der Meer OP. 1985. Male sterility in Cole crops-a serial story. *Eucarpia Cruciferae Newsletter* 10: 58-60.

Van der Meer QP and Van Bennekom JL. 1969. Effect of temperature on the occurrence of male sterility in onion (*Allium cepa* L). *Euphytica* 18: 389-394.

Welch JE and Grimball Jr EL. 1947. Male sterility in the carrot. *Science* 106: 594-597.

Williams ME, Lecmans J and Michiels F. 1997. Male sterility through recombinant DNA technology. In : *Shivanna KR and Sawhney VK (eds), Pollen Biotechnology for Crop Production and Improvement.* Cambridge Univ. Press, pp. 237-257.

Woong Yu I. 1985. *Inheritance of cytoplasmic male sterility in pepper* (*Capsicum annuum* L.) MSc Thesis, Kyung Hee University, South Korea.

Yang CJ, Chen BJ and Liu JG. 2007. A new hot pepper F_1 hybrid "Chanjiaozidantou". *China Vegetables* 3: 30-31.

Yoo Woong IL. 1990. *The inheritance of male sterility and its utilization for breeding in pepper (Capsicum spp.).* Ph.D. Thesis, Kyung Hee University, South Korea.

Yordanov M, Gentschev S and Stoyanova Z. 1971. Morphological changes in male sterile tomato mutants treated with gibberellins. *CR Academy of Science and Agriculture,* Bulgaria 4 (1): 55-61.

Zhang XP, Rhodes BB, Baird WV, Skorupska HT and Bridges WC. 1996. Development of genic male sterile watermelon lines with delayed-green seedling marker. *Horticulture Science* 31: 123-126.

Zhu YY, Wu XG, Gong J, Jin HJ, Yoo WY, Li SJ and Wu QS. 2001. Breeding and utilization of a male sterile line with radish cytoplasm in broccoli. *Journal of Nanjing Agricultural University* 24(3): 15-18.

The Science of Horticulture : Vol. 1, 2011
© K.V. Peter (ed.), pp. 145-163
New India Publishing Agency, New Delhi (India)
E-mail : newindiapublishingagency@gmail.com
Web: www.bookfactoryindia.com

Chapter 6 Polyembryony : The Basics

KUNDAN KISHORE

Polyembryony is the occurrence of more than one embryo in a seed. It includes all those instances in which there is a clear indication of the actual occurrence of two or more embryos in a developing ovule (Tisserat *et al.* 1979). The additional embryos in polyembryony result from the differentiation and development of various maternal and zygotic tissues associated with the ovule of seed. Earlier, polyembryony is said to have an abnormal feature but now it is considered as a desirable character in citrus, mango, jamun, almond, etc. to obtain true to type planting materials. The polyembryony is one of the apomictic processes which have been described to occur in the ovules of angiospermic species (Koltunow, 1993). Polyembryony in horticultural crops occurs as facultative apomixis wherein simultaneous growth of multiple embryos of somatic origin co-exist in the same seed containing sexual embryo resulting from either selfing or cross pollination.

In normal sexual cycle (amphimixis), diploid sporophytic cell (Megaspore Mother Cell) of ovule transforms to haploid gametophytic cells (embryo sac) through meiosis which contain egg cell that forms embryo after fertilization with pollen grain (syngamy). But in some plants meiotic division and syngamy are eliminated and still a viable embryo is formed inside the seed. The formation of embryo with asexual means is called apomixis (Apo=away from+ mixis = act of mixing) and seed is called apomictic seed. Thus apomixis refers to substitution of the usual sexual reproduction by a form of asexual reproduction which does not involve meiosis and syngamy for embryo formation (Bhojwani and Bhatnagar, 1999)

Polyembryony was first reported by Leeuwenhoek in citrus as early as 1719 and the different cases of polyembryony in plants were studied by Braun in 1859. In 1878 Strasburger demonstrated the formation of plural embryos in many genera of angiosperms. In 1901, Earnst summarized the works done in polyembryony and classified the various means by which adventitious embryos may be derived. Following Ernst's work, it soon became apparent that polyembryony is not an abnormal feature but rather a desirable character (Kepiro and Roose, 2007). About 244 species of 144 genera belonging to 59 angiosperm families are reported to exhibit either polyembryony or adventitious embryony (Kumar *et al.* 1991). Despite its wider occurrence, the evolution of polyembryony is poorly understood. Several workers studied the evolution of apomixis and adventitious embryony in the context of their similarity to asexual reproduction (Khoklov, 1976). Because of the lack of clear distinction between polyembryony, apomixix and adventitious embryony, the former is also assumed to be akin to asexual reproduction.

To understand clearly the origination and formation of multiple ovules in angiospermic seed, it will be better to know the anatomical structure of ovule.

Ovule

The female reproductive apparatus in a flower is gynoecium and its functional unit is called carpel. A typical carpel is comprised of swollen ovary, style and stigma. The ovule forms inside an ovary which in turn forms seed. A well developed ovule consists of embryo sac, nucellus enclosed almost completely by one or two integuments leaving a small opening at the apical end and micropyle. The micropyle is the main passage for entry of pollen tubes in embryo sac. The base of the ovule is called funiculus which is attached with ovary wall through placenta. The anatropous type of ovule is common in horticultural crops (Kishore *et al.* 2009), in which micropyle lies close to funiculus and ovule looks inverted (Fig. 1). The central portion of the ovule is occupied by embryo sac which is surrounded by sporophytic nucellar cells. The nucellar region opposite to the micropyle is called chalaza. In horticultural crops, the embryo sac, nucellus and chalaza are surrounded by two integuments thus make the ovule bitegmic (Frost and Soost, 1968; Bhojwani and Bhatnagar, 1999).

The embryo sac, also called female gametophyte, is the most important structure wherein fusion of male gamete takes place and zygotic embryo and endosperm form. Embryo sac is a 7 celled- structure containing two polar nuclei in the centre, egg apparatus at the micropyler region comprising of one egg cell and two synergids and at the chalazal end, three antipodal cells are present (Fig. 2). Cells of egg apparatus and antipodal cells are uninucleate and haploid whereas the central cell (polar nuclei) is binucleate or diploid. In angiosperm, double fertilization is the rule in which one male gamete fuses with egg cell and forms embryo (2n) while other gamete fuses with the polar nuclei and forms endosperm (3n). In case of polyembryonate species, extra embryos are formed by nucellar cells (2n), integuments (2n), synergids (n), antipodal cells (n), etc. The developing embryos get nutrition from endosperm which acts as a conductive tissue and the enlarging embryos gradually replace most of the endosperm (Pullaiah *et al.* 2001).

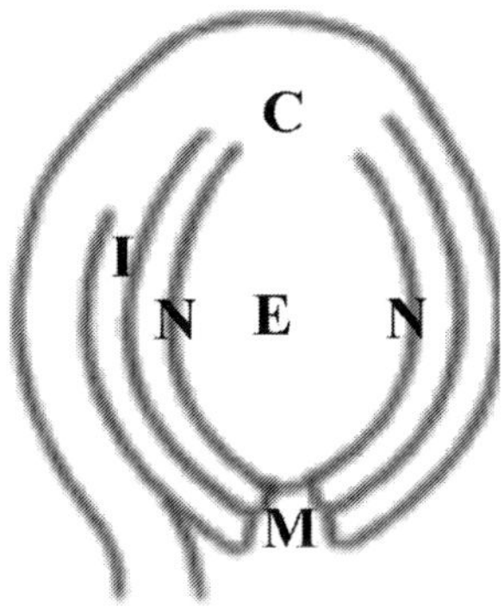

Figure 1. Anatropous ovule of citrus
E- Embryo sac, N- Nucellus
M - Micropyle, I- integument
C- Chalaza, F- Funiculus

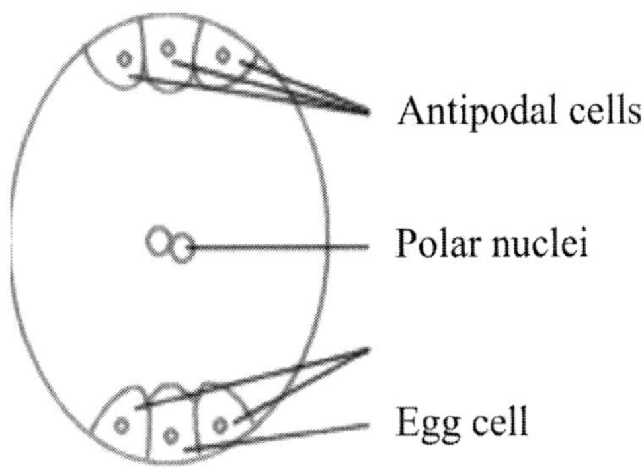

Figure 2. An Embryosac

Different Ways of Polyembryony

The polyembryony is a type of apomixis that may occur by following ways (Bhatnagar and Bhojwani, 1999).

1. Cleavage of proembryos
2. Formation of embryos by cells of embryo sac other than egg
3. Development of more than one embryo sac within the same ovule or multiple polyembryony
4. Formation of embryo by some sporophytic/maternal tissue of ovule.

Cleavage of Proembryo

In this case, the embryo that forms after normal fertilization divides irregularly and forms mass of cells which proliferate and develop into many embryos. The cleavage polyembryony is quite common in Orchidaceae.

Embryos from Cells of the Embryo Sac other than The Egg

In this category, the most common source of additional embryo is the synergids (n). The ploidy of embryo depends on the origin. If it arises from unfertilized synergid, the embryo will be haploid and if it arises from fertilized synergid the resultant embryo will be diploid. The formation of diploid embryo from synergid is brought about by the entry of more than one pollen tube into the embryo sac or by the presence of additional male gamete in the same pollen tube. Embryos arising from unfertilized synergids are known in French bean (*Phaseolus vulgaris*). In this case, the synergid embryo (haploid) and zygotic embryo (diploid) can be distinguished on the basis of ploidy level. The development of embryo from antipodal cells and endosperm cells is rather rare. The embryo from antipodal cell may be haploid or diploid depending up on the fusion with male gamete but generally fails to produce germinating embryo and gets degenerated at an early stage whereas, endosperm produces triploid embryo and seedling.

More than One Embryo Sac in the Same Ovule

Multiple embryo sac may arise in an ovule either from the same megaspore mother cell or from other megaspore mother cell or from any other sporophytic tissue of ovule. Although many embryos develop inside an ovule but hardly one or in few cases two embryos survive and produce seedlings. This condition is also called multiple polyembryony.

Activation of Some Sporophytic Cells of the Ovule

The embryos arising from the maternal sporophytic cells (outside the embryo sac) are called adventive embryos. The maternal cells which involve in embryo development are nucellus and integument. The embryo developed from nucellus is called nucellar embryony which is a common feature in the families of horticultural importance (Kobavashi *et al.* 1979). Polyembryony in horticultural crops is associated with nuceller embryony because it frequently results in polyembryonic seeds from which multiple seedlings germinate.

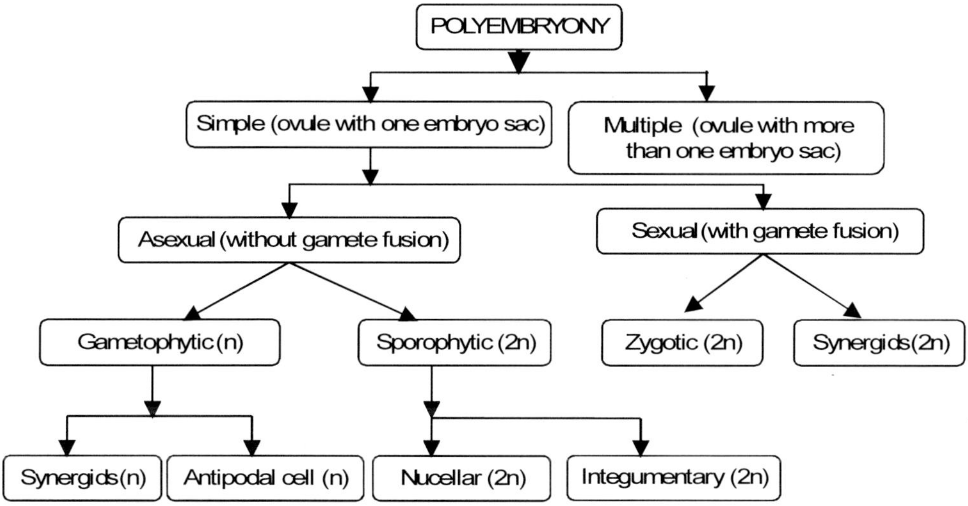

Adventive/Nucellar Embryogenesis

Nucellar embryony is an adventitious form of apomictic reproduction wherein the somatic cells of the nucellus tissue are initialized to enter into an embryonic pathway of development (Koltunow *et al.* 1995). In polyembryonic seed, many nonzygotic, nuceller embryos are initiated directly from the maternal cells surrounding the embryo sac containing a developing zygotic embryo. During embryo sac expansion, embryogenic nuceller cells obtain access to endosperm and develop into embryo along side the zygotic embryo that may or may not develop completely. Nuceller embryo gives rise to seedlings that are of the same genotype as the female parent. In addition to the presence of normal zygotic and nucellar embryos, twin and triplet zygotic embryos were also observed in the seeds of mandarin collected from Darjeeling district (Das *et al.,* 2007). These embryos were developed by fission of the original zygotic embryo.

The adventive embryogenesis is completed in four steps: (I) formation of adventive Embryo Initial Cells (AEICs), (II) differentiation of AEICs (III) division of AEICs and (IV) development of adventives embryos (Wakana and Uemoto, 1988). AEICs generally appear before anthesis and are characterized by their homogenous cytoplasm, a large nucleus and irregular plastids. After pollination, the cell composition is changed and starts synthesizing more energy. In fertilized seeds, the division of the AEICs generally occurs before the division of zygote but after endosperm division (Kobayashi *et al.* 1979; Koltunow *et al.* 1995). The recent studies show that the initiation of adventive embryos in Citrus occurs autonomously and not affected by pollination, fertilization or the development of zygotic embryo or endosperm

(Wilms *et al.* 1983). However the development of adventive embryos is greatly influenced by endosperm development. Lack or poor development of endosperm results in poor development of adventitve embryos. Therefore, in unfertilized seeds, the nucellar embryos fail to develop beyond certain stage and are incapable of normal germination. (Wakana and Umeto, 1987). The presence of endosperm promotes the development of adventives embryos at the micropyler end, but suppresses their development towards chalazal end. The degree of suppression is directly related to the distance of embryo from the micropylar end. In normal seeds, the AEICs at the chalazal end generally do not develop beyond the initial celled stage.

In a genotype that produces polyembryonic seeds by nucellar embryony, normal zygotic embryo is also formed and such a genotype can produce different types of seeds (Wakana and Uemoto, 1988): (i) seeds with one mature zygotic embryo developed by sexual reproduction and in this case seed becomes monoembryonic: (ii) seeds with one mature nucellar embryo only and in this case also seed shows monoembryony: (III) seeds with multiple mature nucellar embryos resulting polyembryonic seed: (IV) seeds with one mature zygotic embryo and one or more mature nucellar embryo and in this case seed becomes polyembryonic.

In citrus, nucellar embryony does not prevent normal sexual reproduction and zygotic embryo is also formed (Esan and Soost, 1977; Wilms *et al.* 1983). Thus citrus seed produces both types of seedlings having zygotic and nucellar origin. The number of nucellar embryo varies with seed to seed and species to species so a seed may contain two or more seedlings but not every seed produced by the plant with nucellar embryony has multiple mature embryos.

Regarding the location of embryos in the seed, the positioning of nucellar embryos is not fixed and they may be arranged near micropyle or away from micropyle. The zygotic embryos are generally present at the micropylar end of embryo (Thakur and Bojwa, 1971).

Carimi *et al.* (1998) reported that zygotic embryo developed at faster rate and was at more advanced stage at the micropyler region of the embryo 105 days after pollination (DAP), while the growth of nucellar embryos was relatively slower and they attained the developmental stage similar to zygotic embryo 150-200 days after pollination.

Classification of Polyembryony

The term monoembryony is used to refer to a single seed that contains one embryo to describe a strictly sexual seed parents. Monoembryonic species do produce seeds with two or more embryos by twining at a low frequency but these embryos will be genetically different from the maternal parent.

Camaroon and Soost (1979) classified the polyembryony on the basis of frequency of polyembryony.

i. *Strictly polyembryonic* – Plant species in which the frequency of multiple embryos is lesser than 7% is described as strictly polyembryonic.

ii. *Nearly monoembryonic* – In case of nearly monoembryonic plant species, the frequency of polyembryony varies 6-10%.

iii. *Polyembryonic* - If the per cent age of multiple seed is more than 10, this condition is called polyembryony and plants are called polyembryonate.

Ernst (1910) divided polyembryony into two categories on the basis of embryogenesis

i. *True polyembryony* – When two or more embryos arise in the same embryo sac from nucellus (citrus, mango, jamun), synergid etc.

ii. *False polyembryony* – In this case, more than one embryo sac is formed in an ovule (*Fragaria sp.*) which is followed by the formation of multiple embryos.

Yakovlev (1967) proposed a genetic basis of classification of polyembryony.

i. *Gametophytic* – Multiple embryos arise from the gametic cells of the embryo sac (synergid, antipodal) after or without fertilization. In this case, haploid embryos are formed.

ii. *Sporophytic* : When multiple embryos arise either from zygote or from sporophytic cells of ovule (nucellus, integument) and the resulting embryo will be diploid. Hence citrus, mango and jamun exhibit true and sporophytic polyembryony.

Polyembryonate Crops

Presence of multiple embryos in the same seed is more frequent in Gymnosperms than in Angiosperm. In fruit tree crops, polyembryony is also common and occurs in many crops; citrus (Frost, 1938), mango (Sachar and Chopra, 1957), Jamun (Narayanswamy and Roy, 1960), Kiwi (Crete, 1944), Almond (Kester and Gordzeil, 1996), *Fragaria sp.* (Lebegue, 1952) mangosteen (Campbell, 1967) and peach (Toyama, 1974). Occurrence of polyembryony depends up on species and varieties, in other words, not all species of a genus exhibit polyembryony and not all varieties of a species show polyembryony and the reason could be genetic. Most of the citrus species show polyembryony, while C. *medica* (citron), C. *grandis* (pummel) C. *latifolia* (Tahiti lime) and C. *nobilis* (King mandarin) are monoembryonic. In mango, most of the varieties grown in coastal areas of India exhibit polyembryony. A list of polyembryonic species and varieties are given below.

List of polyembryonic species and varieties

Crops	Polyembryonate species/ varieties
Citrus and related species	*Citrus reticulata* L. (tangerines), C. *reshni* hort. ex Tanaka (Cleopatra mandarin) *C. deliociosa* Ten. (Willow leaf Mandarin), *C. unshu* Marc. (Satsuma mandarin), *C. medurensis* Lour. (calamondin) *C. sinensis* Osbeck (sweet orange) *C. paradisii* Macfad (grapefruit) *C. aurantium* L. (sour orange) *C. aurantifolia* Swingle (sour lime) *C. limon* Burm. (lemon), *C. jambhiri* lush., (rough lemon), *C. latipes* (Swingle) Tanaka (khasi papeda), *C. depressa* Hayata. (flat lemon), *C. volkarmeriana* (volkamer lemon), *C. tiawanica, C. limettoides* (sweet lime), *C. limetta, Citrus karna* (karna khatta), *C. limonia* (Rangpur lime), *Fortunella sp., Pocirus trifoliate Raf.*, Citrange,
Mango varieties	Bappakai, Chandrakaran, Kensington, Kitchner, Kurukkan, Muvandan, Mylepelian, Nekkare, Olour, Valaikolumban, Goa, Goa Kasargod, , Olour, Nileswar Dwarf, Salem Peach, Prior, Starch, Cambodiana, Carabao, Corazont, Paho, Pahutan, Pico, Senora and Strawberry
Almond varieties	Nonpareil, Mission, Price, Sonora, Carmel

Factors affecting polyembryony

The number and type of embryo produced may vary from tree to tree and also at different positions on a single tree. (Parlevliet and Cameron, 1959). The variation is suggested to be controlled by minor genes, pollen sources and environmental conditions (Khan and Roose, 1988). Occasionally, the zygotic embryo does not survive and multiple. Embryos arise from nucellus and develop into mature embryos. Even if the zygotic embryo does not survive, the development of nucellar embryo is dependant on the initial development of zygotic embryo or on the process of fertilization of ovule (Tisserat *et al.* 1979) or on pollination. Thus nucellar embryo apparently may not develop independently of the process of sexual reproduction and in most of the cases, it requires the development of zygotic embryo. In citrus, neither the seed setting nor the formation of the nucellar embryos takes place without pollination of flower. The development of nucellar embryos is induced by fertilization. The process of formation of these embryos is initiated not only by the pollen transferred to the stigma of the flower, it depends upon the degree of pollen germination and the speed of pollen tube growth (Akira *et al.* 1987; Koltunow,

1993). It was observed that number of nucellar seedlings depends upon the period of fertilization of the ovule. If fertilization is delayed and the rate of fertilization is slower, the development of zygotic embryo is also delayed which stimulates more formation of nucellar embryos (Esen and Soost, 1977). It is established that pollen plays a definite role in the formation of additional embryos. Pollen from different species of citrus decreases or increases the number of nucellar seedlings. The pollen of winter hardy wild species *Citrus latipes* (Khasi papeda) is more viable and remains active during crossing than cultivated species and stimulates formation of nucellar seedlings (Kolelisavili, 1970)

Polyembryony is also affected by the type of pollinator (Soares Filho *et al.*, 1995), amount and pollen viability , plant nutrition, air temperature, environmental and soil humidity, and speed of the wind. Dhillon *et al.* (1993) reported that adventitious embryos developed into globular or early cotyledon stages in the absence of pollination, but the embryos required the endosperm in order to grow, making pollination and fertilization necessary for the development of the polyembryonic seeds. The growth of nucellar embryo depends upon fertilization for endosperm formation. During fertilization, one male gamete meets with egg cell to form zygotic embryo while other male gamete fuses with polar nuclei to form endosperm. Since endosperm supplies food to developing embryos, the fertilization is a prerequisite for the production of matured embryo. The growth of nucellar embryo is arrested at later stage of development if endosperm is not formed (Koltunow *et al.* 1996).

Therefore any factor that affects pollination, fertilization or seed development will also affect the percentage of polyembryony and embryos/ seed. The difference in results obtained in various locations shows the importance of determining the characteristics of polyembryony.

The number of embryos/ seed was observed to be inversely proportional to the length of embryo. The number of embryos / seed was negatively correlated with embryo length, implying that as the number of embryos / seed increased, embryo size decreased. This may be due to competition for nutrients among embryos (Xiang and Roose, 1988).

Causes of Polyembryony

Many theories are proposed to explain the occurrence of polyembryony but most of them are not sufficiently validated. Haberlandt (1921, 1922) proposed the 'necrohormone theory'.The theory advocated that the degenerating cell of the nucellus acts as a source of stimulus for the adjacent cells to divide and form adventives embryo. But the theory could not be validated as adventive embryo could not be induced by damaging the nucellar cells. The monoembryonate conditions in some species of citrus are ascribed to the synthesis and release of certain volatile and non-volatile embryogenic inhibitors in their ovules which do not occur in the ovules of polyembryonate species. Ethanol and ethylene are important volatile inhibitors produced by the ovule of monoembryonate species of citrus (*Citrus medica*). The ethylene among other substances could be repressing development of nucellar embryos through its union with an ethylene receptor protein of nucellar cells. The fewer number of adventives embryos in south side of tree than that of north side, might be due to more synthesis of ethylene, ethanol and abcisic acid due to variation in sunlight and temperature (Garcia *et al.* 1999). The non-volatile component of the inhibitors was auxin, ABA and GA_3. But genetic theory is the most accepted theory as the presence of polyembryony is determined by the gene.

Inheritance of Nuceller Embryony

Polyembryony in citrus *in vivo* is controlled by multiple recessive genes (Maheshwari and Rangaswamy, 1958). In monoembryonic species, these recessive genes may synthesize a potent inhibitor of embryogenesis. Parlevliet and Cameron (1959) studied inheritance of polyembryony in citrus and came out with following findings.

- The polyembryony in citrus is generally controlled by a dominant gene having heterozygous allele (Pp) while homozygous recessive gene (pp) is present in monoembryonic citrus species.
- The hypothesis is supported by the findings obtained by crossing monoembryonic (pp) and polyembryonic (Pp) parents with the resultant ratio of 1:1.
- When polyembryonic parents were crossed, the ratio of polyembryonic and monoembryonic offspring was 3:1.
- A variable degree of polyembryony was recorded in polyembryonate offspring obtained by crossing monoembryonic and polyembryonic and between polyembryonic parents that implied the presence of minor genes affecting degree of polyembryony.
- The presence of modifier or duplicate genes should also be considered as in some crosses between monoembryonic and polyembryonic parents, and between polyembryonic parents, the progeny ratio varied greatly from 1:1 and 3:1 ratios.

According to Sturrock (1968), polyembryony in mango appeared to be inherited by recessive gene, however more recent studies demonstrated that polyembryony in mango is controlled by a single gene (Aron *et al.*, 1998) as in citrus and the ratios obtained after crossing were equal than that of findings of Parlevliet and Cameron (1959).

Thus it can be concluded that mostly polyembryony is controlled by heterozygous dominant gene and the absence of dominant allele leads to monoembryony. But in some of the cases, modifier genes and minor genes are also present tinkering with the ratios and degree of polyembryony respectively.

Organogenesis

A typical dicotyledonous seed of citrus, mango, jamun, etc. comprise of embryonal axis with two cotyledons attached laterally with axis. The portion of embryonal axis above the level of cotyledon is called epicotyle and the portion below the level of cotyledon is called hypocotyls. The epicotyle terminates into plumule (embryonic shoot), and the hypocotyls at its basal end produce radical (embryonic root)

When a citrus seed germinates, the hypocotyle and primary root are the first organs to appear and the primary root grows rapidly and soon becomes yellow-brown. After the elongation of primary root, the epicotyl begins to elongate by intercalary growth and produces shoot (Fig. 3).

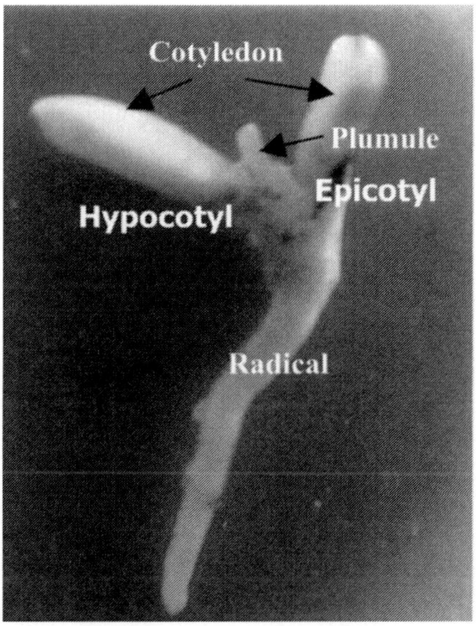

Figure 3. Organogenesis in *Citrus reticulata*

The elongating shoot may lift the seed coat out of the ground if the seed was planted with the chalaza upright. While the epicotyls are elongating, the first foliage leaves expand only slightly, but thereafter they expand and mature over a period of several weeks. The growth of various parts of the citrus seedling occurs in a sequence; first, the hypocotyl-root axis grows; second, the epicotyl; third, the first foliage leaves; and finally, after about two months, stem apical growth commences, and the seedling becomes an established plant (Bhojwani and Bhatanagar, 1999).

Degree of polyembryony

The degree of polyembryony is the frequency of occurring multiple seedlings in a seed. It varies with species and varieties and the variation also occurs with location, environment and position of fruits on tree. On the basis of degree of polyembryony, species/varieties can be divided into three categories; slightly polyembryonic (polyembryony up to 25%), moderately polyembryonic (polyembryony up to 50%) and highly polyembryonic (polyembryony > 50%). Iwamasa *et al.* (1967) studied the embryos in citrus species and reported that the number of embryos in trifoliate orange, Dancy tangerine and sweet orange were 5, 40 and 17 respectively. Although the occurrence of more than one embryo within a mature seed was a common phenomenon but in addition to the presence of zygotic and nucellar embryos, twin and triplet zygotic embryos were also observed and the frequency of occurrence was up to 30% (Das *et al.* 2007).

The study on Sikkim mandarin (*C. reticulate*) showed a polyembryony range of 55.0-67.0%. The maximum number of seedlings/seed was observed to be 6 and the percent occurrence of multiple seedlings decreases with the increase in the seedling number/seed. The percent of one seedling/seed was 51.5 followed by two seedlings/seed (36.1). The % of 3, 4, 5 and 6 seedlings/seed was 7.5, 1.8, 0.6 and 0.1 respectively (Fig. 4, 5, 6). It was also observed that the degree of polyembryony was positively correlated with the occurrence of three or more number of seedlings (Kishore *et al.* 2008).

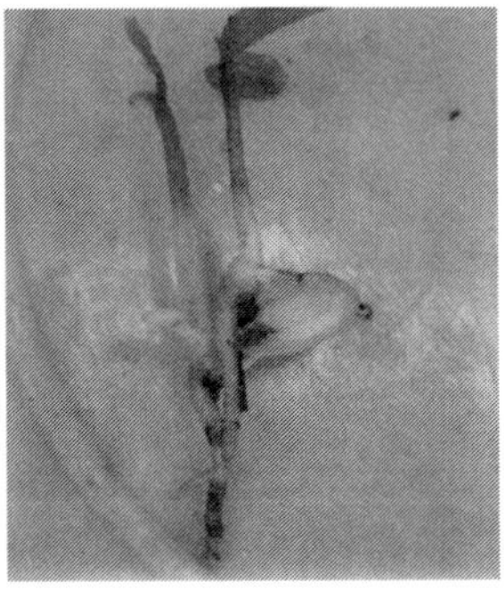

Figure 4. Citrus seed with two seedlings

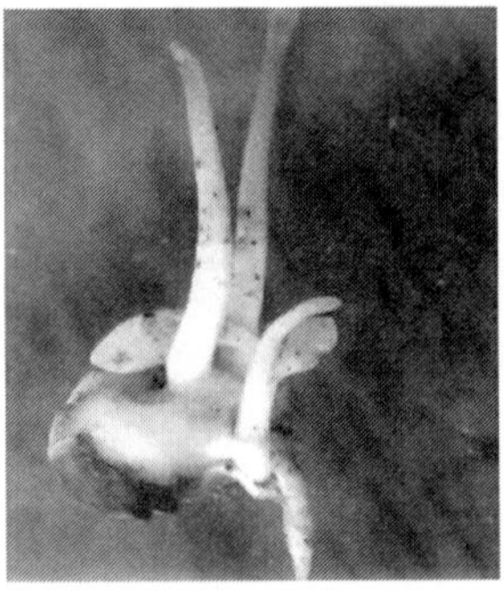

Figure 5. Citrus seed with three seedlings

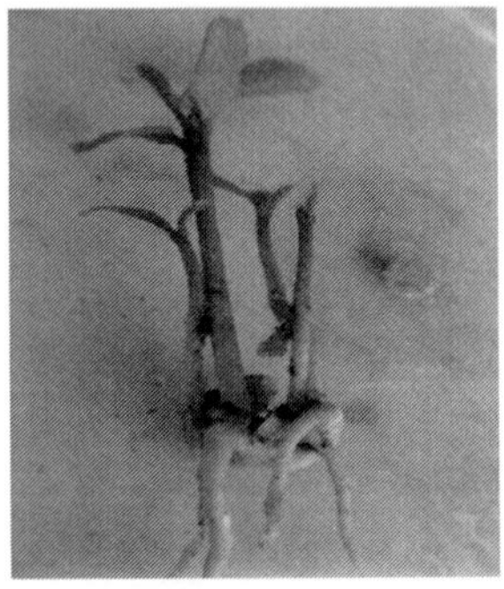

Figure 6. Citrus seed with four seedlings

Identification of Nucellar Seedlings

Morphological identification of nucellar and zygotic seedlings at juvenile stage is practically difficult. Moreover it is not an effective method, as the chance of selecting wrong seedling is more. Although, visual recognition is the simplest method and is effective when the male and female parents differ significantly in their growth habit but when the two parents are similar, as is the case with the most citrus cultivars, separation is more difficult. The difference in zygotic and nucellar seedlings can be made only after fruiting and the plant possessing the characters of mother pant will be nucellar in origin while the plant showing variation in characters is zygotic in origin. On an average, zygotic seedlings are smaller than nucellars, and many, but not all can be distinguished using physical character variations enough in height or size alone could not be used as a criterion for selection. There is also considerable variation among zygotic seedlings and visual recognition on the basis of height, leaf size, thorn length, petiole length and stem diameter and is especially difficult in non-hybrid cultivars. Morphologically, the zygotic seedling can be identified if one variety is crossed with trifoliate orange (*Poncirus trifoliate*) as the resultant zygotic seedlings will have trifoliate leaves as this character is controlled by dominant gene (Khan and Roose, 1988).

Many distinguishing characters for nucellar and zygotic seedlings at seedling stage were proposed by various workers and some of these characters are proved to be a marker. Cytologically the nucellar cells destined to form adventive embryos can be distinguished from other cells of the nucellus by their dense cytoplasm and starchy contents (Wilms *et al.* 1983). The zygotic embryo present at the micropyler region holds two original cotyledons of the seed. Zygotic embryo was usually larger than the other embryos and also took the most space in a seed (Fig. 7), while nuclear embryos were tiny, heart shaped, green and were crowded at the micropyler region and the most of these could be easily separable (Das *et al.* 2005). Nucellar embryos can be distinguished from the zygotic embryo by their lateral position in the embryo sac and irregular shape. Individual nucellar embryos had two cotyledons, and rudimentary radicals and plumules were present in between two cotyledons. Generally, the most vigorous and first germinating seedling of citrus is zygotic as it has the largest cotyledon and occupies most of the space of seed (Fig. 8)

Similarly, in mango, the earlier concept was that the zygotic plantlets are being the weakest in polyembryonic mango seed because it probably degenerates due to competition with nucellar plantlets (Sachar and Chopra, 1957). The recent findings prove that zygotic plantlets are vigorous and positioned near micropyle and possess big cotyledons. The vigorocity of zygotic plants can be explained by a heterotic effect of cross between the female plant and unidentified male plant (Maria *et al.* 2006).

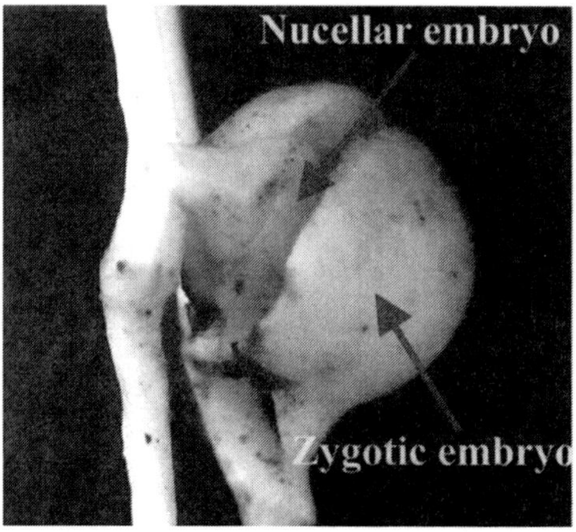

Figure 7. Zygotic and Nucelar embryos of *Citrus reticulata*

Figure 8. Zygotic and Nucelar seedlings of *Citrus reticulata*

Interestingly, in addition to presence of the single normal zygotic embryo, twin and triplet zygotic embryos were also found (Das *et al.* 2007). Some identical twin seedlings showed a common primary root and in this case, seedlings germinate with common hypocotyls and primary root but the two epicotyles are distinctly separated from the cotyledonary node. In some other cases, during germination, the two hypocotyls and also the epicotyls of the identical twins are fused (Fig. 9 and 10). Twins are seemed to develop by fission of the original zygotic embryo and the original two cotyledons of the zygotic embryo are also divided almost equally and shared among twins or triplets. The presence of twins further makes the identification process rather more complex.

The biochemical techniques are effective and expeditious tools to identify zygotic and nucellar seedlings at an early stage. Among biochemical techniques, isozyme markers are commonly used to distinguish nucellar (true-to-type) and zygotic (off-type) citrus seedlings (Moore and Castle, 1988; Anderson *et al.* 1991). This technique is based on the presence or absence of protein bands in the sap of leaf, root or bark sample. Since the results may be influenced by the environment, as well as by the stage plant and its organs, this method is relatively less reliable for seedling identification.

Molecular techniques by using molecular markers (RAPD, RFLP, SSR, ISSR, QTL, etc.) are the most recent, advanced, effective, expeditious and reliable tools to identify zygotic and nucellar seedlings at an early stage as results are not influenced by external factors. Das *et al.* (2007) identified the zygotic seedlings through RAPD markers and reported that zygotic seedlings (twin or triplets) usually had one or more extra band than those of the nucellar seedlings. They also observed the variability in mandarins of North East by RAPD profiling. Maria *et al.* (2006) and Abirami *et al.* (2008) identified the nucellar seedlings of mango through RAPD and also determined the position of zygotic and nucellar embryos in an ovule. Carimi *et al.* (1998) used RFLP analysis of low and high- copy- number DNA to distinguish zygotic and nucellar seedlings and reported that in 82-88 % cases, zygotic embryos develop in early stage of ovule development. Nicasio *et al.* (2002) identified nucellar and zygotic seedlings with markers amplified with SSR (simple sequence repeat) and inter-simple sequence repeat (ISSR) primers.

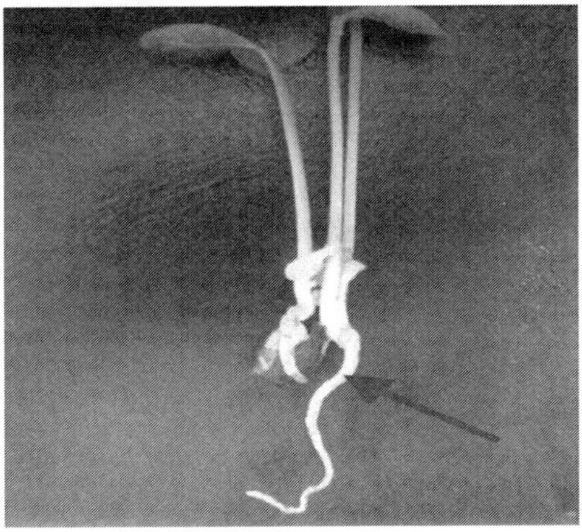

Figure 9. Zygotic twins in *Citrus reticulata*

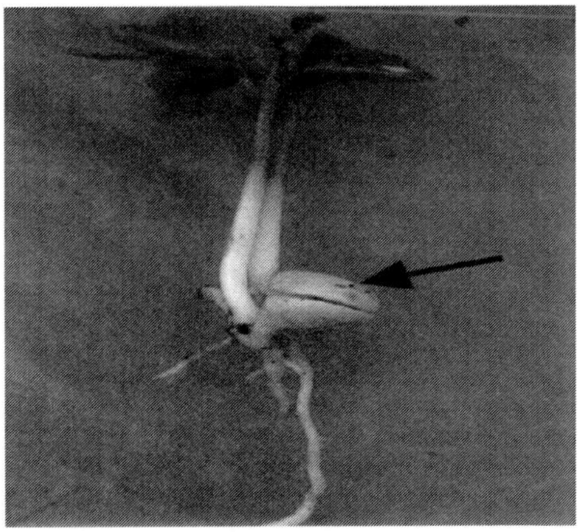

Figure 10. Zygotic twin in *C. reticulate* having equal cotyledonsize

Significance of Polyembryony

Polyembryony or nucellar embryony plays an important role in horticulture. It provides uniform seedlings of the parental type, as obtained through vegetative propagation and because of this the nuclear seedlings of citrus, mango and jamun furnish better clones of orchard stock. Nucellar embryony possesses following advantages:

1. The nucellar embryos are free from diseases and it is the only practical approach to raise disease free clones for root stocks and seedlings.
2. The nucellar seedlings have a tap root and, therefore develop a better root system.

3. The nucellar seedlings show a restoration of the vigour lost after repeated vegetative propagation.
4. Polyembryony helps in the large scale propagation of desired genotype.
5. Avoids the complications associated with sexual reproduction like pollinators and cross incompatibility.

Although nucellar polyembryony has great significance in horticulture for the production of uniform and disease free planting materials but it poses problems for breeders. For the development of new lines/varieties, zygotic embryos/seedlings are required to be raised but if parents are highly polyembryonic, the chance of getting hybrid seedlings will be less.

Conclusion

Polyembryony is an example of facultative apomixis wherein simultaneous development of nucellar and zygotic embryos takes place in the same seed. It is exhibited by many horticultural crops and in most of the crops, it is used for production of true to type plants. It is considered to be an essential trait because it provides a low cost, low-technology method for propagation of genetically uniform plants. The number of nucellar embryo varies with species to species and plant to plant, moreover, a variation in the degree in polyembryony may observe in the same plant. The degree of polyembryony may be more than 100 % in different crops. In highly polyembryonic plants like *Citrus reticulata, Citrus sinensis, Citrus jambhiri,* etc. the maximum number of seedlings/seed may be 6 or more. The visual identification of nucellar seedlings is difficult but some of the indicators are useful in differentiating zygotic and nucellar seedlings. The first germinating and vigorous seedling with big cotyledon is generally considered zygotic. But the occurrence of zygotic twins in citrus further makes the selection process complex. Although, visual recognition is the simplest and effective when the male and female parents differ significantly in their morphological characters but when the two parents are similar, separation is more difficult. Biochemical and molecular techniques are the most effective and expeditious tools to identify zygotic and nucellar seedlings at an early stage.

References

Abirami K, Singh S K, Singh Room, Mohapatra T and Kumar Anand Raj. 2008. Genetic diversity studies on polyembryonic and monoembryonic mango genotypes using molecular markers. *Indian Journal of Horticulture* 65(3): 258-62.

Akira Wakana and Shunpei Uemoto. 1987. Adventive Embryogenesis in Citrus I. The Occurrence of Adventive Embryos without Pollination or Fertilization. *American Journal of Botany* 74(4): 517-30.

Anderson C M, Castle W S and Moore G A. 1991. Isozymic identification of zygotic seedlings in Swingle *Citrus paradisi x Poncirus trifoliata* nursery and field population. *Journal of American Society for Horticultural Sciences* 116; 322-6.

Aron Y, Czosnek H, Gazit S and Degani C. 1998. Polyembryony in mango *(Mangifera indica* L.) is controlled by a single dominant gene. *Hort science* 33(7): 1241-2.

Bhojwani S S and Bhatnagar S P. 1999. Polyembryony. (in). *The Embryology of Angiosperms.* pp 236-253. Vikas Publishing House Pvt. Ltd. New Delhi.

Camaron J W and Soost, R K. 1979. Sexual and nucellar embryony in F1 hybrid and advanced crosses in *Citrus* with *Poncirus. Journal American Society Horticultural Sciences* 104(3): 408-10.

Campbell, C W. 1967. Growing the Mangosteen in Southern Florida. *Proceedings of the Florida State Horticultural Society.* 79: 399-401.

Carimi, F., Tortorici, M.C. Pasquale, F.de. and Crescimanno, F.G. 1998. Somatic embryogenesisand plant regeneration from undeveloped ovules and stigma/style explants of sweet orange naval group (*Citrus sinensis* (L.) Osb.). *Plant Cell, Tissue and Organ Culture* 54(3): 183-9.

Crete P. 1944. Polyembryony in *Actinidia chinensis. Bulletin Society of Botany France* 91: 89-92.

Das A, Mandal Bidisha, Sarkar J and Chaudhuri S. 2005. Variability in multiple embryo formation and seedling development of mandarin orange (*Citrus reticulata* Blanco) of north-eastern Himalayan region of India. *PGR News Letter* 143: 35-9.

Das A, Mandal Bidisha, Sarkar J and Chaudhuri S. 2007. Occurrence of zygotic twins seedlings in mandarin orange plants of northeastern Himalayan Region. *Current Science* 92(11): 1488-9.

Dhillon, R S, Kaundal G S and Cheema S S. 1993. Nucellar embryony for propagating *Citrus. Indian Horticulture* 38: 44- 5.

Esen A and Soost K. 1977. Adventitve embryogenesis in Citrus and its relation to pollination and fertilization. *American Journal Botany* 64: 607-14.

Frost H B . 1938. Nucellar embryony and juvenile characters in clonal varieties of citrus. *Journal of Heredity* 29: 423-32.

Garcia R, Asin M J, Forner J and Carbonell E A. 1999. Genetic analysis of the apomixis in Citrus and Poncirus by molecular markers. *Theoretical and Applied Genetics* 99: 511-18.

Iwamasa, M, Ueno, I and Nishiura, M. 1967. Inheritance of nucellar embryony in citrus. *Bulletin Horticultural Research Station Japan* Series B 7: 1-8.

Kepiro J L and Roose M L. 2007. Nucellar Embryony. (in). Citrus: Genetics, Breeding and Biotechnology pp. 141-49. (Eds) Khan I A. CAB International, UK.

Kester D E and Gradziel T M. 1996. Almond. (in): Fruit Breeding pp1-97. (Eds) Janik J and Moore J D. John Wiley and Sons, New York.

Khan I A and Roose M L. 1988. Frequency and characteristics of nucellar and zygotic seedlings in three cultivars of trifoliate orange. *Journal American Society Horticulture Science* 113:10510.

Khokhlov, S.S. 1976. Evolutionary genetic problems of apomixis in angiosperms. (in): Apomixis and breeding pp 317. Khokhlov, S. S. (Eds). Sharma, B. R. transl. Amerind, New Delhi.

Kishore Kundan, Kumar A, Rahman H, Monika N and Pandey Brijesh. 2008. Studies on degree of polyembryony in Sikkim Mandarin (*Citrus reticulate* Blanco). Abs. p 66. 3rd Indian Horticulture Congress. Nov. 6-9, OUAT, Bhubneshwar.

Kishore Kundan, Pathak K A and Shukla Rohit. 2009. Sequential development of reproductive organs in yellow passion fruit (*Passiflora edulis*). *The Indian Journal of Agricultural Sciences* 79(2): 141-3.

Kobayashi S, Ieda I and Nakantani M. 1979. Studies on the nucellar embryogenesis in citrus. Formation of the primordial cells of the nucellar embryony on the ovule of the flower bud, and its meristematic activity. *Journal of Japanese Society of Horticultural Sciences* 48: 179-85.

Kolelisavili, M V. 1970. Use of polyembryony in the breeding of lemon crop. (in): Apomixis and breeding. pp 71-3. S S Khokhlov (Eds). Amerind Publishing Co. Pvt. Ltd., New Delhi.

Koltunow A.M, Hidaka T and Robinson S P. 1996. Polyembryony in Citrus. *Plant Physiology* 110: 599-609.

Koltunow A M, Soltys K, Nito N and Mcclure S. 1995. Anther, ovule, seed and nucellar embryo development in Citrus sinensis cv. Valencia. *Canadian Journal of Botany* 73: 1567-82.

Koltunow, A. M. 1993. Apomixis: Embryo sacs and embryos formed without meiosis or fertilization in ovules. *Plant Cell* 5: 142537.

Kumar A. Gupta B and Singh Uma 1991. Polyembryony in Acacia nilotica. *Indian Forestry* 117: 124-5.

Lebegue A. 1952. Polyembryony in Angiosperm. *Bulletin Society of Botany France* 99: 329-69.

Maheshwari, P and Ranganswamy, N S. 1958. Polyembryony and in vitro culture of embryos of *Citrus* and *Mangifera. Indian Journal of Horticulture* 15:275-82.

Maria Cristina Rocha, Cordiero Alberto Carlos, Queiroz Pinto, Victor Hugo Vargas, Ramos Fabio Galop Falaro and Lilia Marta Santo Fraga. 2006. Identification of plantlets genetic origin in polyembryonic mango (*Mangifera indica* L.) Rosinha seeds using RAPD markers. *Revista Brasileira de Fruiticulture.* 28(3): 454-7.

Moore G A and Castle W S. 1988. Morphological and isozymic analysis of open pollinated citrus- rootstock population. *Journal of Heredity* 79:59-63.

Narayanswamy S and Roy S K. 1960. Embryo sac development and polyembryony in *Sizygium cumini. Botanical Note* 3: 273-84.

Nicasio Tusa, Loredana Abbate, Sergio Ferrante, Sergio Lucretti and Maria-Teresa Scarano1. 2002. Identification of zygotic and nucellar seedlings in citrus interploid crosses by means of isozymes, flow cytometry and ISSR-PCR. *Cellular and Molecular Biology letters* 7: 703 8.

Parlevliet J E and Camaron, J. E. 1959. Evidence of the inheritance of nucellar embryony in citrus. *Proceedings American Society Horticultural Society* 74: 252-60.

Pullaiah T, Lakshminarayan K and Hanumanth Rao B. 2001. Polyembryony. (In). Textbook of Embryology of Angiosperm. pp 164-77. Regency Publication, New Delhi.

Sachar R C and Chopra R N 1957. A study of endosperm and embryo in *Mangifera. Indian Journal of Agricultural Sciences* 27(1): 219-38.

Soares Filho WS, Lee LM and Cunha Sobrinho AP (1995) Influence of pollinators on polyembryony in citrus. *Acta Horticulture* 403:256-265.

Sturrock T T. 1968. Genetics of mango polyembryony. *Proceedings of Florida State Horticulture Society* 81(1): 311-4.

Thakur D R and Bajwa B S 1971. Extent of polyembryony in some species and varieties of citrus. *Indian Journal of Horticulture* 2: 25-28.

Tisserat B, Esan E B and Murrashige T. 1979. Somatic embryogenesis in angiosperm. Horticultural Review 1: 1-78.

Toyama T K. 1974. Haploidy in peach. *Horticulture Science* 9: 187-8.

Wakana A and Uemoto S. 1987. Adventive embryogenesis in citrus. I. The occurrence of adventives embryo without pollination or fertilization. *American Journal of Botany* 74: 517-30.

Wakana A and Uemoto S. 1988. Adventive embryogenesis in citrus (Rutaceae). II. Post fertilization development. *American Journal of Botany* 75: 1033-47.

Wlims H J, Van Went J L, Cresti, M and Ciampolini F. 1983. Adventive embryogenesis in Citrus. *Caryologia* 36: 65-78.

Yakovlev M S. 1967. Polyembryony in higher plants and principles of its classification. *Phytomorphology* 17 (14): 27882.

Xiang C and Roose M L. 1988. Frequency and characterization of nucellar and zygotic seedlings in 12 citrus rootstocks. *Scientia Horticulture* 37: 47-59.

The Science of Horticulture : Vol. 1, 2011
© K.V. Peter (ed.), pp. 165-178
New India Publishing Agency, New Delhi (India)
E-mail : newindiapublishingagency@gmail.com
Web: www.bookfactoryindia.com

Chapter 7 Principles of Food Irradiation

TRIPATHI, P.C.

Food irradiation involves exposure of food to short wave energy to achieve specific purposes like extension in shelf life, insect disinfestations and elimination of food-borne pathogens and parasites. Compared to heat or chemical treatment, irradiation is more effective and an appropriate technology to destroy food-borne pathogens without much loss to the quality of food. It offers a number of advantages to producers, processors, retailers and consumers. Irradiation is a cold process and can be used to pasteurize and sterilize foods without affecting freshness and texture of food. Unlike chemical fumigants, irradiation does not leave any harmful toxic residues in food and is more effective. It is efficient and can be used to treat prepacked commodities. The FAO estimates that about 25% of world's total food production is lost due to insects, bacteria and rodents after harvesting. Use of radiation can play an important role in cutting these losses and reducing dependence on chemical pesticides. The food borne diseases claim millions of lives all over the world every year. Irradiation of food can reduce these incidences by destroying harmful microorganisms.

Use of irradiation to control spoilage was demonstrated in the early decades of the 20th century. It could not be used for commercial purpose due to inability to obtain required quantities of ionizing radiation. In the 1940s Cobalt-60 and Cesium-137, which emit gamma rays, became available. The availability of these sources stimulated research in food irradiation aimed at development of a commercial process. A lot of research was conducted on food irradiation during 1950's and 1980's to utilize irradiation in food preservation. By the continuous research efforts, the methods were standardized and commercialized for many food products . Today, more than 42 countries worldwide including India have approved the use of irradiation for over 100 food items and about 30 of these are applying the technology on a limited commercial scale. South Africa has cleared more than 40 food items. The volume and number of irradiated food products entering trade have grown steadily in recent years, particularly in China, Belgium, France, The Netherlands, South Africa and the United States.

Some Milestones in the Food Irradiation

1905 U.S. and British patents issued for use of ionizing radiation to kill bacteria in foods

1921 Studies on the beneficial effect of x-rays in spoilage causing organism *Trichinella spiralis* in raw pork

1923 First published results of animal feeding studies to evaluate the wholesomeness of irradiated foods

1940s Beginning of food irradiation in United states

1950s Beginning of food irradiation program in many countries including India

1970s Successful in reduction of losses in onion, potato and many other foods in India

1990s Commercial application of irradiation in potatoes, onion and spices in India

2000s Commercial application of irradiation in fruits in India

2007 United States accepted import of Irradiated Indian Mangoes

1. What is Food Irradiation?

Ionizing radiation comprises high-energy gamma rays, X-rays and accelerated electrons. Their energy is high enough to dislodge electrons from atoms and molecules and to convert them to electrically charged particles called ions. Gamma rays and X-rays, like radio waves, microwaves, ultraviolet and visible light rays, form part of the electromagnetic spectrum and occur in short-wavelength, high-energy region of the spectrum and have the greatest penetrating power (Fig 1).

Figure 1. Electromagnetic spectrum

The low energy rays cannot penetrate the food. Therefore they are not used for irradiation The X- rays are high energy beams but these are generally not used for food irradiation because the efficiency of conversion from electrons to X-rays is generally less than 10%. Gamma rays and X-rays have similar properties and effects on materials but main difference between them is their sources of origin. X-rays with varying energies are generated by machines while gamma rays with specific energies are produced by spontaneous disintegration of radioisotopes. Only certain radiation sources i.e. radio isotopes can be used in food irradiation. The radioisotope almost exclusively used for the irradiation of food by gamma rays is cobalt-60. It is produced by neutron bombardment in a nuclear reactor of the metal cobalt-59. It has a half-life of 5.3 years. The gamma rays produced are highly penetrating and can be used to treat full boxes of fresh or frozen food. Cesium-137 is the only other gamma-emitting isotope suitable for food irradiation but the supply of commercial quantities of cesium-137 is almost negligible. Thus Cobalt-60 is the only choice for gamma radiation source. Over 80% of the global Cobalt-60, is produced in Canada. India also produces some quantity of Cobalt –60.

Quantity of radiation energy absorbed by the food as it passes through the radiation field during processing is called radiation dose. The radiation dose varies for different types of food products. It is measured in Grays (Gy) or kilo Grays (kGy). The earlier unit of radiation was rad. One Gray is equivalent to one joule of energy absorbed by one kilogram of irradiated food. The maximum dose of 10 kGy is recommended by the Codex General Standard for food irradiation. But in some countries up to 50 kGy doses of irradiation are recommended for use.

Gamma irradiation does not increase the temperature of food being irradiated thus is often called as cold pasteurization. Irradiation works by disrupting the biological processes which lead to decay. It prevents the division of food spoilage microorganisms, such as bacteria and moulds by changing their molecular structure. The reactions with the DNA cause the death of microorganisms and insects. It can also slow down ripening or maturation of certain fruits and vegetables by modifying / altering the physiological processes of the plant tissues. Irradiation doses used for food processing can be categorized into three groups.

Low-dose irradiation: The low doses (< 1 kGy) of irradiation are used for disinfestation of insects in stored grain, pulses and food products and for the destruction of parasites in meat and meat products. Irradiation doses of 0.25 to 1 kGy are used for disinfestation including quarantine treatment and elimination of food borne parasites. For delaying of fruit ripening, 0.25 - 0.75 kGy irradiation is used while 0.03 – 0.15 kGy is used for sprout inhibition in bulbs and tubers.

Medium dose irradiation: Medium doses (1-10 kGy) of gamma irradiation are used for eliminating microbes in fresh fruits, meat and poultry products, to destroy food pathogens in meat, and to help in the hygienisation of spices and herbs. The 1.5 to 3.0 kGY radiation is used for reduction of spoilage microbes to improve shelf-life of meat, poultry and sea foods under refrigeration. For elimination of pathogenic microbes in fresh and frozen meat, poultry and seafoods and reducing microorganisms in spices, medium doses of radiation are used.

High dose irradiation: High doses of radiation (above 10 kGy) are used for sterilization of packaged meat, poultry and their products.

Use of different doses of irradiation for food irradiation purpose is approved by different countries but low to medium doses of radiation are accepted by many countries.

Infrastructure for Food Irradiation

The radioisotopes are used for food irradiation. These radioisotopes emit radiation continuously, which are harmful to living organisms. Thus these facilities must be regulated and inspected by national radiological safety and health authorities. In India these are regulated by Atomic Energy Regulatory Board (AERB) and Department of Atomic Energy. The common features of all commercial irradiation facilities are the irradiation room and a system to transport of food into and out of the room. A thick concrete shield (1.5 – 1.8 m. thick) is essential surrounding the irradiation room to make it sure that ionizing radiation does not escape to outside of the room. Since a gamma irradiation is used for most of the food irradiation process, it is more essential because the isotope source continuously emits radiation. The isotope when not in use for food treatment is stored in a 5-6 m. deep well. The water absorbs the radiation energy and protects workers from exposure .A conveyer which carries the products for irradiation is also required to move the food products through the irradiation chamber. Apart from the basic irradiation facilities, other facilities like loading- unloading area, temporary storage for food products are required. In India, these facilities are available with BARC, Mumbai. Other commercial irradiation plants are established in Vashi (Mumbai), Lasalgoan (Nasik), Ahmadnagar for onion, potato and spices and a few others are under construction.

2. Benefits of Food Irradiation

Food irradiation offers a wide range of benefits to food industry and the consumer. Irradiation reduces pathogen load, kills the storage pests, delays ripening and enhances shelf life of various food products.

2.1 Reduction of Pathogenic Microorganisms

There are many microorganisms like *Escherichia coli, Salmonella, Campylobacter jejuni, Listeria monocytogenes* and *Vibrio* which are the primary causes of food borne diseases. Incidences of food contamination by these microorganisms are increasing. These pathogens are of primary concern from a public health point because of the severity of illnesses and higher number of outbreaks. Medium-dose irradiation (1 to 10 kGy) of food kills these bacteria and reduces number of spoilage microorganisms. This results in low number of outbreaks of these diseases.

There are parasites like *Trichinella spiralis* which causes illness and death. The larvae of this parasite can be inactivated by irradiation with a minimum dose of 0.3 kGy.

2.2 Decontamination

Spices and condiments are used in various types of food products for seasoning due to their distinctive colours and aromas. These are often contaminated with microorganisms because of the environmental and processing conditions. The heat treatment cannot be used for these products as this causes adverse affect on colour, aroma and oleoresins. The fumigants like ethylene oxide are often used to kill the biological contaminants of spices and condiments. Many countries have restricted or prohibited use of these chemicals because of certain health hazards. The irradiation does not have any such problems. It does not increase temperature of the product and is ideal for the decontamination of spices. The use of irradiation has since emerged as a viable alternative of these chemicals in spices and condiments. Irradiation of spices is practiced in over 20 countries on a commercial scale.

2.3 Extension of Shelf Life

Shelf life of many fruits and vegetables, meat, poultry, fish and seafood can be considerably prolonged by treatment of good quality fresh food with low-dose irradiation. Irradiation of mushrooms at 2 to 3 kGy inhibits cap opening and stem elongation. Shelf-life extension can be increased at least two-fold by irradiation if they are stored at low temperature after irradiation. All fruits and vegetables are not suitable for irradiation because undesirable changes in colour and texture or both limit their acceptability. Different varieties of a particular fruit or vegetable may respond differently for inhibition of spoilage microorganisms by irradiation. The effectiveness of irradiation is also influenced by maturity of fruits, physiological status of fruits, time of irradiation etc.

2.4 Disinfestation

Disinfestation is aimed at preventing losses caused by insects in stored cereals, pulses, flour, dried fruits, dried nuts and other dried food products including dried fish. The main problem faced in storage of cereals and pulses is incidence of storage pests e.g. beetles, moths, weevils and others. These are not quarantine insects but they cause extensive damage to stored products. Irradiation (1 kGy or less) is an effective pest control method for these commodities and a good alternative to methyl bromide and phosphine, the major fumigants used to control grain pests. Irradiation is a fast treatment and its efficacy is not temperature dependent. The relatively low doses of irradiation (1 kGy or less) are required for insect control. The re-infestation in the irradiated products can be prevented by proper packaging. Disinfestation can also be used for fresh fruits of citrus, mango, and papaya for fulfilling the quarantine requirements. Low dosages of radiation(0.15 to 0.3 kGy) effectively control fruit fly and other insects.

2.5 Inhibition of Sprouting

Sprouting is one of the major storage problems in tuber and bulb crops like potato, onion, yam etc. This is one of the major hindrances in the long-term storage and distant transport of these commodities. Chemicals like maleic hydrazide, propham, or chloropropham, recommended for sprout suppression are not fully effective under all sorts of climatic conditions. Moreover some of these are now restricted or prohibited chemicals due to health reasons. Irradiation is proved as effective measure for control of sprouting. A very low radiation dose of 0.15 kGy or less, inhibits sprouting of products such as potato, yam, onion, garlic, ginger, and chestnut. It leaves no residues and allows storage at higher temperatures.

3. Uses of Irradiation

In India, the major emphasis was given on the use of low doses of irradiation. Elaborate studies were conducted on the use of low dose of gamma irradiation for control of sprouting in onion, garlic and potatoes at BARC,

Mumbai. The results of these studies revealed that 60 to 90 Gy completely inhibited sprouting in onion and it was found better than recommended sprout suppressant maleic hydrazide. There was no internal injury in the bulbs and the bulbs remain healthy for longer period. It was also found that onion can be stored for 8 to 9 months with irradiation combined with low temperature storage. Further low temperature induced sprouting in onion was completely inhibited by gamma irradiation (Fig. 2). The irradiation was also effective in control of sprouting in *kharif* onion, which usually does not have dormancy, and these onions can be stored for 3-4 months as against normal shelf life of 30 to 45 days. Only constraint is that these bulbs can not be used for seed production purpose. In garlic, irradiation treatment (75 gy) increased shelf life and prevented internal sprouting of cloves and these garlic were firm even after 9 months of storage. Similarly sprouting was inhibited by the application of gamma irradiation in potato. This was commercially used by many traders and farmers.

Figure 2: Irradiated and unirradiated onions after 4.5 months of storage

The low dose of gamma irradiation is also used for mangoes for delay in ripening and disinfestation of insects particularly seed weevil. This is also commercialized and irradiated mangoes are being exported. In spices and rice, medium doses of irradiation are used against storage pests.

Globally, gamma irradiation is used for fresh and processed foods to prevent spoilage. The effect of irradiation on food depends on dose of irradiation to which the food is subjected. High-dose levels of irradiation (5 to more than 70 k Gy) can be used to sterilize foods by eliminating all microorganisms and spores in the food. The low dose of irradiation (less than 0.1 kGy) is usually used to inhibit sprouting in potato, onion and garlic. Low doses of irradiation were also effective in disinfesting grain products prior to storage and transshipment. It is also successful in reducing microbial and insect contamination on fresh fruits and vegetables and fruit fly in grapefruit.. Shelf life of mushroom, potato, tomato, onion, mango, papaya, banana, apricot, strawberry and fig can be extended with low-dose irradiation without affecting quality of products. The quality of citrus fruits, avocados, pears, cantaloupes, and plums are degraded by irradiation. The shelf life of banana, mango, and papaya can be increased by irradiation at 0.25 to 1 kGy as it delays the ripening. In South Africa, mango irradiation (750 Gy) retarded maturation, thereby making it possible to transport the fruit over longer distances and simultaneously eliminate plant quarantine problems by killing the mango weevil.

Strawberries are frequently spoiled by *Botrytis* mould. Radiation treatment (2 to 3 kGy) followed by storage at 10°C results in a shelf-life of up to 14 days. The enhancement of shelf life depends on physiological condition of fruits and vegetables. For example, if strawberries are irradiated before they are ripe, the red colour does not develop satisfactorily. It is important to irradiate them at the time of initiation of ripening so that colour can be developed properly and shelf life increased. Higher doses of irradiation (>25 kGy) are applied for sterilization of canned products. These sterilized products can be stored at room temperature for longer period. Though all fresh products are not suitable for irradiation but in majority of cases, it was found effective. The doses recommended for some of products are summarized below.

Use of Irradiation in some food products

Food Products	**Radiation Dose (k Gy)**	**Effect**
Spices and other seasoning products	Up to a maximum of 30	Reduces microorganisms and insects. Replaces chemicals used for this purpose.
Strawberries	1 to 5	Extends shelf life by delaying mould growth.
Grain, fruit, vegetables, and other foods subject to insect infestation	0.1 to 2	Kills insects or prevents them from reproducing, partially replace post-harvest fumigants used for this purpose.
Bananas, avocadoes, mangoes, papayas, guavas, and certain other non-citrus fruits	1.0 (maximum)	Delays ripening.
Potatoes, onions, garlic, ginger	0.05 to 0.15	Inhibits sprouting.

(Source: American council on science and health, 1988)

4. Effect of Irradiation on Food

4.1. Effects on Microorganisms

Effect of irradiation is due to ionizing radiation. The microorganisms are destroyed by the passage of an ionizing particle or quantum of energy produced by them. This effect is less when food is irradiated in frozen state. Irradiation may also induce fatal changes in microorganisms. Bacterial spores are more resistant to ionizing radiation than are vegetative cells. Gram-positive bacteria are more resistant than Gram-negative bacteria. The resistance of yeasts and moulds vary considerably and some are more resistant than bacteria. The bactericidal efficacy of irradiation depends on dose of irradiation, condition, type and number of organism, condition of organism, physical state of the food etc. The dose of irradiation required for killing a few organisms is given below.

Doses of Radiations required for killing organism

Organism	Approximate lethal dose (k Gy)
Insects	0.22 to 0.93
Viruses	10 to 40
Yeasts (fermentative)	4 to 9
Yeasts (film)	3.7 to 18
Moulds (with spores)	1.3 to 11
Bacteria (cells of pathogens)	1.4 to 4.8
Gram-negative	1.8 to 2.3
Gram-positive	0.23 to 8.8
Bacterial spores	3.1 to 50

Source: Frazier W C and West stuff D.C., 1988

4.2. Effect of Radiation on Nutrients in Food

There is no significant effect on the nutritional quality of protein, lipid and carbohydrate constituents by irradiation. There are losses of vitamin E and B but these losses are less than the other methods of food preservation. The nutrient loss by irradiation can improve by keeping the food at low temperature during irradiation. Over all, the nutrient losses in the irradiation are much less than in other methods. The losses in colour, texture and flavour are very low in irradiation while these are very high in traditional methods of preservation.

5. Food Safety of Irradiated Foods

There is concern about safety of irradiated food .The free radicals created by ionizing radiation may be present in the food at the time of ingestion. Free radicals are also formed when food is fried, baked, ground, and dried. In foods with a high moisture content, free radicals disappear rapidly but in dry foods, the free radicals are much more stable and do not dissipate quickly. The assessment of the safety for human consumption of irradiated foods has involved basically four aspects i.e. radiological safety, microbiological safety, nutritional adequacy and toxicological safety. After reviewing an extensive amount of data, the joint FAO/IAEA/WHO Expert Committee in 1981 concluded that irradiated foods are safe and wholesome. The committee's main conclusion was that the irradiation of any food commodity up to an overall average dose of 10 kGy presents no toxicological hazard and can be used for consumption without toxicological test. The irradiation of foods up to an overall average dose of 10 kGy does not produce any special nutritional or microbiological problems and can be used for food preservation. These standards were accepted in 1983 by many countries for trade and marketing of irradiated foods.

5.1 Irradiated Foods and Radioactive Foods

Irradiated foods and radioactive food are often confused. Irradiated foods are those deliberately processed with certain types of radiation energy to increase the shelf life of the foods or to inhibit sprouting or to destroy food-poisoning bacteria. The radioactive foods are those that have become accidentally contaminated by radioactive substances from weapons testing or nuclear explosion or leakage. Irradiated products neither contain any radioactive materials nor they emit radiation, as food exposed to sun light does not emit sun light.

6. Food Irradiation Regulation in India

In India, Bhabha Atomic Research Centre (BARC) has done extensive research and development work on preservation of foods by radiation. BARC is the Nodal agency for assistance and guidance on processed technology. BARC provides information on availability, cost and possible alternative of essential and critical machinery, pollution and other environmental aspects, utilization and disposal of radioisotopes. BARC provides training to the staff on one of the existing plant on various aspects of operation and safety. Board of Radiation and Isotope Technology (BRIT) also provides technological inputs and guidance for formulation of irradiation projects. Board of Radiation and Isotope Technology (BRIT) of the Ministry of Atomic Energy has set up a facility for radiation processing of spices at Navi Mumbai. Several regulations are passed from time to time such as;

The Atomic energy (Control of Irradiation of Foods) Rules, 1996 ,The Prevention of Food Adulteration (Fifth Amendment) Rules, 1994 . The Atomic Energy (Control of Irradiation of Food) Rules, 1996 include the modalities by which food may be irradiated, as well as the procedure for food irradiation licences and their suspensions/revocations. The Department of Atomic Energy (DAE) and Atomic Energy Regulatory Board (AERB) are the regulatory bodies for food irradiation in India. Firstly the inspectors from AERB ensure that the installation satisfies all the prescribed safety requirements. Then AERB issues a certificate of approval and DAE licenses the irradiator. The AERB is empowered to withdraw the certificate of approval. Government of India has permitted use of irradiation technology in preservation of food items like potato, onion, rice, semolina (*sooji or rawas*), Wheat *atta or maida*, mango, raisins, dried dates, ginger, garlic, shallots (small onion) and meat products including chicken.The Ministry of food processing industries (MFPI) has taken interest in radiation processing technology as a means of preservation of perishable fruits and vegetables, pest control and enhancement of quality. The MFPI is now encouraging entrepreneurs for setting up of facilities for radiation processing of foods in private sector.

Food items approved for radiation preservation by the Ministry of Health and Family Welfare under Prevention of Food Adulteration Rules, 1955

Food Items	**Purpose**	**Minimum Dose (kGy)**	**Maximum Dose (kGy)**
Onion	Sprout inhibition	0.03	0.09
Potato		0.06	0.15
Ginger		0.03	0.15
Garlic		0.03	0.15
Shallots (Small onion)		0.03	0.15
Mango	Disinfestation (Quarantine)	0.25	0.75
Rice , Semolina (sooji, rawa), Wheat atta, maida, Raisin, figs, dried dates	Disinfestation	0.25 0.25 0.25	1.0 1.0 0.75
Meat and meat products including chicken	Shelf-life extension and pathogen control	2.5	4.0
Spices	Microbial contamination	6.0	14.0

Source: Minhas, 2002

7. International Trade of Irradiated Foods

Extensive research work is carried out on irradiation of food in many countries which have approved the use of irradiation in several food commodities. More than 19 countries have approved uses of irradiation in some foods for controlling spoilage. The approximate quantity of spice and dry vegetables irradiated was 10,000 tonnes in 1987. This was increased to 89,000 tonnes in 2000. More than 50 % of this quantity was irradiated in Unites States of America. Mangoes from India, Philippines and other countries are accepted in some countries. Still there are countries who have imposed restrictions on the trade and marketing of irradiated foods.

Food products approved by Countries for Irradiation

Country	Food Products
Argentina	Potatoes, strawberries, onions, garlic
Belgium	Potatoes, strawberries, onions, garlic, spices
Bulgaria	Potatoes, onions, garlic, grain, dry food concentrates, dried fruits, fresh fruits
Canada	Potatoes, onions, wheat flour, poultry, spices and some dried vegetables
Finland	Spices, herbs
Chile	Potatoes, papaya, wheat, chicken, onions, rice, fish products, spices
France	Potatoes, onions, garlic, spices, dried fruits and vegetables
India	Potatoes, onions, garlic, spices
Israel	Potatoes, onions, poultry, spices, fresh fruits and vegetables
Czechoslovakia	Potatoes, onions, mushrooms
The Netherlands	Asparagus, strawberries, mushrooms, potatoes, onions, poultry, rice and ground rice products, spices,
Philippines	Potatoes, onions, garlic
South Africa	Potatoes, onions, garlic, chicken, papaya, mangoes, strawberries, dried bananas, avocados, beans
Spain	Potatoes, onions
Thailand	Potatoes, onions, garlic, dates, wheat, rice, fish, chicken
U.S.S.R.	Potatoes, grain, fruits and vegetables, , poultry, onions, meat products
U.S.A.	Wheat and wheat flour, potatoes, spices, pork, fresh fruits and vegetables
Yugoslavia	Cereals, potatoes, onions, garlic, poultry, dried fruits and vegetables

8. Consumer Acceptance

Initially consumers were concerned about irradiated foods. Their concerns include safety, quality and nutritional value of the foods. The higher quality and longer shelf life of irradiated foods have gradually impressed the consumers. But all consumers want that irradiated products should be labeled for various reasons. The supporters of food irradiation firmly believe that these foods will be accepted when offered for sale. There are anti-food radiation activists who oppose use of irradiated foods. They raised their concerns about the use of irradiation. Some of their concerns are i) it will allow for "dirty" meat, poultry, or produce to be processed and sold to consumers, ii) Irradiating micro organisms such as *E. coli* and *Salmonella* may give rise to even more dangerous, radiation-resistant strains of bacteria iii) Irradiation destroys nutritional quality of the products, iv) the chromosomal abnormalities and polyploidy in food products may be dangerous to human safety. But a large group of consumers are still undecided about the use of irradiated foods. Consumer education is important to provide them reliable information on irradiated foods. The concerned agencies are taking up measures to educate the consumers.

Economics of Irradiation

Food processing measures increase the value of the product and also involve additional cost. There are many other factors besides the processing which increase cost of processing. These include availability, transportation cost, loading, unloading, storage charges, treatment charges etc. The processing brings benefits to producers, traders and consumers in terms of lower storage losses, more availability, better quantity, storage life, convenience and improved hygiene of the food. The major factors influencing the cost of food irradiation are design parameters of irradiation unit, handling conditions, capital costs, spare parts, storage capacity, operating costs, utilities, maintenance, etc. Irradiation cost may range from Rs. 0.25 to 0.50 / kg. for a low dose application such as sprout inhibition of potato and onion and insect disinfestations in cereals and pulses to Rs. 1-3 / kg for high dose application such as treatment of spices for microbial decontamination. The irradiation charge for onion at Lasalgoan (Nasik) plant is Rs.1500 / tonne. The cost of treatment of spices is higher but these are competitive with alternative treatments. The alternate methods particular fumigants have side effects and they also affect the sensory and other quality parameters. The cost of phyto-sanitary treatments like hot water, vapour treatment and cold treatments is 2 to 10 times higher than irradiation. Cost of irradiation can be further brought down by a multipurpose facility treating a variety of products around the year. Considering the prevention of losses by irradiation and retention of quality characteristics, the irradiation is the most economical and viable method of extension of storage life of food product.

Irradiation provides considerable advantages in preservation of food to a varying extent as determined by the treatment. Now it is an established , versatile and environmentally-friendly method of treatment of foods for sanitary, phytosanitary and shelf-life extension purposes. Food irradiation is effective in controlling microorganisms. It decontaminates food of bacteria, yeast, molds and insects. The decontamination can improve the hygienic quality of the foods and prevents potential health hazards. It also controls maturity, senescence and sprouting of fresh fruits and vegetables. The nutritive value of foods remains unchanged except insignificant change in some vitamins. In a nutshell, irradiation has proved to be one of the best methods of food preservation. It can reduce the post harvest losses and facilitate greater supply of safe food.

References

The Gazette of India: June 22, 1996. Asadha 1, 1918 [Part II -Sec 3(I)]

Anonymous 1999. Facts about Food irradiation A series of Fact Sheets from the International Consultative Group on Food Irradiation (ICGFI), Vienna, Austria

Anonymous 2009. Food irradiation: A guideline for consumer, policy makers and the media. Grocery Manufacturers Association, Washington DC.

Frazier, W.C., and West Hoff, D.C. 1988. Chapter 10. Preservation by radiation. In: Food Microbiology. Fourth edition. McGraw-Hill. New York, NY.

Karoly Vas 2002. Recent Advances in the Preservation of Food by Irradiation IAEA Bulletin- Vol. 18, NO. ¾..

Minhas, B.S. 2002. Thematic planning meting on irradiation as a sanitary and phytosanitary treatment for food and agricultural commodities for the new millennium. Country Paper –India at Vienna, Austria, May 6-10, 2002.

Snyder, O.P. and Poland, D.M. 1995. Food irradiation Today. Hospitality Institute of Technology and Management, St. Paul, Minnesota, USA.

Tripathi P.C. and Lawande, K. E. 2007. Effect of cold storage and gamma-irradiation on storage losses in Onion. *Indian J. Hortic.*64 (3):340-344.

Tripathi, P.C. and Lawande,K.E. 2007. Effect of cold storage and gamma-irradiation on storage losses in garlic. *J. Spices and aromatic plants.* 17 (1):

Tripathi, P.C. and Lawande, K.E. 2006. Comparison of chemical sprout suppressant and irradiation on storage life of onion under different storage conditions- Final Report of CSS Project. NRC Onion and Garlic, Rajgurunagar (Pune).

Venugopal, V. Warrier, S.B. and Sharma, Arun 2002. Food Irradiation: Recent Global Trends, Food Technology Division, BARC Newsletter.

The Science of Horticulture : Vol. 1, 2011
© K.V. Peter (ed.), pp. 179-198
New India Publishing Agency, New Delhi (India)
E-mail : newindiapublishingagency@gmail.com
Web: www.bookfactoryindia.com

Chapter 8 History of Plant Pathology of Horticultural Crops

VEENA ASHOK

Plant diseases are mentioned in some of the oldest books available (Homer, C - 1000 B.C. mentioned about the therapeutic properties of sulphur and in Old testament - 750 B.C.) and were feared as much as human diseases and war.

Major Events in History of Plant Pathology

470 B.C.	Democritus noted plant blights and described a way to control them using olive grounds left after extraction of olive oil.
300 B.C.	Theophrastus made plant disease as an object of a systematic study. He wrote two books a) The Nature of Plants and b) Reasons of vegetable growth. He is considered as "The father of Botany". Theophrastus and his contemporaries believed that plant diseases were manifestations of the wrath of God.
Around 1200 A.D.	Mistletoe was recognized as the first pathogen for which a cultural control was recommended (Albertus Magnus).

1670	Robert Hooke in England invented the double lensed (compound) microscope. Antonius van Leeuwenhoek improved significantly the lenses and the structure of the microscope and began to examine the anatomy of plants, body of filamentous fungi and even bacteria.
1729	Pier Antonio Micheli described many new genera of fungi and illustrated their reproductive structures and proposed fungi arose from their own spores rather than spontaneously.
1735	Swedish philosopher – botanist Carl von Linne published his main work "Systema Naturae" by which he established the diagnosis of plant species and the binomial nomenclature of plants.
1743	Needham observed nematodes inside wheat kernels but failed to suggest they were the causes of the problem.
1845	Irish famine- Early 1840's potato crops began to fail to varying extents in several areas of Europe and Ireland. In 1845, due to the huge loss of potato crop caused by late blight of potato (*Phytophthora infestans*), hunger was quickly followed by starvation, which resulted in the death of many Irish. In 1846 season, once again they lost the crop. Hunger, dysentery and typhus spread among the farmers again, and more of the survivors emigrated to North America. It is estimated that one and a half million Irish died from hunger, and about as many left Ireland.
1845-1885	Coffee rust fungus, *Hemileia vastatrix* attacked the coffee-trees in Ceylon (presently Sri Lanka). The Coffee Rust fungus was first recorded by the Reverend Miles Berkeley, who was sent a collection from the Royal Botanical Gardens at Peradenija. Berkeley's collection of fungi from Sri Lanka was already quite large, containing 1100 accessions and this collection was different from all of them. Berkeley believed this to be a new species and named it *Hemileia vastatrix* and published it in November 1869. *Hemileia* was first found on coffee, in 1875, in the Madulsima district where it was associated with premature leaf drop. Even at this time, it was already of concern since it had affected three acres of coffee plants. When Berkeley learned of this

disease and the fear associated with it, he thought these fears to be well grounded since he had already seen what the Late Blight had done to potatoes. In his publication concerning this species, Berkeley even suggested immediate application of sulphur because this disease would be difficult to combat once it had been allowed to spread. Unfortunately, no one paid attention to the fears of some of the growers except Berkely. That was in 1869. In five years, the coffee-leaf disease had spread over the whole island and no plantation was free of the disease. A commission was set up to study the disease. By this time, the disease had spread to southern India and Malaysia and was menacing the entire coffee industry in the East. Prior to the Coffee Rust, coffee was an important part of the social and political environment in England. By 1675 there were approximately 3000 coffee houses, in London, which became gathering places for social, political as well as religious discussions. It was a place where the daily papers, desks and writing material were available. When the Coffee Rust destroyed the coffee plants in Sri Lanka, Java, Sumatra and eventually spread to Arabia, Liberia and Africa. The British, no longer able to grow coffee, simply gave it up, and began growing tea in Sri Lanka.

1854 French wine production was reduced by 80% due to powdery mildew disease caused by *Uncinula necator*. It was noted in England that when a mixture of powdered lime and sulfur was dusted on the vines, it significantly protected the leaves and berries from powdery mildew.

1859 Charles Darwin published the book "The origin of species by means of Natural Selection" and showed that species of all organisms, plants and animals, evolve over time and adapt to changes in their environment for survival.

1861-63 Anton de Bary proved that potato late blight was caused by a fungus, *Phytophthora infestans*. De Bary also showed that the fungus did not just reappear from nowhere the following growing season but instead survived the winter in partially infected potato tubers in the field or storage. He disproved the theory of spontaneous generation.

Louis Pasteur proved that microorganisms were produced from preexisting microorganisms and that most infectious diseases were caused by germs.

Robert Petri developed artificial nutrient media for culturing the microorganisms (Petri dishes).

Robert Koch established that for proving that a certain microorganism was the cause of a particular infectious disease, certain necessary steps (Koch's postulates) must be carried out and certain conditions satisfied.

1864 Pasteur invented pasteurization.

1866 Anton deBary discovered that some fungi induce rotting of vegetables by secreting substances, enzymes that diffuse into plant tissues in advance of the pathogen.

1878 Grape leaves in some French vineyards began to show whitish downy spots on their undersides, while the upper sides of such leaves corresponding to the underside downy spots became yellow at first and then turned brownish black and died. This condition became known as downy mildew and was caused by the fungus, *Plasmopara viticola*. Within 5 years of its appearance in France, it spread to all the vineyards of that country and into those of adjacent countries. French botany professor Pierre Alexis Millardet while walking among vineyards, noticed that in some of them , the vines of a few rows along the dirt road had a bluish film on their leaves. What was the most noteworthy was that these vines seemed to still have all their leaves healthy, whereas vines in rows that did not have the bluish film, the leaves, young twigs, and berry clusters were affected severely by downy mildew. The owner of the vineyard told him that the bluish film was actually copper sulfate mixed with some hydrated lime to better stick on the leaves. The mixture was sprayed on the vines to create the impression that it was poisonous in that way to keep passerby from going into his vineyard and taking his grapes. With that information in hand, Millardet went back to his laboratory where he mixed copper sulfate and hydrated lime in various proportions and tried them on disease affected vines. Finally in 1885, he found the best combination for the control of downy mildew. This solution became known as Bordeaux mixture.

Woronin in Russia was the first to show that a plant disease, the clubroot disease of cabbage was caused by a protozoan Plasmodiophoromycete.

Burrill showed that bacteria caused the fire blight disease of pear and apple.

1886 Adolph Mayer injected juice obtained from tobacco plant leaves showing various patterns of greenish yellow mosaic into healthy tobacco plants and later developed similar mosaic patterns. Mayer concluded that the disease was probably caused by bacteria.

1887 Koch formulated "Koch's postulates". These rules became the standard procedure for proving that disease is caused by a bacterium or any other kind of pathogen.

1892 Ivanowski. Whatever caused the tobacco mosaic disease could pass through a filter that retains bacteria and he concluded the disease was caused by a toxin secreted by bacteria or by unusually small bacteria that passed through the pores of the filter.

1898 Beijerinck found Tobacco mosaic disease was caused not by a microorganism, but by a "contagious living fluid" that he called a virus.

1935 Stanley concluded that the virus was an autocatalytic protein that could multiply within living cells.

1936 Bawden and colleagues demonstrated that virus consisted not only of protein, but also a small amount of RNA.

1955-81 Lethal yellowing of coconut palms

In 1955, coconut palm trees in the Key West islands of Florida were noticed to drop their coconuts prematurely. First the lower, older leaves and then the next younger leaves turned yellow and then brown and died. Finally, all the leaves and the vegetative bud died and the entire top of the tree fell off, leaving the tall palm trunk looking like a telephone pole. The lethal yellowing disease was first found in mainland Florida in 1971 and killed 15,000 coconut palm trees by 1973, 40000by 1974, and, by 1975, 75% of the coconut palm trees in Dade Country were dead or dying from the disease. In Jamaica, of six million

	trees counted in 1961, 90% were killed by lethal yellowing by 1981. More than a million coconut palm trees were killed in Ghana within 30 years.
1956	Gierrer and Schramm showed that the protein could be removed from the virus and that ribonucleic acid carried all the genetic information that enabled it to cause infection and to reproduce the complete virus.
1963	Vermeulen presented convincing evidence of the pathogenicity of flagellates to coffee trees and in 1976 flagellates are reported associated with several diseases of coconut and oil palm trees in South America and in Africa.
1967	Doi *et al.*, observed walless mycoplasma like bodies in the phloem of plants exhibiting yellows and witches broom symptoms.
1971	Diener and colleagues discovered viroids while they were studying the potato spindle tuber disease. Viroids are the smallest infectious agents which multiply autonomously in plant cells.

Plant Pathology in the 20th Century

Early Developments

The Descriptive Phase

Agriculturists, botanists, naturalists and other scientists became aware of and familiar with the existence of plant diseases and with some of the causes of plant disease. Reports began to be published in scientific, popular, and semi popular journals describing numerous plant diseases on a variety of agricultural and ornamental plants. The availability of improved magnifying lenses and microscopes made possible the detection and description of many fungi and nematodes and later bacteria associated with diseased plants. Development and introduction of techniques for growing microorganisms in pure culture by Brefeld, Koch, Petri, and others (1875 – 1912) contributed greatly to Plant Pathology. Koch's postulates (1887) had a profound effect on plant pathology. The invention of electron microscope made it possible to visualize and describe most viruses and, after 1970, helped detect and describe the molecules and viroids.

During the phase, many observations were reported concerning the biology of microorganisms involved. These observations were correlated with the prevailing environmental conditions and with differences in disease severity among the various hosts. Different types of control practices, mostly cultural but also some chemical ones were tried for various diseases. Discovery of Bordeaux mixture could control many diseases on almost all crops.

The Experimental Phase

Scientists (USDA and state agricultural stations) began to experiment in all areas of Plant Pathology. Although new diseases and pathogens continued to be discovered and described, plant pathologists began to ask questions and to design experiments to answer about how pathogens enter their host plants, multiply, and spread within the plant; the mechanism of host plant cell death and breakdown; pathogen sporulation; spore dispersal, over-wintering, over-summering, and germination; and the effect of environment on disease development, among others. They also studied variability among plant species and varieties in disease expression and loss. Experimentation also grew rapidly to control plant diseases and to avoid or reduce the losses from them.

The Etiological Phase

The etiological phase of plant pathology involved observations and experiments aimed at proving the causes of specific plant diseases. Although the etiological phase began with the proof of pathogenicity of late blight fungus, rust and smut fungi, etiological studies were facilitated and accelerated by the development of techniques for the pure culture of fungi and bacteria and by the necessity to satisfy Koch's postulates for every disease. The etiological phase resumed, continued and accelerated as new types of pathogens, such as viruses, phytoplasmas, fastidious bacteria, protozoa, and viroids were discovered. The etiological phase often depended on improvements in methodology and instrumentation, such as electron microscope, special nutrient media, density gradient centrifugation, electrophoresis, the development of serological techniques, the polymerase chain reaction and the development of DNA probes and other nucleic acid tests and tools.

The Search for Control of Plant Diseases

The earliest use of chemicals for the control of plant diseases probably began in the late 1600s when some farmers in Southern England planted wheat seed that had been salvaged from a ship wreck. They noticed that far fewer wheat plants produced from such seed were infected with smut than wheat plants produced from other seed. This led some farmers to treat wheat seed with brine to control bunt. In the mid- 1700s, copper sulfate was substituted for sodium chloride, and bunt control improved significantly.

It was noted round A. D. 1200 that a tree can be cured from mistletoe infections if the branch carrying the mistletoe is pruned out. In the mid 1700s, recommendations for the control of cankers included excisions of the canker and the application of grafting wax on the cut area. In the early 1800s, lime sulphur and aqueous suspensions of sulphur were recommended for the control of mildew of fruit trees.

The Main Areas of Progress

Chemical Control of Plant Diseases

1885 The discovery of Bordeaux mixture proved that the plant diseases can be controlled chemically and gave great encouragement and stimulus to the study of nature and control of plant diseases.

1913 Organic mercury compounds were introduced as seed treatments, and such treatments were routine until the 1960s when all mercury containing pesticides were banned because of their toxicity.

1928 Alexander Fleming discovered the antibiotic penicillin. It opened a new area for research in the control of plant diseases.

1934 First dithiocarbamate fungicide, thiram was discovered, which led to the development of a series of effective and widely used fungicides, including ferbam, zineb and maneb.

1950 Antibiotics, primarily streptomycin was used to control plant diseases. Soon after the antibiotic, cycloheximide was shown to be effective against several plant pathogenic fungi.

1965 First systemic fungicide, carboxin was discovered, and it was soon followed by the introduction of several other systemic fungicides, such as benomyl.

1967 Tetracycline was shown to control plant diseases caused by mollicutes; a few years later, tetracycline was shown to control plant diseases caused by fastidious bacteria that live in the xylem of their host plants.

Appearance of Pathogen Race Resistant to Bacteriocides and Fungicides

1954 Strains of bacteria causing disease in plants were resistant to certain antibiotics.

1963 Strains of plant pathogens were found resistant to certain protective fungicides.

1970 New isolates/ strains of numerous fungal plant pathogens appeared that were resistant to a fungicide that was previously effective. This leads to the use of mixtures of fungicides, alternating compounds in successive sprays and spraying with a systemic compound in the early stages of the disease and with a broad spectrum compound in the later stages of the disease.

Public Concern about Chemical Pesticides

1950 A few voices of concern about using pesticides were beginning to be heard.

1962 Rachel Carson's book "Silent Spring' was published. It created greater awareness of the possible adverse effects of pesticides. After a period, scientists agreed to do research on the issue of safety of pesticides and began testing insects, earthworms, birds, fish, plants, animals, water streams, lakes and even soil and underground water reservoirs for pesticides. To the surprise of many scientists, pesticides, particularly the persistent types, were found in many of these bodies, sometimes in fairly high concentrations.

Mid 1960s All pesticides containing mercury were banned by the US government and soon afterward DDT and chlorinated hydrocarbons were also banned.

Mid 1980s Approximately 85-90% of the pesticides or pesticides used previously for plant disease control are banned by US government or discontinued by the manufacturers. Much stricter rules are imposed on pesticide applicators, pesticide applications, and handlers of products treated with pesticides, with each restriction making it safer, but more expensive, to apply pesticides.

Alternative Controls for Plant Diseases

Concern over the toxicity of pesticides led to the reexamination and improvement of many old practices and to the development of some new cultural practices for use in controlling plant diseases. The modification of cultural practices, use of resistant varieties and monitoring of the appearance and development of plant disease epidemics that allow for a reduced use of pesticides have become the basis of "integrated management" of plant diseases.

Numerous nonpathogenic microorganisms, mostly fungi and bacteria antagonized various plant pathogenic fungi, bacteria and nematodes and some of them protected the host plant from infection by the pathogen.

1963 First biological control was obtained. (Fungus against fungus)

1972 Control of crown gall bacterium with non– pathogenic bacterium.

1980s Introduction of one or several appropriate viral genes into host plants through genetic engineering and expression of these genes by the host.

1996 Use of pathogenic microorganisms or chemicals which cause tiny necrotic lesions in the treated plant and, by so doing, activate the defense of the whole plant against subsequent infections by pathogens of the same or different types. Actigard was market tested.

Mechanisms by Which Pathogens Cause Disease

Once it became apparent that fungi and other microorganisms were the causes rather than the results of plant disease, efforts began to understand the mechanisms by which the microorganisms cause disease.

1886 deBary working with *Sclerotinia* rot disease of carrots, concluded that the pathogen produces enzymes and toxins that degrade and kill plant cells from which the fungus can then obtain nutrients.

1905 I. R. Jones- cytolic enzymes were reported involved in several soft rot diseases of vegetables caused by bacteria.

1915 Pectic enzymes play a significant role in their ability to cause diseases on plants.

1925 A suggestion was made that the bacterium *Pseudomonas tabaci,* the cause of the wildfire disease of tobacco, produces a toxin that is responsible for the bacteria - free chlorotic zone (halo) surrounding the bacteria - containing necrotic leaf spots was confirmed in 1934.

1940 Cellulases are involved in plant disease development.

1950's Wildfire toxin was the first toxin to be isolated in pure form.

Early 1970s onwards many studies are devoted to the elucidation of numerous metabolic changes associated with the hypersensitive response i.e, the localized defense reaction of a resistant plant to a pathogen. In the hypersensitive response, numerous enzymes, known as plant pathogenesis related (PR) proteins are activated.

Genetic Inheritance of Resistance and Pathogenicity

1894 Erikson showed biological races of pathogens.

1909 Orton working with *Fusarium* wilt of water melon distinguished among disease resistance, disease escapes and disease endurance.

1914 Stakman and colleagues established that morphologically indistinguishable races of a pathogen within a pathogen species differ in their ability to attack certain varieties.

1946 Flor- Gene for Gene relationship- for each gene for resistance in the host, there was a corresponding gene for avirulence in the pathogen and for each gene for virulence in the pathogen there was a gene for susceptibility in the host plant.

1963 Vanderplank- suggested two kinds of resistance, vertical resistance, which is controlled by a few "major" resistance genes and horizontal resistance is determined by many "minor" resistance genes.

1964 Z. Klement and colleagues recognized that the hypersensitive response also operates against bacterial plant pathogens.

Epidemiology of Plant Diseases

1944 Mills developed a table listing the duration of rain required at each temperature for apple buds, leaves and fruits to become infected by apple scab fungi. He and others used this information to predict the disease.

1963 Vanderplank through the book "Plant Diseases, Epidemics and control" established epidemiology as an important and interesting field of plant pathology. In his book, he discovered the principles and variables in plant disease epidemics, stated the difference in the development and control of monocyclic and polycyclic pathogens and described the general structure and patterns of epidemics.

1969 First computer simulation program of plant disease epidemics was published for the fungal induced early blight disease of tomato and potato. Since mid 1970s, disease modeling and computer simulation of epidemics were developed for many diseases and together with newly developed disease monitoring instrumentations, were used in plant disease forecasting systems. Disease forecasting has become an important component of Integrated Pest Management (IPM) and has helped to reduce the amount of pesticides applied to crops without reducing yields.

Biological Control

1914 C. F. Von Tubeuf used the term "biological control" for the first time in relation to plant pathogens.

1921 C. Hartley First attempt at control of a root disease with microorganism introduced into the soil.

1926 G.B. Sanford observed that potato scab was suppressed by green manuring through activities of antagonists.

1927 W.A. Millard and C.B. Taylor controlled scab of potato by adding *S. praecox* plus green grass clippings to sterilized soil.

Molecular Plant Pathology

The beginning of Molecular Plant pathology can probably be traced to the isolation by W. Stanley in 1935 of the tobacco mosaic virus as a crystalline protein, which he believed infectious.

1941 Beadle and Tatum-one gene codes for one enzyme.

1942 H. H. Flor, gene for gene concept

1953 Watson and Crick showed that DNA exists in a double helix and their discovery impacted greatly all fields of Biology.

1956 Gierrer and Schramm showed that it was RNA, not the protein of Tobacco Mosaic Virus was responsible for the infection of plant cells.

Mid 1970's Studies on *Agrobacterium tumefaciens* revealed that the T-DNA of its Ti-plasmid contained several genes of which two, coding for growth regulators were responsible for the production of tumours (galls) by the infected plants. It was later showed that the two genes could be removed and replaced with one or more genes from other organism such as plants, other bacteria, viruses and even animals, genes that could be transferred into and expressed (translated) by the plant cells.

1986 The first practical results of molecular plant pathology in improving disease resistance came in 1986 when R. Beachy and colleagues obtained tobacco plants resistant to TMV by transforming them ie, introducing into them the coat protein gene of the virus in a way that the plants could express the gene and produce the virus protein. Such transformed plants are called transgenic and the resistance they acquire is called pathogen derived resistance.

1989 M.B. Dickman and P.E. Kolattukudi transformed a fungus that normally could enter host plants only through wounds, with a cloned gene coding for the enzyme cutinase. That enzyme enabled the fungus to penetrate host plants directly through the cuticle, thereby proving that cutinases play a role in the direct penetration of some plants by fungi.

1990 R. Cheim and colleagues obtained transgenic tobacco plants that expressed increased disease resistance by transforming them with the gene for stilbene synthetase, the enzymes which synthesize a phytoalexin.

Discoveries in molecular plant pathology came fast in the 1990's. The concept of systemic acquired resistance (SAR) burst onto the scene through the discovery of D.F. Klessig and colleagues and J. Ryals and co-workers that salicylic acid; a relative of aspirin is associated with SAR.

1991 R. Broglie and co- workers showed that plants transformed with the gene that codes for cutinase exhibit enhanced resistance to diseases by fungi that contain chitin in their cell walls.

1993 G.B. Martin isolated the only resistance gene conferring resistance in tomato to a bacterial pathogen through the hypersensitive response. Tomato plants transformed by

B. Baker and co workers with the tobacco plant resistance gene, which makes tobacco resistant to tobacco mosaic virus, were also made resistant to the virus, proving that atleast some resistance genes may function in species other than the one in which they normally occur.

1998 V.M. Williamson and colleagues found that a single cloned disease-resistance gene from tomato can confer resistance to both a nematode pathogen and an insect.

Advances in molecular plant pathology have also provided a new set of diagnostic tools and techniques that are used to detect and identify pathogens even when they are present in very small numbers or in mixtures with other closely related pathogens. Such tools include detection with monoclonal antibodies, analysis of isozymes or fatty acid profiles of pathogens, analysis of fragments of other nucleic acids produced by specific enzymes, calculation of percentages of hybridization of their nucleic acids and determination of nucleotide sequences of the nucleic acids of the pathogens.

Since 1980's Segments of DNA (probes), complimentary to specific segments of the nucleic acid of the microorganisms, have been labeled with radioactive isotopes or with colour producing compounds and are used extensively for the detection and identification of plant pathogens.

History of Plant Pathology in India

Plant pathological developments in India have two main stages.

1. Formative period of plant pathology in India (1905-1950): Work was mainly confined to mycology with special emphasis in the early period on symptomatology and description of the pathogen. Later stress was laid on the physiology of disease, development of resistant varieties and chemical control.

2. Modern period (1950-1970) when the new techniques evolved was applied in the fields of mycology, virology, and bacteriology.

Plant pathology in India began with the establishment in 1905 of the Indian (then Imperial) Agricultural Research Institute at Pusa, Bihar and the appointment of E.J. Butler as the first Imperial Mycologist. The credit for paving foundation of plant pathology goes to him and he is known as "Father of Indian Plant Pathology".

The first Indian Universities established in 1857 at Calcutta, Madras and Bombay, emphasized taxonomy of fungi.

Plant Pathology as a University Science became established at Lucknow (1921), Allahabad (1887) and Madras (1857) Universities.

1947 Indian Phytopathological society by B. B. Mundkur under the chairmanship of S. R. Bose was established.

Diseases Caused by Fungi

1903 E.J. Butler wrote a monograph on potato diseases.

1915 J.F. Dastur reported that potato blight due to *Phytophthora infestans*, though uncommon in the Indian plains is found on the hills, and caused severe damage to crops grown at an elevation of 2000m and above. Dastur described two new diseases of potato: leaf rot caused by a species resembling *P. parasitica* and a tuber rot caused by a new species, *P. himalyensis*.

1918 E.J. Butler published the book "Fungi and Diseases in Plants" which served as the major source of literature and inspiration to budding plant pathologists.

Strong schools for fundamental pathology, especially the biochemistry of host - parasite interaction, were established at Lucknow and Madras Universities under the leadership of S.N. Gupta and T.N. Sadasivan.

Dasgupta studied the role of enzymes in pathogenicity. A general high metabolic rate and higher level of several enzymes were shown in virulent strains.

Diseases Caused by Bacteria

1827-1831 Bacterial canker (*Xanthomonas citri)* affected citrus specimens were collected from Dehra Dun in Himalayas suggesting the origin of disease as India

Although bacterial diseases were important and widespread, very little work was done until 1950, when M.K. Patel established a school of bacteriology at the College of Agriculture, Pune. He and his associates established more than 40 new species of plant pathogenic bacteria, mostly belonging to the genus *Xanthomonas.*

The first disease suspected to be caused by bacteria was 'bangle blight' of potato in Bombay. Cappel reported it in 1892 and was prevalent in Pune and other places in Bombay. Coleman (1909) was the first to establish the bacterial association with bangle blight. Mann and Nagpurkar isolated the organism, established its pathogenicity and studied its biochemical and physiological properties.

1928 Patwardhan reported black rot of cabbage in India.

Non Parasitic Diseases

1940 A school for the study of deficiency diseases was established by S.N. Gupta at Lucknow University.

1954 Vasudeva and Raychaudhari reported a serious disease of guava from Rajasthan, characterized by severe reduction in leaf size, interveinal chlorosis, suppression of growth and cracking of fruits.

History of fungicides Use in India

1885 Ozanne was the first to try copper sulphate for the control of fungal diseases

1906	Sulphur was used on grapes in Bombay. Butler recommended Bordeux mixture brushing for the control of Koleroga of Arecanut.
1909	McRae claimed control of *Phytophthora omnivore* var.*arecae* on arecanut with Bordeaux mixture in which resin was added as a sticker.
1911	McRae advised to spray potato plants with Bordeaux mixture against early and late blights.
1927	Anstead claimed protection of rubber from *Phytophthora* with Bordeaux mixture. Kamat found Bordeaux paste quite satisfactory against gummosis of citrus.
1930	Narasimhan suggested the use of linseed oil in Bordeaux mixture for improving coverage and tenacity.
1931	Dastur claimed success in controlling foot rot of betelvine by soil drenching with 2-2-50 Bordeaux mixture.
	Mayne obtained control of coffee rust by Bordeaux spray.
	Murray controlled powdery mildew of rubber by sulphur dusting.
	Uppal *et al.,* obtained control of powdery mildew of vine by sulphur dustings.
	Uppal could control fig rust by dusting 300 mesh sulphur applied at monthly intervals.
1935	Uppal tried colloidal copper for the control of fig and obtained good results.
	Dastur stated that citrus canker could be controlled by Bordeaux.
	The use of organic sulphur fungicides and others was effected in India mostly after 1950s

Occurrence of Major diseases in India

Mango malformation

Malformation is a devastating and somewhat mysterious disease of mango. The disease was first observed at Darbhanga (Bihar) in 1891. It occurs in Maharashtra (Burns, 1910; Narasimhan, 1954), Uttar Pradesh, Punjab, Haryana, Bihar and other places. Studies attempted to determine the nature of the disease since its recognition in India in 1891. Numerous workers attributed it to different causes: nutritional imbalance, virus, eriophyid mites etc.

1891 Maries postulated that mango malformation is caused by excessive soil moisture.

1910 Burner suggested that because of the vegetative activity during the fruiting period, inflorescences produced in April or later were changed into leafy structures. Therefore the disease was considered as a physiological disorder.

1944 Hassan hypothesized the causal organism to be an eriophyid mite, *Aceria mangiferae*.

1946 Sattar suggested its viral origin because malformation occurred both in grafted and seedling plants and incidence increased slowly.

Several reports were published during 1950's and 1960's affirming viral transmission. Later results of electron microscopy, transmission, cultural and serological studies in India and Egypt convincingly disproved the involvement of virus and MLO.

1966 Summanwar *et al.*, reported that *Fusarium. moniliformae* was associated with malformed tissues.

Apple Scab

1930 The disease was first noticed in Ambri variety in Kashmir valley where it remained static until a severe epidemic appeared in 1973

1977 It was noticed in Himachal Pradesh and by 1983 it quickly spread to almost entire state of H.P when there was an epidemic resulting in huge losses to the public exchequer. It rendered 10% apple crop unfit for consumption in HP which had to be destroyed and as a result the state suffered a loss of Rs.15.0 million.

Bunchy Top of Banana

The disease was recorded as causing serious damage to bananas in Fiji as early as 1829. In India, the first occurrence of the disease was reported from Kerala and spread to Tamil Nadu, Andhra Pradesh, Karnataka and west coast of Maharashtra.

Late Blight of Potato

It was recorded from Nilgiri hills sometimes between 1870 and 1880. Soon after, it was noticed in Darjeeling in 1883 and spread rapidly to other areas viz., Khasi hills, Kumaon hills and Shimla hills. In the plains, it was first observed in 1898 - 1900, in Hoogly district of West Bengal. Since 1913, it was recorded at several places in Assam and Bihar.

Cassava Mosaic

African cassava mosaic, although noticed in some parts of India since 1942, was authentically reported during 1966 only. Malathi *et al.,* 1987 found that the disease occurring in India is caused by a distinct virus "Indian cassava mosaic virus".

Root (Wilt) of Coconut

First reported after the great floods in 1882 in three independent locations each about 50 km apart in the erstwhile state of Travancore (Kerala). Since then it spread from the original foci of infection and now occupies a contiguous area covering 9/14 districts of Kerala.

1908 Butler isolated *Botryodiplodia* sp from the roots of disease palms and suggested that root rot alone could be sufficient to cause the disease syndrome.

1951 Menon and Nair isolated two bacteria from roots of diseased palms.

1954 Nagaraj *et al.,* based on systematic nature and resemblance of symptoms proposed viral etiology.

1966 Davis and Pillai ruled out the role of micronutrients in the incidence of disease.

1969 Srivastava observed characteristic streaming movement of bacteria in vascular tissues of roots of diseased palms and isolated bacterium tentatively identified as *Pseudomonas* sp.

Summanwar *et al.,* reported the isolation of a virus from roots and leaves of diseased coconut palm.

1975 Cecil observed Ca and Mg contents of healthy palms significantly higher than those of apparently healthy and diseased palms. He also observed that palms in the root (wilt) affected areas are in a state of unbalanced nutrition as compared to healthy palms.

1980 Solomon and Sasikala ruled out the association of TMV through serological studies.

1983 Solomon *et al.,* observed presence of Mycopasma Like Organisms in sieve tubes of juvenile tissues such as sub-meristem, petioleof developing leaves etc.

Koleroga of Arecanut

First reported from the erstwhile Mysore state in South India by Butler, (1906) and subsequently from areas of present Uttara and Dakshina Kannada of Karnataka and also from certain parts of Malabar and Cochin by Coleman, (1910). The fungus was first described as *Phytophthora omnivora* and Co leman, (1910) named it as *P.omnivora* v ar.*arecae*.

Foot rot of Black Pepper

Severe incidence of black pepper vine death was reported in Wynad region of Kerala as early as 1902 and was investigated by Barber and later by Butler. Though *Phytophthora* isolation from black pepper was reported earlier in 1929 by Venkata Rao, the first authentic report was by Samraj and Jose in 1966 that established the pathogenicity of *Phytophthora* in black pepper and adopted the identification of Muller (1936) as *P.palmivora var.piperina.* The disease was loosely referred to as quick wilt disease of black pepper based on sudden wilting and death of the vine. However, the terminology of the disease was changed to *Phytophthora* foot rot by Nair and Sarma in 1988.

Katte Disease of Cardamom

This is the first reported disease of small cardamom found to occur in almost all cardamom growing areas. Mollison, (1900) reported its occurrence for the first time from North Canara area in Karnataka state.

1914 Van Bundren referred this as unknown disease while in 1914 contented that the disease was due to addition of inferior quality of leaf manure to the crop.

1929 Sahasrabudhe and Bapat concluded that the mosaic like symptoms were due to soil sickness caused by the soil protozoan, *Olpidium striatus.*

1938 Thomas suggested viral nature of mosaic disease for the first time.

1945 Uppal *et al.,* conclusively proved viral etiology of the disease and successfully transmitted the disease with banana aphid, *Pentalonia nigronervosa.*

1985 Naidu *et al.,* purified and characterized the virus.

References

Annual Review of Phytopathology 1972, Annual Reviews INC, 4139, El Camino Way, Palo Alto, California, USA.

Chadha, K.L.(Ed) 1993.Advances in Horticulture(13 volume series).Malhotra Publishing House,New Delhi-64

George, N.2005.Plant Pathology(Fifth Edition)Agrios-Academic Press,30 Corporate Drive,Suite 400,Burlington,M A 01803,USA

Nene,Y L 1971. Fungicides in Plant Disease Control .Oxford and IBH Publishing Co.New Delhi.

Vegetable Science

The Science of Horticulture : Vol. 1, 2011
© K.V. Peter (ed.), pp. 199-210
New India Publishing Agency, New Delhi (India)
E-mail : newindiapublishingagency@gmail.com
Web: www.bookfactoryindia.com

Chapter 9 History of Plant Breeding with Emphasis on Vegetables

PETER, K.V. AND PRADEEP KUMAR, T.

The history of plant breeding is divided into 4 periods- knowledge of plant breeding before 1900, application of genetic principles (1900-1940), plant breeding research after 1940 and the recent developments in plant breeding. Emphasis is given on the land marks in genetic improvement of vegetable crops.

The major land marks in genetics and breeding of crop plants are listed below.

AD 1 - 1900

1665 Hooke - described the cell

1675 Malpighi - conducted anatomical investigations in plants; repeated the observations of Hooke

1676 Millington - anthers function as male organs

1694 *Camerarius* - first to demonstrate sex in plants. Suggested crossing as a method to get new types. Worked with mulberry and maize. Systematic plant breeding was established at this point. Slow progress because workers did not recognize difference between genetic and environmental variation

1716	Mather - observed natural crossing in maize. Noted xenia, where ears of yellow maize planted next to red and blue maize had red and blue kernels in them in the first year
1719	Fairchild - created first artificial hybrid of Carnation x Sweet William, commonly known as Fairchild's mule
1753	Linnaeus - published "Species Plantarum". Binomial nomenclature of plant taxonomy officially begins with his general list of plant species
1761-66	Koelreuter - demonstrated that hybrid offspring received traits from both parents (pollen and ovule transmit genetic information), and was intermediate in most traits. First scientific hybrid was produced (tobacco). Demonstrated the identity of reciprocal crosses. Noted hybrid vigor, segregation of offspring (parental and non-parental types) from a hybrid
1778	Herbert - suggested winter hardiness was inherited after noting the research of Koelreuter
1779	Knight - emphasized the practical aspects of hybrids. Worked on crop improvement of garden pea. Noted the advantage of outcrossing to produce new forms
1785	Van Mons - started systematic selection of horticultural plants. Seed catalog in 1825 listed 1050 cultivars
1793	Sprengel - classic publication on the role of insects in pollination of angiosperms. Studied flower structure and color in relation to pollination and insect habits
1801	Lamarck - theory of evolution through inheritance of acquired characters
1812	Gauss - suggested the theory of least squares. Had begun the idea of "probable error" by 1820
1819	Shirreff - utilized pure line selection (progeny test) to develop a new oat cultivar (released in 1824), and a new wheat cultivar (released in 1832). Had first nursery plots, with 70 families by 1857. Suggested crossing parents with desirable characteristics to obtain progeny of value
1820	Goss - reported dominance and recessiveness in peas
1820	Knight - noted that male and female parents contribute equally to the F_1, with segregation following in the F_2. Made first wheat variety cross in Europe

1827-49	Von Gartner - made the most extensive crossing experiments to date, with 10,000 crosses in 700 species from 80 genera
1831	Brown - discovered the eukaryotic cell nucleus
1837-38	Schleiden and Schwann - developed the cell theory of higher plant organization
1844-63	Godron - used sterility of interspecific hybrids as evidence that the parents were distinct species
1844	Unger - showed that cells come from the division of preexisting cells
1846	Quetelet - described biological traits in quantitative terms
1847	Sutton - developed cultivars of beets, peas and potatoes for production in Great Britain
1848	Hofmeister - observed dark staining bodies (chromosomes) in the cell nucleus
1855	Virchow - established that the egg of one generation comes from the egg of previous generation; the "continuity of heredity"
1858	Hofmeister - discovered the female gametophyte
1859-89	Darwin - published "Origin of Species"; noted inbreeding, sterility and differences in reciprocal crosses
1860	Louis Vilmorin - stressed the value of self-pollination in breeding stable cultivars of wheat and sugar beets
1860	Naudin - did research much like Mendel, describing the separation of essences in various proportions in germ cells and recombination in the offspring of a cross. Used *Datura* for his rather complex experiments. Thought the whole plant was a mosaic of the 2 parents.
1861	Pasteur - used heat to kill bacteria
1861	Schultze - found that protoplasm was an essential element in the cells of all organisms
1866	Mendel - published "Experiments in plant hybridization"; discovered unit factors (genes), segregation of F2, recombination of 2 or more genes, and dominance of one allele over another; formulated the laws of inheritance

1875-77 Strasburger - first adequate description and drawing of chromosomes; nuclei arise only from nuclei; suggested the terms "gamete" and "chromosome "

1878 Henry Vilmorin - did the first extensive work with interspecific crosses: *Triticum vulgare* x *T. polonicum* and *T.durum*

1885-87 Weismann - wrote 12 essays on heredity and evolution, important in directing the trend of biology and thought; summarized current knowledge and pointed out the broad implications: continuity of germplasm, equal inheritance from parents, and non-heritability of acquired characters

1884 Strasburger - demonstrated fertilization and showed the fusion of the 2 nuclei to form the zygote

1888 Strasburger - demonstrated reduction division in plants

1889 Galton – wrote a book on natural inheritance

1894 Pearson - made contribution to mathematical aspects of testing; introduced the term "standard deviation"

1900 Correns (Germany) and DeVries (Holland) and von Tschermak (Austria) - independently rediscovered Mendelían laws of heredity

1900 Bateson - introduced the terms "allelomorph", "homozygote", "heterozygote", "F_1" and "F_2".

AD 1901 - 1940

1901-02 Bateson and Punnett - called public attention to the importance of Mendel's work; first report of linkage (using garden pea)

1902 DeVries - proposed the mutation theory of evolution from evidence in *Oenothera*

1902 Biffen - first inheritance studies on disease resistance; found that stripe rust resistance was due to a single gene

1903 Johannsen - developed the pure line theory of selection

1904 Hannig - contributed to the idea of embryo culture

1904-05 East and Shull - began inbreeding experiments with maize

1905 Shamel - reported yields of maize inbreds for 3 generations and their hybrids; first report of hybrid made by crossing inbred lines

1905-07	Williams - developed the remnant seed breeding plan for maize
1906	Bateson - introduced the term "genetics"
1908-09	Hardy and Weinberg - developed law of equilibrium for populations
1908-10	East - published his work on inbreeding and proposed the idea ofmultiple alleles at a locus
1910	Morgan - discovered genes on chromosomes
1910	Bruce, Keeble and Pellew - proposed theory for hybrid vigor due to favorable dominant genes
1910	Wood and Stratton - suggested check plots to correct for soil variability in yield trials
1911	Collins and Kempton - demonstrated linkage in maize
1911	Mercer and Hall - introduced questions about number, plot shape and size, and replications in field trials
1912	Jennings - selfing reduces heterozygosity by one-half each generation
1912	Harris - suggested the use of Chi-square for testing the goodness of fit for segregation ratios
1913	Sturtevant - provided the experimental basis for the concept of linkage vs. map distance
1916	Jones and Gilman - released a yellows resistant cabbage
1916	Winkler - introduced the term "genome"
1916	Shull - introduced the term "heterosis" to replace "heterozygosis"
1917	McFadden - described wheat-rye hybrids (triticale)
1917	Jones - proposed that hybrid vigor was due to linkage of favorable dominant alleles; developed the first commercial hybrid of maize, ëBurr-Leamingí (4-way hybrid)
1917	Winge - proposed the theory of polyploid origin of plant species
1918	Sakamura - classified chromosome numbers in wheat species
1918	Fisher - introduced the ideas of quantitative inheritance and correlation

1920	East and Jones - produced the first tobacco cultivar by hybridization
1920	Jones - proposed the 4-way hybrid for commercial cultivars
1920	Blakeslee - discovered trisomics in *Datura* and coined the term " monosomic" for diploid missing one chromosome
1920	Wright - published "Biometrical relations between parent and offspring"
1920	Harland and Pope - described the backcross breeding technique for small grains
1920	Bridges - discovered duplications, deficiencies and translocations in chromosomes
1925	Sturtevant - found first chromosome inversion
1925	Fisher - introduced analysis of variance and other statistical methods
1926	Sumner - isolated the first enzyme in crystalline form (protein)
1926	Muller - reported artificial mutations in animals by X-rays
1928	Griffin - discovered genetic transformations in bacteria
1928	Stadler - described the mutagenic effects of X-rays in barley
1929	McClintock - first to report and number 10 chromosomes in maize
1931	Stern, Creighton and McClintock - provided the cytological proof of crossing-over
1932	Davis - suggested the inbred-variety cross method for testing maize inbreds
1932	Jenkins and Brunson - used the topcross method to give comparative tests for combining ability
1933	Rhoades - discovered cytoplasmic sterility in maize
1933	Morgan - awarded Nobel Prize for theory of the gene
1934	Dustin - discovered colchicine
1935	Vavilov - contributed much to knowledge of species origin in "The scientific bases of plant breeding"
1936	Fisher - published the book "Statistical methods"

1937 Nebel and Ruttle - used colchicine to double chromosome number

1937 Dobzhansky - published book "Genetics and the origin of species"

1939 Harvey - showed that different genotypes of maize and tomato have different nutrient requirements.

Genetics and Plant Breeding Research After 1940

1941 Beadle and Tatum - published their classic study on the biochemical genetics of *Neurospora*

1942 Atkins and Mangelsdorf - suggested the use of isogenic lines for comparative studies of plant characters

1943 Jones and Clarke - described the inheritance of male sterility in onion; first use of male sterility in commercial hybrid production

1944 Avery, MacLeod and McCarty - described the transforming principle and suggested that DNA, not protein, is the hereditary material

1945 Hull - proposed recurrent selection

1946 Muller - received Nobel Prize for his contribution to radiation genetics

1946 Delbruck and Bailey - demonstrated genetic recombination in bacteriophage

1946 Lederberg and Tatum - demonstrated genetic recombination in bacteria

1946 Auerbach and Robson - showed that chemicals could induce mutations

1948 Borvin and Vendrely - showed that the quantity of DNA and chromosome number was related

1948 Frandsen and Frandsen - described the polycross method in detail (the method had been used before)

1948 Comstock - suggested reciprocal (half-sib) recurrent selection

1948 Green and Green - described crossing over between alleles at the *lz* locus of *Drosophila*

1951 Kihara - reported production of seedless watermelon fruits through triploidy

1952	Zinder and Lederberg - described transduction in *Salmonella*
1953	Watson, Crick and Wilkins - proposed a model for the structure of DNA
1955	Benzer - demonstrated fine structure of the genetic material of phage T4 of *E. coli*; coined the terms "cistron", "recon" and "muton"
1955	Mather proposed disruptive selection.
1956	Demerec and Hartman - showed that gene order and biochemical sequence were related in certain instances
1956	Ochoa, Kornberg - synthesized RNA and DNA polymers
1957	Taylor, Woods and Hughes - made the first successful attempt to determine the mechanism of chromosomal duplication using tritiated thymidine and autoradiography
1957	York- exploited self incompatibility for commercial production of F_1 hybrid seeds in cabbage.
1958	Beadle, Tatum and Lederberg - received the Nobel Prize for their work in biochemical genetics
1958	Peterson and Weigle – advocated use of gynomonoecious lines for the production of hybrid seeds in cucumber.
1958	Sanger - received the Nobel Prize for being the first to determine the sequence of amino acids in a protein molecule (insulin)
1959	Nieuwhof- proposed family selection method for cabbage
1959	Ochoa and Kornberg - received the Nobel Prize for their research on the *in vitro* synthesis of nucleic acids
1960	Peterson and Anhder – suggested use of gynoecious lines for the production of hybrid seeds in cucumber
1966	Davis - suggested utilization of marker genes for hybrid seed production in onion.
1968	Shimotsuma – developed reciprocal translocation stocks in water melon for production of seedless fruits.
1968	Ogura - reported male sterility in radish
1970	Berg, Cohen and Boyer - introduced recombinant DNA techniques

1970	Cross - reported male sterility in carrot
1970	Pochard - In bell pepper, *ms-509* male sterile line was developed
1973	Stanley N. Cohen and Herbert Boyer-created first GMO
1975	Casali and Tigchelaar – proposed single seed descent method (SSD)
1976	Rick- suggested single cell culture and cell hybridization for tomato improvement
1981	Tanskley - demonstrated usefulness of isozyme markers in vegetable breeding
1983	Karry Mullis-invented polymerase chain reaction
1991	Georgiev – successfully exploited *ms-10* gene for hybrid seed production in tomato
1994	Flavr savr tomato introduced
1995	Bt corn introduced
1996	Monsanto - commercially launched glyphosate resistant transgenic soyabean.
2002	More - developed tropical gynoecious parthenocarpic lines in cucumber.

Development of Transgenic Technology

Transgenic technology provides the means to make "crosses" , unthinkable earlier. Organisms which have until now been completely outside the realm of possibility as gene donors can be used to donate desirable traits to crop plants. These organisms do not provide their complete set of genes, but rather donate only one or a few genes to the recipient plant. A single insect-resistance gene from the bacterium *Bacillus thuringiensis* can be transferred to a tomato or cabbage plant to make Bt tomato or bt cabbage. Transgenic plants were first created in early 1980s by four groups working independently at Washington University in St. Louis Missouri, the Rijksuniversiteit in Ghent, Belgium, Monsanto Company in St. Louis, Missouri, and the University of Wisconsin. On the same day in January 1983, the first three groups announced at a conference in Miami, Florida, that they inserted bacterial genes into plants. The fourth group announced at a conference in Los Angeles, California, in April 1983 that they inserted a plant gene from one species into another species.

The Washington University group, headed by Mary-Dell Chilton, produced cells of *Nicotiana plumbaginifolia*, a close relative of ordinary tobacco, which were resistant to the antibiotic kanamycin. Schell and Marc Van Montagu, working in Belgium produced tobacco plants resistant to kanamycin and to methotrexate, a drug used to treat cancer and rheumatoid arthritis. Robert Fraley, Stephen Rogers, and Robert Horsch at Monsanto produced petunia plants resistant to kanamycin. The Wisconsin group, headed by John Kemp and Timothy Hall, inserted a bean gene into a sunflower plant. These early transgenic plants were laboratory specimens, but subsequent research developed transgenic plants with commercially useful traits like resistance to herbicides, insects and viruses.

Vegetable Improvement in India

Research on vegetable crops was started by ICAR,New Delhi with the sanctioning of a nucleus 'Plant introduction scheme' at the IARI, New Delhi in 1947-48. Simultaneously, the ICAR also started such ad-hoc schemes in Punjab, Uttar Pradesh, West Bengal, Maharashtra, Himachal Pradesh, Jammu and Kashmir and Tamil Nadu. Entire seeds of temperate vegetables were imported from European countries before World War II. When supply of vegetable seeds from Europe was affected during the War, authorities felt the need for production of these seeds within the country. Thus, a vegetable Breeding Station at Katrain in Kullu valley, Himachal Pradesh was established in 1949. This station was transferred to Indian Agricultural Research Institute in 1955 for undertaking research on genetic improvement and standardization of seed production technology of temperate vegetables. In fact, the Vegetable research in India got momentum in fifties with the sanctioning of a nucleus plant introduction scheme at the IARI, New Delhi. Evaluation of thousands of foreign and indigenous collections resulted in selection of 30 good varieties. Late Dr Harbhajan Singh, Head, Division of Plant Introduction, deserves acclaim for his well foresighted and acumen mind which gave varieties like Pusa Sawani in bhindi, Pusa Purple Long, Pusa Purple Round and Pusa Purple Cluster in brinjal , still held in esteem by the Indian vegetable growers.

Establishment of the Indian Institute of Horticultural Research (IIHR) at Bangalore with a full fledged division of vegetable crops in 1968 is another landmark in the history of genetic improvement of vegetable crops in India. The vegetable research in India received further boost when a separate Division of Vegetable Crops and Floriculture was started at the IARI, New Delhi in 1970. The All-India Co-ordinated Research Project of the ICAR, conceptualized during reorganization of the Council, is important in the history of agricultural research in the country. The All-India Co-ordinated Research Project for Improvement of Vegetable Crops was started in 1970-71, during the Fourth Five-Year plan, to provide a national grid for multilocation testing of vegetable production technologies developed by various research institutes and state agricultural universities.

The first major task of the All-India Co-ordinated Vegetable Improvement Project was multilocational testing of varieties developed by different research institutes and agricultural universities at centres under the project located in different agroclimatic conditions and to identify varieties suitable for different crop growing areas. Later, research was undertaken on agro-techniques, like plant population density, fertilizer requirements and chemical weed control. Research programmes were also undertaken on chemical control of diseases and insect pests. To start with, only 14 vegetables were covered under the programme. Later, with addition of a few more crops, number of crops covered under the project were brought to 24. However, the major emphasis was laid on 15 important crops: tomato, brinjal, chillies, cauliflower, cabbage, peas, onion, muskmelon, watermelon, okra, Frenchbean, pumpkin, carrot, radish and garlic. Workshops/ group meetings were organized in which active workers associated with different research projects participated and very vital information regarding identification of varieties and improved package of practices emerged. The Project Directorate of Vegetable Research was started in 1987 during the Seventh Five-year Plan by upgrading erstwhile All-India Co-ordinated Vegetable Improvement Project. This Directorate operated from the original headquarters of the All-India Co-ordinated Vegetable Project located at the IARI, New Delhi from 1987 to 1992. This was shifted from New Delhi to Varanasi in 1992 and upgraded to the status of Indian Institute of Vegetable Research.

The National Bureau of Plant Genetic Resources, New Delhi started in 1976 by upgrading the erstwhile Division of Plant Introduction at Indian Agricultural Research Institute.The NBPGR is doing intensive work on exploration, collection, evaluation, conservation and exchange of germplasm of vegetable crops besides work on crop improvement. The National Research Centre (Arid Horticulture), Bikaner, (Now Central Institute of Arid Horticulture) is also working on vegetables under the arid areas. A National Research Centre on Onion and Garlic was also established at Pune during VIII Plan. Besides ICAR, The National Horticultural Research and Development Foundation (NHRDF) is also conducting research on onion and garlic at its headquarters at Nasik and other regional stations. A number of short-term, time- bound and result-oriented adhoc schemes to solve area-specific problems of selected vegetable crops are also being supported by the ICAR at various central institutes and State Agricultural Universities (SAUs) (Chadha, 1991). There are now four institutes and National Research Centres working on vegetables in India.During 11th plan all the National Research Centres are strengthened into Directorates. The AICVIP is operating in 26 centres and 42 voluntary centres spread over different agroclimatic zones of India. The AICRP has made significant contributions to vegetable production by developing 311 varieties in 22 vegetable crops, identified for cultivation in different climatic zones (Rai and Pandey, 2005). Out of these, 195 are high-yielding open- pollinated, 35 are resistant to diseases and 81 are hybrids. As the result of multidisciplinary and multilocational testing of new research materials during the last three decades, improved varieties/hybrids in 21 major vegetable crops were identified and recommended for cultivation. These include 67 varieties of tomato, 77 of brinjal, 15 of onion, 19 each of okra and cauliflower, 33 of gardenpea, 26 of chillies, 12 of muskmelon, 8 each of cowpea and cabbage, 7 each of frenchbean, garlic and bottle-gourd, 6 each of bitter-gourd, pumpkin and cucumber, 5 of watermelon, 3 each of carrot, ridge-gourd and sponge-gourd and 2 for dolichos-bean. By and large, vegetable research was carried out in India by public institutions . From 1969 onwards, there had been some effort to start R & D activities by private companies with foreign collaboration. Private companies conducting research include Indo-American Hybrid seed Co., Bangalore ; Mahyco, Jalna, Maharashtra; Sutton & Sons, Calcutta, West Bengal; Bejo Sheetal Hybrid Seeds, Jalna, Maharashtra; Biogene, Bangalore and Novartis India Ltd., Pune. A number of open pollinated varieties and hybrids in vegetable crops were developed by different private seed companies which performed well under the multilocation testing of All-India Co-ordinated Vegetable Improvement Project.

Reference

Peter, K V and Pradeep Kumar, T 2008. Genetics and Breeding of Vegetables. DIPA, ICAR, New Delhi. 12, 364p.

The Science of Horticulture : Vol. 1, 2011
© K.V. Peter (ed.), pp. 211-267
New India Publishing Agency, New Delhi (India)
E-mail : newindiapublishingagency@gmail.com
Web: www.bookfactoryindia.com

Chapter 10 Basics in Vegetable Seed Technology

KANWAR, J.S. AND GEETA BASSI

India is a vast country with varied soil and agro-climatic conditions and therefore is able to grow a wide range of vegetables. In view of their productivity, diversification, nutrition, value addition and export, vegetables are an important component in Indian agriculture. Up to 1939, vegetable seeds were used to be imported from Australia, U.K., U.S.A., but after World War I and II, India started seed production of vegetables. In the recent past, the seed industry has made rapid strides over the past three decades and has contributed significantly to the quantum jump in agricultural production and India emerged as the second largest producer of vegetables, next to China in the world. During 2005-2006, India produced 113.5 million t of vegetables from 7.2 million ha with a productivity of 15.76 t/ha. The seed industry has increased certified and quality seed production and distribution by the organized sector to about 60 lakh quintal from 300q in 1964-65. India can be proud of having launched and managed both crop improvement and seed improvement as simultaneous and parallel programmes playing complimentary and synergistic roles.

Background

Seed is the basic the cheapest, critical but the most effective and vital agricultural input for attaining sustained growth in agricultural production and productivity, which is dependent on the development of new and improved varieties of crops and an efficient system for supply of their quality seed. Seed is the starting point of most of our vegetable crops although some are propagated vegetatively from cuttings or tubers. The vegetable seed is, therefore, basic for all inputs such as fertilizers, crop protection, irrigation, harvesting and marketing. India has made substantial significant advances in the seed sector in the last four decades. However, per capita consumption of vegetables in our country remained 130 g as against a minimum 300 g recommended. In order to bridge the gap of consumption and demand of increasing population, the production of vegetables needs to be increased many folds. Moreover, large scale vegetable production for processing and export can also earn foreign exchange for the country besides generating employment. Thus, it is essential that seed of the highest possible quality may be made available to the farmers. Quality seed ensures better germination, rapid and uniform emergence and vigorous growth. These attributes result in a good stand and increased yield and quality. Poor quality seed results in "skips" and yield reduction, all of which decrease profitability of growers. The quality seed development starts from selection of land, agronomy, plant protection, harvesting, threshing, seed processing and storage. The quality seed production demands utmost care and vigilance at each and every stage. It is an established fact that good quality seed is a prerequisite for optimum return from the crop.

Categories of Seeds

In the Indian generation system of seed multiplication, there are primarily three categories of seed i.e. breeder seed followed by foundation seed and certified seed. Certified seed is actually sold to the farmers to raise the commercial crop. There is another category of seed known as Truthfully Labeled Seed in which case, there is no certification but labeling is compulsory. Label colour and size have also been specified for each category of seed.

Nucleus seed: This seed is produced by the concerned breeder. It is a group of progenies of individual plants taken at random from a variety for the purpose of purifying that variety of mixtures and off- types. There are two classes of nucleus seed *viz* nucleus seed stage I and nucleus seed stage II. The stage I seed is produced by following plant to progeny rows but when the demand of breeder seed is more in case of popular and high yielding varieties, stage II multiplication procedure is adopted. In this case, progeny to plot method is followed. Purity and quality of nucleus seed plot are monitored by a team of experts. It is generally produced as plant/head-to-row and serves as the stock seed for breeder seed plot. This seed is not certified but at Punjab Agricultural University, its production is monitored by a team of experts.

Breeder Seed: Breeder seed is the seed or vegetatively propagating material directly controlled by the originating or sponsoring plant breeder of the breeding programme of institution and/or seed whose production is personally supervised by a qualified plant breeder , that provides the source for the initial and recurring multiplication of foundation seed. Breeder seed is genetically so pure as to guarantee that in the subsequent generations, it conforms to the prescribed standards of genetic purity. The breeder seed production is undertaken by a qualified plant breeder/sponsored plant breeder. Normally, the breeder seed is produced by State Agricultural Universities (SAUs) and Indian Council of Agriculture Research (ICAR) institutes. Breeder seed in small quantity is produced in progeny-row methods and large scale in open fields in proper isolated field. The minimum number of single plant progenies, isolation distance and minimum number of field inspections are given elsewhere. The breeder seed plot is jointly inspected by the monitoring team consisting of concerned breeder and representatives of ICAR, National Seed Corporation and state seed certification authority. The BSP III certificate is issued after the inspection of seed plot.

Foundation seed: Foundation seed is also known as certified foundation seed. It is the progeny of breeder seed or it is produced from foundation seed which can be clearly traced to breeder seed. Production of foundation seed is supervised and approved by the certification agency and is so handled as to maintain specific genetic identity and genetic purity and is required to conform to certification standards specific for the vegetable crop/variety being certified.

Certified foundation seed can be multiplied up to 2 stages from breeder or foundation seed itself. During the production and multiplication of certified foundation seed, the following guidelines are to be observed as per Indian Minimum Seed Certification Standards (1988):

(1) Foundation seed produced directly from the breeder seed is designated as foundation seed stage-I.

(2) Certified foundation seed produced from foundation seed stage-I is designated as foundation seed stage-II.

(3) Certified foundation seed stage-II cannot be used for further increase of foundation seed. It is only used for production of certified seed. Production of foundation seed stage-II is ordinarily adopted when the breeder seed is in short supply, *e.g.*, when vegetatively propagated crops, apomictically reproduced crops and self-pollinated crops, often cross pollinated crops, these being gene-pools, should not lose their genetic identity and purity. Measures to safeguard the same are adequately taken. Minimum seed certification standards are same for both foundation seed stage-I and II.

Requirement of plant progenies/isolation distance and field inspections for Nucleus and Breeder Seed production of vegetable crops.

Crops	Minimum single plant progeny	Minimum plants / progeny	Isolation distance (m)	Minimum field inspections	Check row after progeny row	Remarks
Brinjal	100	15	200	3	25	-
Chilli	100-150	20	400	3	25	-
Okra	100	50	400	3	60	-
Cucurbits	200-300	10	1000	3	20	Muskmelon seed field should be isolated all around from snap melon, long melon and wild melons, whereas pumpkin seed field should be isolated from summer and winter squashes. Similarly, watermelon should be isolated from citron.
Cauliflower	200	10	1600	3	20-30	Plants with loose or flat curds or leaves protruding through them and extra early or extra late flowering plant should be rogued out.
Cabbage	200	10	1600	3	20-30	-
Carrot	200-215	40-50	1000	3	40	Extra early and extra late bolting plants should also be removed at early bolting stage and again when about 90% of the plants are bolted.
Radish	200-215	40-50	1600	3	40	-do-
Turnip	200-215	40-50	1600	3	40	-do-
Onion	300	70-80	1000	4	50	-do-
Tomato	75-100	20-25	50	3	25	The plant showing early blight leaf spot and TMV should be rogued off.
Pea	200	40-50	20	3	50	Rabit ear and mosaic affected plants should be removed.
Methi	300	20	50	2	50	-
Garlic	200		5	2	30	-

(4) Certification tag for both foundation seed stage-I and II is of white colour and contains the information as to its stage (I or II).

The foundation seed is produced by State Agricultural Universities, National Seed Corporation (NSC), State Seed Corporations (SSC) and recognized private seed companies.

Certified Seed: Certified seed is the progeny of foundation seed and its production is so handled as to maintain specific genetic identity and purity according to standards prescribed for the crop being certified. Certified seed can only be multiplied if there is a genuine shortage of foundation seed, but it does not exceed three generations beyond foundation seed stage-I and it is determined by the certification agency that genetic identity and genetic purity are not significantly altered. Certified seed produced from certified seed is not eligible for further seed increase under certification. Certification tag seed is subscribed with 'not eligible for further seed increase. Certified seed is produced by NSC, SFCI, SSCs and private seed companies.

Class of seed	Label Colour	Label Size
Breeder	Golden Yellow	12 cm x 6 cm
Foundation	Both side White	15 cm x 7.5 cm
Certified	Both side Azure Blue	15 cm x 7.5 cm
T.L. Seed	Opal Green	15 cm x 10 cm

Standards

In India, seed certification is not compulsory, it is a voluntary process. Seed certification standards have been prescribed for foundation and certified seed. There are two types of standards: field standards, which apply to standing crop and seed standards applicable at seed level. Field standards include isolation requirement, maximum permissible level of off types, inseparable other crop plants, objectionable weed plants, pollen shedders (in male-sterile or A lines), plants infected by seed borne diseases etc. Seed standards relate to genetic purity, physical purity, germination, other crop seeds, weed seeds, moisture content etc. Standards for minimum genetic purity and germination for some important vegetable crops are given elsewhere. It is a fact that such standards have not been prescribed for breeder seed. Desirably the standards for breeder seed should be more strict than prescribed for foundation seed. However, the breeder seed should be 100% genetically pure.

Procedure for Seed Certification: Purpose of seed certification is to maintain and make available to the public high quality seeds and propagating materials of superior vegetable varieties/hybrids so grown and distributed as to maintain maximum genetic quality, identity and purity. Seed certification is a systematically designed process to secure, maintain, multiply and make available seeds of superior vegetable crop varieties having both genetic and physical purity. Certified seeds fetch premium in the market as it possesses genetic authenticity of the given cultivar as well as analytical purity.

Certified seed production is completed in six broad phases, namely receipt and scrutiny of application, verification of seed source, class and other requirements, field inspections, supervision of processing and packing, seed sampling and analysis and grant of certificate

Receipt and verification of application: The individuals desirous of producing seed under certification have to pay Rs. 25 as registration fee for any number of varieties/crops but for one season and additional charges @ Rs 750/ ha are charged for conducting seed certification.

Verification of source of seed: Any individual intending to produce the seed under certificate shall submit to the Certification Authority one or more relevant evidences such as certification tag, seed container, purchase record, sale record, progeny certificate etc. as may be demanded by the Certification Authority, in order to confirm if the seed used for raising the crop has been obtained from the source approved by it and conforms to the class of seed required for seed production.

Field inspections: The staff of the Certification Authority organizes their schedules with the complete knowledge of the fields in their area of jurisdiction so as to visit the fields systematically and in time. A record of inspections made is maintained. The number of inspections and stages of crop growth at which they should be conducted, vary from crop to crop depending on the crop duration, pollination behaviour, susceptibility to contamination, etc. Generally, field inspections of seed plot under certification are conducted at various stages as explained below:

(a) *Vegetative or pre-flowering stage*: This inspection is conducted to verify the source of seed, actual area planted, isolation distance from contaminants, planting ratio of parents in case of hybrid seed production and removal of off-types.

(b) *Flowering stage*: Inspection during flowering is done to check the occurrence and removal of pollen shedders in male-sterile lines, synchrony of parental lines, off-types, objectionable weed plants, inseparable other crop plants and seed-borne diseases.

(c) *Post-flowering and pre-harvest stage*: During this inspection, it is confirmed that the grower has continued thorough roguing of off-types, objectionable weed plants, inseparable other crop plants and designated diseased plants. Assessment of raw seed production is also calculated.

Minimum seed certification standards of some vegetable seed crops

S No.	Name of vegetable crop	Submitted sample size (g)	Working sample (g)	Pure seed (%)	Germination (%)	Moisture content (%)
1.	Bitter gourd	1000	200	98	60	7
2.	Bottle gourd	700	70	98	60	7
3.	Cucumber	150	15	98	60	7
4.	Tinda	1000	250	98	60	7
5.	Long melon	150	15	98	60	7
6.	Muskmelon	150	15	98	60	7
7.	Pumpkin	350	100	98	60	7
8.	Ridge gourd	100	40	98	60	7
9.	Snake gourd	1000	40	98	60	7
10.	Snake melon	150	40	98	60	7
11.	Sponge gourd	1000	40	98	60	7
12	Summer squash	1000	40	98	60	7
13.	Winter squash	1000	700	98	60	7
14.	Water melon	1000	250	98	60	7
15.	Brinjal	150	15	98	70	8
16.	Capsicum	150	15	98	60	8
17.	Okra	1000	140	98	65	10
18.	Radish	300	70	98	70	6
19.	Tomato	70	7	98	70	8
20.	Amaranths	70	10	95	70	8
21.	Celery	25	5	97	70	8
22.	Methi	40	4	98	70	8
23.	Lettuce	30	3	98	70	8
24.	Spinach	250	25	96	60	9
25.	Cabbage	100	10	98	70	7
26.	Cauliflower	100	10	98	65	7
27.	Chinese cabbage	40	4	98	70	7
28.	Knol Khol	100	10	98	70	7
29.	Onion	80	8	98	70	8
30.	Carrot	30	3	95	60	8
31.	Turnip	70	7	98	70	6
32.	Sugar beet	250	25	96	60	9

Principles of Seed Production

To produce seed with all quality attributes, a systematic seed production is must. Raising a seed crop differs from raising a commercial crop in several respects. Special methods and precautions are needed to produce seeds of desired quality. It needs good knowledge of crop management, preventing out crossing, curing, post harvest seed handling like processing and storage of seed, etc. and some elementary requirements as following, are to be taken special care of.

Genetic Principles : Genetic purity of a variety can be deteriorated due to:

- *Mechanical admixtures*: It is the most important source of variety deterioration during seed production.
- *Natural crossing* : In sexually propagated especially cross pollinated crops, natural crossing is another important source of variety deterioration. it depends upon the breeding system of the crop, isolation distance and pollinating agent/s.
- *Developmental variations* : When the seed crop is grown out of its area of adaptation for several consecutive generations.
- *Minor genetic variations*: Such variations may exist even in the varieties appearing phenotypically uniform and homogenous at the time of their release.
- *Selective influence of disease*: New crop varieties often become susceptible to new races of diseases and are out of seed program.

Maintenance of genetic purity: Various steps to maintain the genetic purity are:

- Use of approved seed for seed multiplication.
- Inspection and approval of field prior to planting.
- Providing adequate isolation.
- Field inspections at critical stages.
- Sampling and sealing of seed lots.
- Grow out testing.
- Adoption of generation system.
- Avoiding genetic shift by growing crops in areas of their adaptation.

Agronomic Principles

1. *Environmental Requirements* : Vegetable crops for seed production should be grown under the best possible conducive conditions. Best seed yield and quality are obtained when crop is grown in appropriate season and in an area where the crops' reaction to right environment can be observed. The regions with abundant sunshine should be selected for seed production. The flowering in many crops is controlled by the light duration. For example, lettuce and spinach require long day conditions for flowering and seed setting. Some species require a low temperature stimulus to initiate flowering. Many temperate vegetables like cabbage, cauliflower, beet root, European type radish and carrot also require vernalization. Areas of high rainfall should be avoided as it will reduce seed viability and increase need for artificial drying. Excessive wind will increase water loss, prevent activity of pollinators, carry wind borne pollen over long distances and increase seed shattering.
2. *Land Requirements* : The field selected for raising a vegetable seed crop should be free from 'volunteer' plants. Volunteers mean the plants originating from the seed/ plant material of the previous crop. The land should be levelled with proper drainage and should have sufficient organic matter. The cultural operations of vegetables for seed production are similar to the operation on market crop but seed crop is to be maintained beyond the stage of commercial market harvest. Therefore, the control of diseases, pests and weeds should be continued over a longer period.
3. *Pollination Requirements* : Vegetable crops like tomato, garden pea, fenugreek and cowpea are self pollinated and majority of other vegetable crops like okra, chillies, cucurbits and brassicas are cross pollinated. The natural insect population is normally sufficient under open conditions to ensure satisfactory pollination but high plant population grown for seed in a concentrated area, there is a possibility that natural insect population may be insufficient to ensure proper seed set. Therefore, the introduction of supplementary bee hives will improve pollination and seed set, especially in cucurbits. However, care must be taken to ensure that pest protection chemicals are not used in a way to harm useful pollinating insects. Spray of chemicals should be avoided at peak pollination insect activity and should be done in the late afternoon.
4. *Isolation Requirements*: Satisfactory isolation of seed crop helps to maintain purity in three ways:
   (a) Cross-pollination does not occur between cross compatible crops.
   (b) During harvesting, seeds of different varieties of same crop are not mixed

(c) The transmission of pest and diseases from alternative host crops are minimized.

Proper isolation is thus essential to maintain genetic purity and health of a variety. Isolation between cross compatible varieties is achieved in three different ways.

Time Isolation: Isolation by time will allow seeds of different varieties of the same crop to be produced at the same station each year. If the season is too long enough to allow two production cycles of the cross compatible crops then they too are isolated by time. For example, early and mid maturity group of cauliflower grown for seed production can be isolated by time.

Isolation By Distance : The mode of pollination is related to the degree of isolation necessary. In case of self pollinated varieties, the isolation distance is relatively short but, in case of cross pollinated varieties, the distance from other variety should be relatively wide. The isolation distance also depends on the direction of insect flight (in case of insect pollinated varieties) or the direction of winds (in case of wind-pollinated varieties).

Field or Plot Shape : The plots for seed production should be as near to square as possible because the majority of the pollen contamination occurs around the perimeter of a seed crop. Thus by adopting square shape of seed plot, the perimeter of any given area is minimized and minimum amount of seed is discarded while applying the discard strip technique. This technique may be used when seed of high genetic purity is needed or there is evidence of pollen contamination from other cross compatible crops. A 5m wide strip around the fields' perimeter is harvested separately and discarded. The bulk harvest from the inner area forms the main harvested seed lot.

Roguing : The vegetable varieties, which are produced for seed show genetic change over several generations. It is therefore, necessary to exert control and keep the natural variation within the acceptable limits. This is achieved by inspecting the crops at various growth stages and removing individual plant which do not conform to the defined limits of that variety. Thus roguing is a technique that is used in seed production to maintain genetic purity of the variety. Rogues or off-types may occur in a crop due to any of the following reasons.

- The diversity of the morphological types within a crop may be wide. This tendency is greater in predominantly cross-pollinated (e.g. cauliflower, cabbage, *cucurbits* and onion) than self- pollinated (e.g. pea, tomato, fenugreek) crops. This is why varieties of self- pollinated crops are generally more uniform and stable than varieties of cross- pollinated crops.

- The seeds result from cross pollination between the crop for seed production and other compatible varieties or wild plants. These are not always identified in the first generation and show up in the second.
- Some plants may display deviation from the normal type due to mutation.
- Seeds of other varieties may have been accidentally mixed in the seed stock during its production, processing or admixture via seed drill.
- Volunteer plants may arise from vegetative pieces or dormant seed of the previous crop grown in the same field. It is always easier to conduct intensive roguing in breeder seed plots than in large commercial seed production plots. To obtain maximum benefits from roguing operation, following practical points should be adopted:
- The crop should be grown in such away that plants can be seen individually. Paired row system of planting may be followed so that it is easy to walk between rows. This practice facilitates detection of dwarf undesirable plants.
- Walk systematically through the crop so that each plant is seen. Remove the whole off-type plants. Do not simply remove the fruits showing undesirable characters because the remaining flowers on the off-type plant will still contribute to the material in the next generation.
- Inspect the crop with the sun behind as it is difficult to examine plants with the sun on the eyes.
- Do not delay field inspections. The undesirable plants should be removed before flowering as far as possible. Remove cross- compatible weeds and wild relatives. Remove all diseased plants and also related infected weeds. Variety description based on morphological characters like leaf shape, flower colour, fruit shape and colour generally form a good basis for roguing but some characters like leaf colour, plant height, earliness of flower are affected by environment.

From seed production point of view, vegetable crops are divided into two groups *i.e.* Annual crops and Biennial crops.

I. *Annual crops* : The crop plants that complete their life cycle from germination to seed formation in one season only. In this group of vegetable crops, the seed production techniques are same as those for table crop. But special attention is paid to crop inspections, removal of off-types, disease control, harvesting, extraction and processing of seed. This group includes crops like tomato, brinjal, capsicum, pea, cucurbits, okra, etc. Quality seed of these vegetable crops can be produced by the following ways.

(i) By maintaining proper isolation distance according to mode of pollination.
(ii) Roguing off and removing other undesirable plants.
(iii) Selection of plants for seed.

II. *Biennial crops*: The crop plants that complete their life cycle in two seasons *i.e.* during first season, plants go through vegetative phase only and store food. In the second season, the plants grow reproductive parts (essential organs), use the stored food and produce flowers and the seed. This group includes knol khol, cauliflower, cabbage, turnip, carrot, radish and sugar beet. In the first phase, the seed crop is raised as that of the table crop. Then it is either left *in situ* or uprooted and replanted for seed production. According to the temperature requirement, mostly the crops of this group are again divided into two types:

(i) *Temperate or European types* : The seeds of this group can be produced in hills at an altitude of 1500 m and above. It cannot produce seed in plains or tropics since low temperature requirements of bolting for seed setting are not met with.

(ii) *Tropical or Asiatic types*: The seeds of this group can be produced in plains or Tropics.

In biennial group of vegetable crops, there are two distinct methods of seed production:

(a) Seed-to-seed or *in-situ* method
(b) Head/root/bulb-to-seed or transplant method.

In seed-to-seed method, seed crop is raised on the same location where it was sown originally without any disturbance, whereas, in transplant method, the plants or roots or bulbs when mature are uprooted and transplanted in a new location without or after storage for some time as the case may be. Former method is preferred as it avoids extra cost of cultivation in lifting and transplanting. Also it gives higher seed yield as compared to transplant method. It imparts early maturity, thus providing ample time for the cultivation of next crop. Lack of systematic roguing is the major disadvantage of seed-to-seed method. Because true-to-type of the root crops cannot be accurately determined without uprooting. Therefore for raising high quality stock seed it is desirable to raise seed of all biennial crops by transplant method.

The vegetable crops are also classified on the basis of climatic conditions. *i.e.* winter (Pea, Cole crops, leafy vegetables (methi, spinach), onion, radish, turnip, tomato, chilli, etc.) and summer vegetables (okra, cucurbits, cow pea, beans, etc.).

Annual Crops

The seed production of this group of vegetable crops is easier than biennial crops. In most of the vegetables, harvesting for the market and seed purposes coincides, e.g., muskmelon, watermelon, pumpkin, dry pea, beans and chilli. The other vegetable crops have mature seed stage quite distinct from the green marketable stage, e.g., okra, brinjal, cucurbits, etc. The general cultural practices followed for seed production in these crops are similar to that of a market or table crop.

Tomato

General cultural practices for growing tomato for seed are similar to those of the market crop. It is a warm season crop, sensitive to frost and requires an average temperature of 18-27°C for proper growth, development and fruiting. Fruit and seed setting are adversely affected under high temperature and low humidity.

Isolation: Tomato is predominantly a self-pollinated crop but out crossing does take place to a negligible extent. Seed field of tomato is isolated from the fields of other varieties and fields of the same variety not conforming to varietal purity requirements for certification by 50 and 25 m for foundation and certified seed production, respectively.

Field Inspection and Roguing: Seed crop is inspected at least three times during the crop season. The first inspection is conducted during the vegetative stage on the basis of foliage and plant type characters. Second inspection is conducted at flowering and fruiting stage. Shape and colour of the immature fruits is the basis for identification of true-to-type plants. At the third inspection, the colour of the mature fruits, shape and size form the basis of selection. Plants showing early blight, leaf spot and mosaic symptoms are removed.

Picking of fruits : In tomato, germination of seed is affected by stage of fruit maturity at the time of fruit picking for seed extraction. Fruits on turning red, ripe red, and over ripe stages are good for extracting good quality seed. Picking at mature green stage and fruits of last picking should be avoided for seed extraction.

Seed extraction : Tomato seed can be extracted manually using acid or fermentation methods or mechanically using axial-flow vegetable seed extracting machine.

Manual Seed Extraction: In case of manual seed extracting, fermentation and acid treatment methods are generally followed in India. Fermentation of pulped material should not be prolonged beyond 72 h to get seed of high germination. Preferably the seed should be washed free of pulp after 48 h.

The seed should not be dried during noon hours on the seed drying floor since the seed loses its germination due to high temperature injury. After that the seed should be washed as quickly as possible.

Mechanical Seed Extraction: Manual extraction of seeds is time consuming, labour intensive and unhygienic. An axial-flow vegetable seed extracting machine has been designed and developed at Punjab Agricultural University, Ludhiana. The machine is capable of extracting seeds of vegetable fruits like tomato, brinjal, chillies, cucumber, summer squash, squash melon, and watermelon. It is quite efficient, economical and hygenic. Seed loss and damage to the seed coat are negligible.

Seed should be dried as rapidly as possible by spreading in thin layers on the drying floor. In case temperature exceeds 40°C, the seed drying under partial shade is desirable. Under artificial conditions, the seed should first be dried at low temperature and then the temperature could be gradually increased to 40°C. The seed is dried to a moisture level of 6 to 8% and packed in cloth bags or moisture proof/resistant containers and kept in a cool and dry place.

Okra

In plains, two crops of okra, i.e., spring summer and rainy season crops are grown. From the seed production point of view , the rainy season crop is considered the best. First fortnight of June is the best date for okra seed production in North India. In South India, March to May is reported the best planting time.

Isolation: Okra is an often cross-pollinated crop. The extent of out crossing in okra varies from 0 to 60 % depending upon the variety, cropping season, population of pollinators and location . Okra seed field should be isolated from fields of other varieties, fields of the same variety not conforming to varietal purity requirements and wild okra species by 500 m for foundation and 250 m for certified seed.

Field Inspection and Roguing: A minimum of three inspections should be conducted, the first before flowering, the second during peak flowering and fruiting stage and the third at mature fruit stage and prior to harvesting. Characters like growth habit, leaf size and shape, fruit shape, extent of hairiness and pigmentation are some of the useful characters to identify the off type plants.

Harvesting : The ripe pods shed their green colour and become hard when allowed to remain on the plant. If the pods are not harvested at proper time, the longitudinal splitting takes place with the result that shedding of seed starts. Pods should be harvested before bursting and kept for drying. The harvesting of pod should possibly be completed in two to three pickings.

Brinjal and Chillies

Sowing time of these crops vary from region to region. Although in northern India, three or four crops of brinjal are grown but from the seed production point of view, rainy season crop is the best. It is sown in June and transplanted in July. In case of chillies, nursery is sown in end of October and early November and transplanted in the field in February- March in Northern plains. The nursery is protected from frost in December- January.

Isolation : These are often cross-pollinated crops. Therefore, the seed fields should be isolated from the fields of other varieties and fields of the same variety not conforming to varietal purity requirements by 300 and 150 m in case of brinjal, and 500 and 250 m in case of chillies for foundation and certified seed production, respectively.

Field Inspection and Roguing : Removal of the off-type plants should be followed during the various stages of plant growth. A minimum of three field inspections are required for these crops. In the earlier period of growth, roguing should be based on leaf and stem characters. Second roguing should be based on fruiting habit, fruit size, shape and colour. The last inspection is conducted at full fruit maturity stage and prior to harvesting. True-to-type fruits, depending upon shape, size, colour, free from diseases and pests are kept for seed extraction.

Harvesting and Seed Extraction : Brinjal fruits are allowed to mature beyond the edible stage before harvesting for the seed. The stage of harvest of brinjal fruit is very important. Delayed harvest reduce the yield and quality of seed. The seeds are scooped out manually from the cut or crushed fruits. Axial-flow vegetable seed extracting machine is also used for extracting the seeds from brinjal and Chilli fruits. Chilli fruits are picked at red ripe stage. The fruits are either dried and crushed and the seed is separated by winnowing or the seed is extracted manually from the freshly harvested fruits. Seed should be dried in shade.

Cucurbits

Cucurbits form a distinct group of species which are similar in botany, agronomy, ecological requirements and susceptibility to pests and diseases. Cucurbits grow well at day temperatures between 20 - 35^{0}C and cannot tolerate frost and strong winds. Low temperature and short days stimulate development of female flowers. Nearly all are monoecious plants with unisexual flowers which are pollinated by insects. The cucurbits require comparatively dry, warm and long growing season for seed production. The various cucurbits grow very well in plains and seed production of these crops is very successful in plains but variety of cucurbits like cucumber, gourds and pumpkins can be grown for seed production in temperate regions also.

Muskmelon and Water Melon

In North India, the crop is sown in January-February but in South India, it is sown during November. Germination of seed depends upon the soil temperature. At about 25⁰C, the seed germinates in about 4 to 5 days. To get early and uniform crop of melons, transplant method is devised for melon and other cucurbits. The polythene bags of 15 x 10 cm are filled with a mixture of farm yard manure (FYM), soil and silt in equal proportions. The seed is sown in these bags in a protected place. As soon as the risk of frost is over, these seedlings are transplanted in the fields.

Isolation : Cucurbit plants are monoecious and andromonoecious type. In order to produce genetically pure seed of muskmelon, the seed plot should be isolated from the other varieties, fields of the same variety not conforming to varietal purity requirements and from the following and other non-dessert forms.

Longmelon (*Cucumis melo* L. var./ *utilissimus*, Duth Full)

Picking melon " " " conomon, (Makino)

Snapmelon " " " *momordica* (Makino)

Snapmelon " " " *momordica* (Duth. & full)

Weedmelon " " " *agrestis* (Naud)

The isolation distance is kept at 1000 and 500 m for foundation and certified seed fields, respectively.

In case of watermelon, the isolation distance is same, but it should be isolated from the fields of other varieties and wild watermelon.

Field inspection and roguing : Systematic and timely field inspections at different stages of plant growth are essential to ensure production of genetically pure seed. A minimum of three fields inspections, viz., before flowering, during flowering, fruiting and finally at fruit maturity are required to identify and remove the off-type plants. All the plants which do not conform to the varietal purity standards should be rogued out. In most of the cucurbits, identification of off-type plants can be established only after fruit setting. Removal of the off-types even at this stage is helpful in preventing further genetic contamination. Edible fruits are examined for internal fruit characters and sweetness.

Harvesting and seed extraction : Muskmelon fruit is ready to harvest with change in rind colour from green to yellow /cream. In most of the cultivars, a crack develops at the point of attachment of the fruit to the stem. The fruit gets easily detached from the vine. This is called full slip stage. But change in its colour is the only index to fruit maturity.

Unlike muskmelon, watermelon fruit does not change rind colour or develop a slip stage. A dull sound while thumping the fruit with a finger indicates fruit maturity, whereas sharp sound denotes an unripe watermelon fruit. Dryness of the tendrils at the point of attachment to the vine is another indication of watermelon fruit maturity. The third sign of maturity of watermelon fruit is when it changes its belly colour from green to yellow touching the ground. Seed extraction of melons on small scale is done by hand, but for large scale extraction, flesh is allowed to ferment. Seed extracted after 24 to 48 h for fermentation is superior in germination as compared to washing the seed with water immediately after scooping out the seed ball from the fruits. The germination of the immature seeds improves during fermentation process.

Watermelon seeds is also extracted mechanically by using the axial flow vegetable seed extracting machine as described earlier. After extraction, the seeds are dried immediately. The seeds lose its colour and viability if not dried properly. For safe storage, the seed should be dried to a moisture level of 6 to 7%.

Squashes, Gourds and Pumpkin

The seed crops are grown in the same way as the table crops. The only difference is that instead of picking fruits at vegetable maturity, the fruits are allowed to mature fully on the plant itself. This group also requires long dry and warm season for seed production. General agronomic practices are almost similar to that of muskmelon and watermelon. For getting early crop, the seeds are sown in polythene bags and transplanted later in the fields.

Isolation: The isolation distance required for all these crops is 1000 m for foundation and 500 m for certified seed plots. The seed field should be separated from the fields of other varieties and from the same variety not conforming to varietal purity requirements for certification. In addition to that, bitter gourd should also be isolated from Balsam apple, *Momordica balsamina*, L.,Bhat Karela, *Momordica cochinchinensis*, Spreng, Jangli Karela, *Momordica dioica* Roxb. ex. Willd. Pumpkin should also be isolated from winter squash, *Cucurbita maxima,* Duch, Summer squash, *Cucurbita mixta,* pang and vice-versa, and ridge ground from the varieties of sponge gourd.

Field inspection and roguing: Field inspections are conducted the same way as described in case of melons.

Harvesting and seed extraction: The fruits of Indian squash (Tinda), bitter gourd, and pumpkin are ready for seed harvest beyond the edible stage of fruit maturity. At ripening, the tinda fruit sheds its green colour and becomes hard. The colour of entire bitter gourd fruit is changed to red from green. At seed maturity, the skin of bottle gourd becomes tough and dull in colour. On ripening, summer squash turns into deep golden colour.

The large sized fruits give higher seed yield than the smaller ones. Similarly, the stage of ripeness in bitter gourd is also important. The seeds extracted from the fully ripe fruits are of the best quality. Seed of these cucurbits is generally extracted by hand. The seeds of summer squash and squash melon can be extracted with axial-flow vegetable seed extracting machine.

Pea

It is a cool season crop and high temperature adversely affects the pod setting thus reducing the seed yield. The best time for planting the pea seed crop is from mid-October to mid-November for getting optimum seed yield.

Isolation: Being a self-pollinated crop, it requires less isolation. For the production of foundation and certified seed, the pea seed field should be isolated from fields of other varieties and field of the same variety not conforming to varietal purity requirements by 10 m and 5 m respectively.

Field Inspection and Roguing : A minimum of three inspections are required, first before flowering, second at flowering and third at the edible pod stage.

Harvesting : In order to avoid loss of seed shattering, the crop should be harvested when first 25% pods and majority of pods are mature. Then vines are cut and piled up for curing for 3-5 days. The seed is separated by threshing and cleaned free of dust and inert matter.

Leaf Vegetables

Spinach (*Beta vulgaris* L.) and fenugreek (*Trigonella corniculata* L.) are the most commonly grown leaf vegetables in India. Their leaves are rich source of vitamins and minerals, especially iron. These can be grown under a wide range of conditions but the winter season is the best for their seed production. These crops are sown in the month of September-October.

Isolation: Spinach is a highly cross-pollinated vagetable, the pollen grain is very small and easily carried away by wind. Therefore, the isolation distance between the other varieties and fields of the same variety not conforming to varietal purity requirements and from the wild Spinach which grows as a weed should be kept as 1600 m for foundation and 1000 m for certified seed production, whereas Kasuri methi is a self-pollinated and requires only 50 and 25 m of isolation distance respectively.

Field Inspection and Roguing : Leaf colour, size and shape are the most important characters in these crops. A minimum of two inspections are required, one before flowering and the second at flowering and fruit stage.

Harvesting : Harvesting is done when the majority of the fruit balls in spinach and pods in case of fenugreek change colour. Fenugreek is more prone to shattering as compared to spinach, so the harvesting should not be delayed. After harvesting, the seed stalks are piled for curing for 3-4 days and after that the seed is extracted by threshing and cleaned.

Biennial Vegetable Crops

Cole Crops

The important crops in this winter group grown in India are cabbage (*Brassica oleracea var. capitata*), knol khol (*Brassica oleracea var. caulorapa*) and cauliflower (*Brassica oleracea var. botrytis*). Both seed-to-seed and head-to-seed methods are practiced for seed production but the latter is essential for production of pure stock or nucleus seed.

Cabbage and Knol Khol

The cabbage seed production is carried out in temperate regions where chilling requirement for seed production is essential. Both head-to-seed and seed-to-seed methods are followed in cabbage seed production. The sowing time for different varieties is adjusted as to complete head formation before the onset of winter, whether the plants are to be stored in cellars or allowed to be over-wintered *in situ*.

In head-to-seed method, true to type plants are selected and uprooted. The non-wrapper leaves are removed and plants are immediately replanted in such a way that the whole stem below the head goes under ground and the head rests at the surface. At certain places like Kalpa or Lahaul Spiti valley of Himachal Pradesh, the plants after removing the non-wrapper leaves are immediately stored in trenches for over wintering. A single layer of heads is placed in trenches that are 0.75 m deep, 0.9 m wide and 3.3m long 6 to 9 cm deep in the soil. These are covered with wooden planks and about 18cm layer of soil is spread over them. On the both ends of trenches, small holes are provided for ventilation.

There are three different methods for seed production *i.e.*

(i) *Late planting* : When seedlings are transplanted in late September, plants over-winter without forming typical heads and bolt to produce seed in the following spring.

(ii) *Stump method* : The fully mature heads after selection for their trueness to type are cut off just below the base. The beheaded portion of the plant is called a stump. The heads are marketed and the stumps either are left *in situ* or replanted during autumn. The following spring, after the dormancy is broken, the buds sprout from the axils of all the leaves and leaf scars. This method has proved to be slightly superior to head intact method.

(iii) *Stump with central core intact method* : Instead of removing the whole heads, they are chopped off on all sides with downward perpendicular cuts in such a way that the central core is not damaged. This is an improvement over the stump method. It is important to rogue out inferior plants and only well developed true-to-type knobs be selected.

Isolation: Cabbage and knol khol being highly cross pollinated crops, require isolation distance of 1600 m for foundation and 1000 m for certified seed production from the fields of other varieties or of the same variety not conforming to varietal purity requirements and from the following varieties of *Brassica oleracea* L.:

Brassica oleracea (L) var. *oleracea*: Wild cabbage
B. oleracea (L) var.*millecacapitata*(Lov). Helm. Thousand heads kale
B. oleracea (L) var.*gemmifera* DC: Brussels sprouts
B. oleracea (L) var.*acephala* DC: Fodder kale (Karam Sag)
B. oleracea (L) var.*viridis* L: Collards, Tree kale
B. oleracea (L) var.*gongylodes* L: Knol Khol
B. oleracea (L) var.*costata* DC: Portugal cabbage
B. oleracea (L) var.*sabauda* L: Savoy cabbage
B. oleracea (L) var.*italica* (Plenck): Broccoli
B. oleracea (L) var.*botrytis* L: Cauliflower

Inspection and roguing : The crop is inspected before the marketable stage. A minimum of three inspections are conducted. Off type plants are rogued out on the basis of head shape, colour of leaves and maturity time of heads. The plants are selected for head shape (round, flat or conical), free from seed borne diseases and infestation of pests and for lower number of non-wrapper leaves etc.

Harvesting and threshing : When the siliques turn yellow and the seed becomes brown, harvesting is done to avoid shattering. The seed crop is harvested in two to three lots, early plants are harvested first and the remaining crop is cut when about 75% of the pods turn yellowish brown. Harvested crop is collected on threshing floor in heaps for curing for 4 to 7 days. After curing, the seed is threshed and separated from straw. The seed is thoroughly dried in the sun and processed.

Cauliflower

Early autumn varieties of cauliflower set seeds freely in the plains but late varieties produce best quality seed in semi-temperate conditions. Late cauliflower, unlike other cole crops, does not require over-wintering to break dormancy rather it is susceptible to low temperature injury after the curd formation.

Both in situ and transplant methods are followed for cauliflower seed production. Among the Asiatic group of cauliflower, the response of early and mid-season varieties to transplant method is satisfactory but late varieties do not respond well to this method, therefore, seed-to-seed method is always preferred. In order to induce uniform seed stalk emergence, central portion about 5 cm in diameter of the curd is scooped out at the time of full maturity. However, commercial seed production of in-situ method gives higher seed yield in cauliflowers adopted to warm weather conditions.

Isolation : Same isolation distance is to be followed in cauliflower as suggested for cabbage.

Inspection and roguing : In order to maintain purity of the variety, a minimum of three inspections are required in cauliflower. The first inspection is conducted before the marketable stage. The important characters to be noted are leaf shape and size, pigmentation, stem length, number of leaves, branching habit, etc. The second inspection is conducted at the marketable stage and the characters considered are curd colour, shape, size and compactness. The third inspection is executed at flowering and pod formation, keeping in view the number of primary and secondary branches, flower colour, siliqua shape and size. In addition, removal of blind, buttons, fuzzy plants, early bolters and those bolting from center are removed for producing quality seed.

Harvesting and threshing : Seed maturity varies from region to region. The crop is harvested when 60-70% pods turn yellow to brown in colour. The harvested crop is stacked or piled and covered with tarpaulin for 2-3 days for curing. Seeds are threshed, separated, cleaned and sun-dried to 6.5% moisture, graded and packed.

Root Crops

The most commonly grown root crops in India are carrot, radish and turnip. These crops are classified into two distinct groups from seed production point of view *i.e.* annuals or tropicals and European or temperates. The former can be grown for seed in the plains but the latter produces seeds only under the temperate climate. Both the types are adapted to cool weather and are highly resistant to frost. Temperate types undergo dormancy which can only be broken when the roots are subjected to a low temperature, i.e. less than 10°C for a period of 4 to 6 weeks at any time either in storage or under field conditions. It is for this reason that the seeds of latter can only be produced in places like Kashmir, Kullu, Katrain, Darjeeling and Nilgiri hills where winters are severe.

Both seed-to-seed and root-to-seed methods are employed for seed production. The latter method is more acceptable producing seed of high quality. True-to-type roots are selected, leaf tops are cut without damaging the crown leaves and root or only the tip cut roots give higher yield of seed than the smaller stecklings. Large sized roots, *i.e.* more than 7.5 cm diameter give higher seed yield.

Isolation : Carrot seed field should be isolated from fields of other varieties and the fields of the same varieties not conforming to varietal purity requirement. The isolation distance should be 1000and 800 m for foundation and certified seed plots respectively.

Field inspection : A minimum of two inspections should be conducted at the mother root production stage as follows:

(i) The first inspection should be made after 20-30 days of seed sowing and the plants are identified and removed on the basis of foliage characters.

(ii) The second inspection should be made after the mother roots have been lifted to ascertain the varietal purity on the basis of internal and external characteristics of the roots. Any plant not conforming to the varietal purity characters should be rejected.

At seed production stage, a minimum of four inspections should be conducted as follows:

(a) The first inspection should be executed before flowering in order to determine isolation, volunteer plants, outcrosses and other relevant factors. Off- types and extra early bolters should be removed.

(b) The second and the third inspections should be done during flowering.

(c) Extra early or late bolters, the off-types should be removed.

(d) The fourth inspection should be made at maturity to verify the uniformity in crop maturity.

Isolation distance for foundation and certified seed plots should be 1600 and 1000m, respectively, for radish and turnip. However, the turnip should also be isolated from Chinese cabbage, rutabaga, mustard, brown and yellow sarson and sugar beet.

Off-type plants on the basis of foliage characters are removed in the field before the roots are pulled out from the nursery. True-to-type roots are identified after lifting the roots from the field. Roots not conforming to the varietal purity characters for colour, shape, size, flesh colour, etc. are rejected. In case of carrot, central core colour and its size are the important characters representing a variety. Small, mis-shaped, forked, diseased, and other undesirable roots are discarded. A thorough and rigorous selection of true-to-type roots is vital for the production of good quality seed.

Harvesting and Threshing : The turnip pods shatter very badly when fully ripe, therefore, the whole crop is cut when 60-70% of the pods turns yellow brown in colour. The crop is piled in wind-rows to allow it for 5-7 days to dry and cure properly.

Unlike turnip, shattering is no problem in radish. So harvesting is done when the pods are fully ripe and the harvested crop is left in the field for drying and curing.

In carrots, the primary umbels mature first therefore the crop is harvested in two or three pickings. Though shattering is not a problem, but if there are repeated showers after the umbels are ripe, the viability of the seed is affected. The seed after curing are threshed cleaned and graded.

Onion

Onions require cool weather during early development of bulb crop and again during early growth of seed stalk. At maturity, relatively high temperature and dry atmosphere are favourable. The onion is a biennial crop which forms the bulb in the first year and the seed crop in the second year.

The onion seed can be produced by bulb-to-seed method or seed to seed method. The most commonly used method in India is bulb-to-seed method. The seed is produced from the bulbs that have been raised during the preceding year.

For nucleus and foundation seed production, bulb-to-seed method is followed in which storing and transplanting of bulbs provide the opportunity for selection of true to type bulbs. There are some varieties which do not store well or sometimes the seed is required to be produced within the same year, in such cases the seed-to-seed method is practiced. Seed yield in onion is greatly influenced by bulb size, spacing and time of planting. Higher seed yield has been obtained by planting larger size of bulbs of greater than 6 cm diameter.

In case of seed-to-seed method, sowing of nursery in August and then transplanting in the first fortnight of October is suggested. The bulbs for seed production are selected at the time of harvest of bulb crop. The bulb of medium size, true-to-type and free from diseases and injuries are selected and stored under proper storage for seed production. At the planting time, the bulbs are taken out of storage and selection is again done to remove the rotten or undesirable bulbs and are planted in the field during mid-October.

Isolation: The seed plot should be isolated from other varieties of onion by 1000 m for foundation and 500 m for certified seed production.

Inspections

(i) The first inspection is done after transplanting of seedlings to determine isolation, volunteer plants, off types including bolters and other relevant factors.

(ii) During second inspection, the bulbs are lifted to verify the true characteristics of bulbs and again the bulbs are inspected for varietal purity at the time of planting in the main field.

Seed Production stage: A minimum of four inspections are executed as follows:

(i) The first inspection is made before flowering to determine isolation, volunteer plants, off-types including bolters and other relevant factors.

(ii) The second and third inspections are executed during flowering to check isolation, off-types and other relevant factors.

(iii) The fourth inspection is done at maturity to verify the true nature of plant and other relevant factors.

Harvesting : The onion crop is ready for harvesting in May. All the umbels do not mature at the same time and therefore, it is desirable to harvest the crop in two or three pickings. The harvested umbels are ready for picking when about 20% black seed becomes visible. The harvested umbels are kept for curing for 2-3 days. Then the seed is threshed and cleaned. Onion seeds are poor storer and lose viability in less than a year in tropical and sub-tropical climate. The seed needs careful drying and proper storage.

Seed Production of Vegetatively Propagated Vegetables

Potato, ginger, turmeric, garlic, sweet potato, colocasia, etc are some of the vegetable crops which are propagated vegetatively by tubers, rhizomes, cloves, root tubers or the corms. These being short duration crops, fit better in the cropping system, respond well to inputs and produce high yield per unit area and time, have high employment generation potential and are able to meet the increasing food requirements of the increasing population of the country. Potato is cultivated in 1.12 million ha with a production of 18.5 million t and accounts for 0.67% of total cropped area in India. Himachal Pradesh was the first state to organize potato seed production programme on massive scale. Similarly ginger and turmeric are very important spices grown in the country and are cultivated in topical and subtropical parts of the world for their rhizomes used as spices, medicine, dye, snuff and cosmetics, etc. Sweet potato is known as poor man's crop, is the cheapest source of carbohydrates, besides high yield potential and wider adaptability under varying agro climatic conditions. It is an important crop of tropical and sub-tropical region. Last but not the least, garlic is again one of the important bulb crops grown and used as a spice or a condiment throughout India. It has higher nutritive value than other bulb crops.

Potato

Potato is propagated vegetatively by planting its tubers. In the plots, selected for potato seed tuber production, crops like tomato, brinjal, chilli, potato etc. should not be grown in the previous year. Land to be used for seed production of potato should be free of volunteer plants. The soil should be free from soil-borne diseases like common scab, nematodes and bacterial wilt. The area should be free from aphids. Seed tubers should be free from seed-borne diseases and pest attack. The seed materials should be clean, healthy, firm and conforming the variety characteristics. The tubers not conforming trueness should not exceed 0.05% and 0.1% (by number) for foundation and certified seed classes, respectively. Bruised, cut, unshaped, cracked tubers or damaged by insects, worms should not exceed more than 1% (by weight).

Isolation : Potato being vegetatively multiplied, there is no problem of crossing. Therefore an isolation distance of 5 m is kept between different varieties and same variety not conforming to the requirements for seed production to avoid mechanical admixtures.

Roguing and inspections: A minimum of four inspections are required at pre-flowering, fruit ripening and seed maturity stages. First inspection should be done soon after emergence *i.e.* 20-25 days after planting to check isolation, off-types and extent of disease infestations. The second inspection is conducted after 45-50 days of planting or appropriate growth stage depending upon crop duration of concerned variety to check isolation, off-types and disease infection with particular reference to mild and severe mosaics, leaf roll, yellow, brown rot and other relevant factors. Third inspection should be executed immediately after haulm cutting and fourth inspection is conducted 10 days after cutting the haulms and before harvesting to assure that there is no regrowth of haulms. In potato seed tuber production programme, aphids are the main problems. So the critical level of aphid has been fixed (20 aphids/ 100 compound leaves) in seed production areas.

Harvesting : The crop should be harvested when the tuber's skin has matured. After harvesting, tubers are to be stored in a cool place.

Sweet Potato

It is propagated vegetatively by planting root tubers. Land to be used for seed production of sweet potato should be free of volunteer plants. The clean, healthy, firm root tubers which conform to the characteristics of the variety are selected as the seed material.

Isolation : An isolation distance of 5 m is kept for foundation as well as certified seed, between different varieties and same variety not conforming to varietal purity.

Rouging and inspections : Generally two to three inspections are carried out at sprouting, vegetative phase, pre-harvesting and harvesting stages. All the off-type plants are removed keeping in view the plant characteristics *viz.*, leaf colour and shape. Diseased and insect-pest attacked plants are also removed. The bruised, ill-shaped, cracked roots or damaged by insects (except sweet potato weevil) or worms should not exceed more than 1% (by weight). The maximum tolerance limit of roots showing visible symptoms caused by diseases and sweet potato weevil is negligible for both classes of seeds.

Harvesting: The tubers are harvested when the leaves turn pale and later yellow. The immediate drying of cut tuber after its exposure to the air is an indication of maturity while the immature ones will remain moist and turn dark in colour. After harvesting, the tubers are cleaned, graded and properly cured and stored.

Ginger

Ginger is propagated through rhizomes which can be carried easily in living state from one place to the other. Land to be used for seed production of ginger should be free of volunteer plants and should be loose, friable and offer minimum resistance to rhizome development. The rhizomes used for seed should be clean, healthy, bold, firm and should conform to the characteristics of the variety. The seed rhizomes not conforming to the varietal characteristics should not exceed 0.5% and 1% (by number-maximum) for foundation and certified seed classes, respectively. Moreover, the cut, bruised, infested with *Pythium sp., Pseudomonas solanacearum* and *Meloidogyne incognita* or injured rhizomes should be avoided.

Isolation : There is no problem of crossing in ginger as it is vegetatively multiplied. Therefore, an isolation distance of 3 m is kept for foundation as well as certified seed, between different varieties and same variety not conforming to varietal purity, to avoid chances of mechanical admixture.

Rouging and inspections : A minimum of four inspections are required for roguing before the crop maturity. The first inspection should be made at the time of planting to verify the isolation , rhizome rot, seed piece weight and spacing. The second roguing should be conducted about 45 -125 days after planting to check germination, sprouting, rhizome rot and shoot borer incidence. The third roguing and inspection should be conducted after 180-190 days after planting to check off types, rhizome rot, insect infestation and *Phyliosticta* leaf spot disease, etc. Fourth inspection should be made before harvest or between 240-250 days after planting to verify rhizome rot, scale insect and mealy bug infestation. The rhizomes rot infestation can be visually seen by breaking the rhizomes. If there is yellowish ring formation below the corky cells (skin), such rhizomes are discarded.

Harvesting : Yellowish and drooping down foliage indicate maturity of the crop. The rhizomes are uprooted carefully by spade without any injury. On large scale, the ginger fields are ploughed with furrow turning plough and then healthy rhizomes are collected. The mother and daughter rhizomes are kept separate as the former are sold and the latter are kept for seed after thorough screening. The harvested rhizomes are kept under shade in thin layers for 5-6 days and are cleaned.

Turmeric

It is also propagated by rhizomes. It requires a warm and humid climate and is very sensitive to water stagnation and alkalinity. Land to be used for seed production of turmeric should be free of volunteer plants and *Pythium sp.,Ralstonea solanacearum* and *Meloidogyne incognita*. The soil should be loose, arable and offer minimum resistance to rhizome development.

Isolation : As in ginger there is no problem of crossing in turmeric as it is vegetatively multiplied. Therefore an isolation distance of 3 m is kept for foundation as well as certified seed, between different varieties and same variety not conforming to varietal purity, to avoid chances of mechanical admixture.

Rouging and inspections : Inspections and roguing schedule are similar to that in ginger. However, minimum four inspections are carried out at planting, sprouting, vegetative phase, pre-harvesting and harvesting stages.

Harvesting : Complete drying of leaves and pseudo stem indicate the maturity of the crop. Leaves are cut close to the ground and fields are irrigated two days before harvest. Underground rhizomes are dug out with pick axe by manual labour. For preservation of seed rhizomes, the material is stored by heaping them under the shade of trees. Heaps are covered with turmeric leaf and plastered with cow dung mixture. It can be left undisturbed for two or three months until sowing.

Colocasia (Arvi)

Colocasia is grown in sub-tropical and tropical parts of the country and all its parts i.e. corms, petioles and enlarged leaves are used in various ways as food. It is propagated through corms.

It is a warm season crop and is grown as a summer as well as rainy season crop. Land to be used for seed production of colocasia should be free of volunteer plants. The seed material (corms) should be clean, healthy, firm and conform to the characteristics of the variety. The corms not conforming to varietal characteristics should not exceed 0.10% and 0.05% (by number) for foundation and certified seed classes, respectively. The bruised, cracked, cut, corms or damaged by insects or worms should not exceed more than 1.0% (by weight). Specification in respect of size of seed corm is 4-6 cm x 2.5-3.5 cm and corresponding weight 20-40 g for foundation and certified seed. In a lot, corms not conforming to specific size of seed should not exceed more than 5% (by number).

Isolation : There is no problem of crossing in colocasia as it is vegetatively multiplied. Therefore an isolation distance of 5 m is kept for foundation as well as certified seed, between different varieties and same variety not conforming to varietal purity, to avoid chances of mechanical admixture.

Rouging and inspection : The field inspections should be done at sprouting vegetative stage, pre-harvesting and harvesting stage. Therefore a minimum of three inspections are required, first and second about 60 and 90 days respectively after planting and third about 160 days after planting and prior to harvesting or at appropriate stage depending upon the the crop duration of the variety.

Harvesting : The corms are taken out by digging carefully without any injury. These are kept under shade for proper curing for 10-15 days and then stored in trenches/pits for future sowings. In some places, the crop is left as such in the field and covered with grass or plant leaves till the sowing of future crop. For sowing, these corms are uprooted and only selected ones are used for planting.

Garlic

Garlic is propagated vegetatively by planting cloves (sections of the bulb). Each clove develops into a bulb. For the best production, cloves are planted directly into the soil. The well developed, uniform bulbs having appropriate shape, size and compactness are selected. Individual cloves with the pointed end up are planted, 3 to 6 cm deep in well-worked beds. Longer the day length given to the variety, higher is the yield. Bulbing takes place during longer days and at high temperature.

Isolation : Land to be used for seed production of garlic should be free of volunteer plants. Being vegetatively multiplied, there is no problem of crossing in garlic. Therefore an isolation distance of 5 m is kept for foundation as well as certified seed, between different varieties and same variety not conforming to varietal purity, to avoid chances of mechanical admixture.

Rouging and inspections : The rouging and inspections are conducted at vegetative phase, bulb formation and bulb maturity stages. All certified seeds are produced from the cloves obtained from the bulbs whose source and identity are assured and approved by the Certification Agency. A minimum of two inspections are made first when plants are large enough to verify isolation, off-types including bolters and second when leaves begin to fall and before lifting of bulbs to verify off types. The average diameter of each bulb should not be less than 2.5 cm of 25 g in weight. The seed material should be clean, healthy and firm and conform to the varietal characteristics which do not exceed 0.1% and 0.2% (by number) for foundation and certified seed classes, respectively. Also cut, bruised, cracked, immature bulbs or those damaged by insects, slugs, worms should not exceed more than 2.0% (by weight).

Harvesting : The bulbs are harvested when the plants start to die back and 1/2 to 2/3 of the leaf material is yellow or brown, turning from a succulent, fleshy texture to a dry, papery sheath. The bulbs can be dug with a digging fork or a spade to prevent breaking off the stems.

Bulbs are dried for two to four weeks on racks or by tying the bulbs into small bunches of ten to fifteen bulbs and hanging in shady place with good air circulation. Then the bulbs are cleaned by peeling off a minimal amount of outer layers to expose a clean bulb. For storage, additional curing is done at 80 to 100°C for four to six weeks, or until the papery skin changes from pure white to off-white and becomes crisp.

Seed Production Areas

The main areas of vegetable seed production are based on climatic factors which ensure optimum temperature, day length, rainfall, humidity and enable the seed to mature till the harvesting. Low relative humidity and moderate temperature minimize the dispersal of many pathogens. Because of diversity of climatic conditions in different parts of India, the seed production can be organized successfully in several parts of the country. The regions which are the best known for the production of particular vegetable seed are given elsewhere.

Along with the climate, the success of vegetable seed production depends upon the technology and economic factors. During the recent years, some pockets in Maharashtra, Karnataka, Andhra Pradesh and Rajasthan have emerged as the major seed producing areas for most of the vegetables and hybrids in particular.

Vegetable seed production regions in India

Regions	Name of the crops
Deccan: Plateau of North Andhra Pradesh (Telengana), North Karnataka, Eastern Maharashtra (Marathwada and Vidharba)	Okra, tomato, brinjal chilli, onion, cucumber, gourds, hybrid tomato
Coastal districts of Andhra Pradesh	Watermelon, bottle gourd and bitter gourd
Tamil Nadu (small areas only)	Gourds, cucumber and cowpea
Eastern Rajasthan (Kota), Madhaya Pradesh (Nimar Valley and Indore)	Pea, muskmelon, onion, methi and chilli
Gujarat and Rajasthan (Sri Ganganagar)	Pea, cauliflower, onion and okra
Eastern districts of Uttar Pradesh (Etawah, Kanpur, Aligarh, Mainpuri and Bareilly)	Mid season cauliflower, bitter gourd, cucumber
North Bihar (Hazipur) and Himachal Pradesh (Saproon valley, Mandi) and Jammu & Kashmir	Early cauliflower
Himachal Pradesh (Lower hills) and Kashmir Valley	Radish (temperate), cabbage, carrot, beet, knol khol and Capsicum
Punjab and Haryana (small areas)	Mid season cauliflower, Asiatic radish, turnip, carrot and okra
West Bengal	Pumpkin, tomato and leafy vegetables
North Bengal (Kalimpong)	Pumpkin, tomato, leafy vegetables, cauliflower, capsicum, French bean
Arunachal Pradesh (Bomdila district)	Radish (temperate) and French beans, cauliflower

Harvesting

The best time of harvesting vegetable seed crops is at a stage when the highest yield of best quality seed will be obtained. Seeds, apart from using as a propagating media, are largely used for genetic manipulation and improvement of vegetable crops. Therefore curing and seed extraction is an important event to be taken due care of. Longevity of seed depends on a number of factors *viz* kind of seed, proper maturity (both stage and standards), curing to a low moisture content and the storage conditions. The anthesis period in many vegetables is comparatively long with successive flowers on complex inflorescences resulting in a long period of fruit ripening and seed maturation. In many species e.g. lettuce, *Brassica spp.* and okra, there is a strong tendency for the earlier maturity, seeds to drop before the later ones have developed. This loss of seeds before harvesting is often known as shedding or shattering of seeds. However, timely harvesting of vegetable seed crops can minimize shattering problem as the maturity indices can act as a better guide to the vegetable seed producer.

Maturity indices : It is an indication of the proper stage of seed maturation leading to right time of harvesting to obtain higher yields of good quality seed. The effects of environment before harvesting on seed viability are also very important. The maturity standards vary from crop to crop. For example in garden pea, seed moisture content should be between 30-40% at threshing. In carrots, there is a relationship between embryo size with umbel position on the plant and seed maturity. The ripening process is interrupted if the seeds are harvested too early and seed quality is adversely affected. Once the optimum % of seeds reaches maturity, delayed harvesting beyond maturity causes seed losses. The incidence of shedding from ripe material is increased during dry weather. Seed crops which are particularly prone to shedding while being cut should be handled at times of comparatively high relative humidity. In arid areas this can include early in the day when effects of overnight dew are still effective. Although seeds which are produced in wet or fleshy fruits such as cucurbits peppers and tomatoes are not normally subjected to all the effects of environment as are the dry seeded vegetable species, it is generally accepted that the seeds should be allowed to develop fully in the fruit before extraction. Ideally the fruits should remain attached to the mother plant

The number of harvestings will depend upon the vegetable varieties and the agro-ecological conditions of the seed producing area. The appropriate stages of harvest maturity indices in vegetable crops are presented elsewhere.

Post Harvest Handling

The crop should be harvested at proper moisture content so as to minimise damage to the seed. In hybrid seed production, where two parents are involved, the male parent rows are harvested first and moved to a distant place. The whole field is then inspected and the broken or lodged male parent plants are removed from the female parent rows. The hybrid seed on the female parent rows is then harvested. All the efforts in roguing and care during earlier stages will go waste if proper care is not taken during harvesting, threshing, processing, seed treatment, packaging etc. for avoiding mechanical mixing. Correct labeling is very important. The threshers, threshing floors, processing machinery, etc. should be thoroughly cleaned in between handling of different varieties.

Maturity indices, number and of harvesting time for vegetable crops

Seed crop	Maturity indices	Harvestings		Remarks
		No.	Time	
A.				
Tomato	Skin colour change to red and softness of fruits	2-3	Day time	
Okra	Pods become gray or brown, hard according to cultivar	1-2	do	Sequential ripening of pods and a tendency to split on ripening
Onion	Seeds become black on ripening in silvery coloured capsules. Ten percent heads expose black seeds.	1-2	Morning	Prone to shattering
Carrot	Secondary and 3rd order head turn brown	1-2	Morning	Shattering on delayed harvesting
Spinach	Later ripening plants start to become yellow	1	Morning	-do-
Cole group	On ripening, plants start dry out and become orange brown in colour. Oldest pod will become brown first	2-3	-do-	Prone to shattering
Chinese Cabbage	On ripening plants become dry and turn brown/light yellow	1-2	-do-	Prone to shattering
Radish	Pods turn brown, parchment like when the seeds reach maturity	1	Daytime	Do not shatter easily
Turnip	Plants turn to brown parchment colour	1-2	Morning	Prone to shattering
Garden pea	Majority of pods turn parchment like	1	Day-time	Do not shatter easily
Methi	Pods turn brown and leaves get dry	1	-do-	Delay in harvesting causes shattering

Contd...

Beans	Earliest pods turn dry and parchment like and rest turn yellow	1	-do-	Over maturity leads to shattering and cotyledon cracking
B. Fleshy fruits which are dried before seed extraction				
Chillies	Turning of fruit colour green to red, yellow or brown	2-3	Day-time	Dry methods of seed extraction
Bottle Gourd	Rind becomes hard and colour changes to light brown or yellow	1	-do-	
C. Wet fleshy fruits				
Capsicum	Green coloured fruit changes to red or yellow depending on variety	1-2	-do-	Wet method of seed extraction
Tinda	Rind become very hard and fibrous	1	-do-	
Cucumber	Fruit develops external ripening colour, stalk adjacent to the fruit withers. Fruits are cut longitudinally and mature seeds separate easily from the interior flesh	1	-do-	Seed extraction by acid/alkali and fermentation methods
Watermelon	Tendrils wither on shoot bearing fruit, skin colour underside the fruit surface resting on the soil turns pale yellow, dull sound on thumping fruit	1-2	Day time	
Muskmelon	Fruits tend to separate (full slip) from stem. In winter melons, seed maturity is indicated by rind colour change from green to yellow or yellow to white, blossom end of fruit softens, skin becomes waxy and its aroma increases. Easy separation by abscission layer.	1-2	Day time	
Squashes, pumpkins and Marrows	Rind becomes hard and its colour changes from green to yellow golden to straw colour	1-2	-do-	

Contd...

Brinjal	Turning normal fruits colour to yellow/straw.	2-3	-do-	Seed is extracted by fermentation, acid, alkali
Bitter gourd	Fruit and seed becomes red	1	-do-	Hard seeds separated and washed
Summer Squash	Fruits become hard, its colour turns deep yellow or red	1	-do-	
D. Vegetatively propagated				
Seed potato	Haulms get dry, droop down turn dark brown in colour	1	-do-	Delay leads to spoilage of seed tubers
Ginger	Drying and falling down of pseudo-stem turning brown	1	-do-	
Turmeric	Drying and falling down of stem turning brown	1	-do-	
Garlic	The stem get dry and change in colour from green to brown	1	-do-	
Colocasia	Drying and dying of petiole and leaves	1	-do-	Skin becomes tough

Threshing and Seed Extraction

Seed has to be extracted from dry fruits/pods/fruits in which the seeds are wet at the time of extraction. Threshing of seed can be done by hand or by machines. Care should be taken while transportation of material from the field to threshing floor. Thresher or seed extraction machines should be properly cleaned to avoid admixture. Vegetables like tomatoes, cucumber and muskmelon bear ripe seed in the fleshy fruits and seeds surrounded by a gelatinous layer. The seed is separated from this gelatinous material by any of the following methods.

Fermentation

The crushed seed and pulp mass are taken in a container and left to ferment for 1 to 3 days depending upon the ripeness and temperature. The mixture should be stirred daily to avoid discolouration of the seed and to keep uniform rate of fermentation. After the completion of fermentation process, the seed is washed by displacement method. The good seed settles down while the pulp and other debris float on the surface. Dry the seed in shade so that the germination may not be affected.

Acid Treatment

The seed may also be separated from the gelatinous coating by acid treatment. About 10 litres of commercial grade hydrochloric acid are stirred into one tonne pulp after for one hour and then seed is washed out with water. It is important not to allow seed to remain for longer duration in acid as it will reduce seed germination. Also avoid iron containers for fermentation as it may lead to seed discoloration.

Alkali Treatment

Equal volume of alkali (420 g ordinary washing soda is added to 5 litres of boiling water) is added to pulp and mixed thoroughly. The alkali mixture is cooled and allowed to stand over night in an earthen pot. Next day the seed will settle down which can be washed thoroughly.

Mechanical Seed Extraction: The seeds of vegetable fruits like tomato, brinjal, chillies, cucumber, summer squash, squash, melon and watermelon can be extracted mechanically with seed extraction machine. It is quite efficient, economical and hygienic. Seed loss and damage to the seed coat are negligible. Manual extraction of seeds is time consuming, labour intensive and unhygienic.

Seed should be dried as rapidly as possible by spreading in thin layers on the drying floor. When the temperature exceeds 40°C, the seed should be dried under shade. Under artificial conditions, the seed should first be dried at low temperature and then the temperature could gradually be increased to 40°C.

Seed Processing

Vegetable seed after harvest taken for processing have some percentage of undesirable material which has to be removed by seed processing operations. Seed processing helps to improve physical purity of seed lot. In vegetable seed processing, air screen machines are generally used. Air screen machines have at least two vibrating screens, the upper screen removes impurities larger than the seed while the lower screen separates out any seed or other impurities smaller than the optimum seed size of crop. Seed processing is mechanical upgrading the physical quality and improving the planting condition of the seed by applying chemical protectants. During this, the seed mass passes through a number of operations such as cleaning, grading, separation and drying. All unit operations are not necessary for each seed lot. The processor should decide the requirement of unit operation for the particular seed lot, after keen examination.

Cleaning and Grading : The seed processing machines are:

- Air Screen Cleaner- Grader (ASG)
- Length Separator (LS)
- Specific Gravity Separator (SGS)
- Seed Treater

Air Screen Cleaner- Grader : This is a basic and essential machine for seed processing. It removes major dockage and under-sized seeds from the seed lot, which improves the flow of material in subsequent operations. The screens of the air screen cleaner separate seeds mainly according to width and thickness. During processing, smooth flow is advantageous as it increases the processing efficiency as well as enhances recovery by reducing product loss in rejection. The screens that are to be used with the screen grader are recommended on the basis of shape and size of seeds.

Length Separator: The indented cylinder is the second most important cleaning machine. The indented cylinder separates according to the length. It separates the impurities especially broken seeds and weed seeds which are either longer or shorter than the crop seed.

Needle Indented Cylinder Separator: This is special type of indented cylinder in which the cylinder has large number of needles. Insect infested seed will be lifted in the tray because the needle fits in the hole made by the insects.

Specific Gravity Separator: After cleaning of seed by air screen cleaner and length separator, it may be necessary to use a gravity separator, if the germination capacity of seed is not yet above the minimum standards of seed certification, or if a very high germination capacity required. In some cases, impurities (stone, earth balls, etc.) or seed component (immature, insect damaged, diseased seed) of equal dimension may also remain in seed mass, which are impossible to remove with normal set of machines.

Spiral Separator: The spiral separator classifies seed according to its shape and rolling ability. It is used to remove damaged seed from brassica, pea, soybean and lentil. The spiral separator requires floor space and is light in weight. Generally, spiral separator is very useful to remove broken seeds from the round seeded crops.

Drying : At the time of harvest, the seed contains frequently higher moisture content than the optimum for better germination and storability of seed. Seed from pulpy fruits like tomato, watermelon, muskmelon, cucumber and brinjal have high moisture content at the time of harvest and absorb more moisture during wet extraction. Other vegetable seeds like onion, fenugreek and pea have relatively low moisture at harvest. Sometimes due to adverse climatic conditions, seed may also have high moisture content. The seed moisture should be reduced to optimum level (6-8%) before storage.

In order to perform the operation efficiently, it is important to optimize drying method, drying rate, drying temperature and airflow rate. These parameters are important in view of quality preservation as well as energy conservation. Natural drying is a very slow process and also depends on the weather condition. However, it is simpler, cheaper in installation and operation. The artificial drying is reasonably costly but having advantage of faster controlled operation.

Among all artificial drying systems, hot air drying is the most common system. At higher temperature, the operation progresses faster. But high temperature can injure the seed quality. In general, the drying temperature should not be higher than 43.5°C. It is advisable to dry seed at following temperature corresponding to its initial moisture level.

Seed Moisture Content (%)	Drying Temperature (°C)
Over 18	32
10-18	37
Under 10	43

The drying rate is a very important factor for safe drying of seed. Faster drying tends to combat with moulds and drop in vigour, but too fast drying has its danger to drying injuries, hardening or cracking of seed coat. The rate should be less if initial moisture is low and if temperature is below 43.5°C. After completion of drying, the seed lot should be cooled with dry and cool air to get uniform distribution of moisture in the seed. It shows actual final moisture content of seed. Lettuce and cucurbits are considered to be relatively quick driers; carrot, beet and tomato are medium driers while legumes, brassicas, onion are slow driers. Most of the vegetables are dried to 6 to 7 % moisture and are listed below :

Optimum moisture content for different vegetable crops

Crops	Moisture content (%)	Crops	Moisture content (%)
Bean	7.0	Muskmelon	6.0
Beet	7.5	Onion	6.5
Brinjal	6.0	Okra	5.0
Broccoli	5.0	Pea	7.5

Contd...

Cabbage	5.0	Pepper	5.0
Carrot	7.0	Pumpkin	6.0
Cauliflower	5.0	Radish	5.0
Cowpea	7.0	Tomato	6.5
Cucumber	6.0	Turnip	5.0
Others	6.0	Watermelon	6.0

Essential Guidelines in Seed Processing

- Only one crop/variety/lot should be handled in the processing area at one time.
- The machine and its parts should be cleaned thoroughly at the time of changing the variety/lot to avoid mechanical mixing.
- The total processing area should be cleaned with special reference to seed as well as impurity.
- The screens of air screen machines should be examined for clogging and other crop seed before fitting them to the machine.
- The size of grading screen should be selected as per test result of sieve analysis for a particular lot.
- The air velocity of air screen machine should be selected in accordance with the terminal velocity of good seed.

Seed Enhancement

Following seed processing, seed quality can further be enhanced using different techniques. Seed enhancement is a commercially viable technology for seed industry that improves the performance of seed through seed treatments and coatings. These are "value added" techniques which comprise of post harvest treatments which improve germination or seedling growth or facilitate the delivery of seeds and other materials promoting crop establishment. The various seed enhancement techniques are :

Seed treatment: It refers to the application of fungicide, insecticide or a combination of both, to seed so as to disinfect and disinfest from seed borne or soil borne pathogenic organisms and storage insects. It also acts as seed protectants by protecting the seed as well as young seedling from organisms in the soil, which might otherwise cause decay of the seed before germination.

Seed dressing: It is the application of finely ground solids (usually a fungicide or insecticide) dusted onto the surface of seeds in small quantities to protect seeds from diseases and/or insect-pests.

Seed coating: When finely ground solids or liquids, containing dissolved or suspended solids, form a more or less continuous layer covering the natural seed coat. It includes pelletting and many other seed treatments.

Seed Pelletting: The application of solid materials to seeds in sufficient quantity to make the pelletted seed subsequently larger and/or heavier and approach a spherical or elliptical shape.

Film coating : In contrast to pelletting, film coating is the application of a very thin polymer layer that seals precise amounts of chemical protectants to the seed.

Seed soaking : A process by which seeds can be led to absorb nutrients, protectants, growth regulators etc. by immersing them in appropriate solutions for extended periods.

Some practical applications of seed coatings are briefly highlighted below:

1. *Enhanced seed quality*: Stimulation of germination and improved seedling growth by influencing the micro environment of seed with the application of micro-nutrients/fertilizers/chemical protectant coatings.
2. *Crop establishment*: Better crop stand and uniform establishment in the field.
3. *Sizing and shaping of seeds*: Uniform coverage between and on individual seeds results in uniform size and weight, thereby improves plantability.
4. *Aesthetics/identification*: Cosmetic coloration helps in seed identification/ product branding and improved aesthetics.
5. *Precision planting*: This is made feasible by increasing the seed size and weight of small seeds. This is particularly useful for high value seeds of vegetables like tomato, brinjal, capsicum, etc and flowers. The tentative increase in seed weight/size ranges from 0.2 % to 400 fold depending upon the type of seed coating.
6. *Economics*: It requires low doses of chemical protectants as against blanket applications of these chemicals for broad area applications, resulting in savings of chemical protectants (fungicides, insecticides) and fertilizers, as against broad-acre applications.
7. *Eco-friendly*: Because low and effective dosage of chemical protectants/ fertilizers is used thus reducing the risk of run off of the chemicals to the soil/environment.

8. *Reduced seed rate*: This is primarily because of precision planting. Seed coating significantly reduces the seed rate per unit area.
9. *Flowability*: Non clumping of seeds particularly chaffy seeds.
10. *Controlled delay in germination*: Hydrophobic polymer seed coatings are used to obtain synchronous flowering of male and female inbred parents in F1 hybrid seed production.

Seed Priming: It is another approach to improve seed quality of the poorer seeds in the seed lot by physiological treatments like seed priming. Heydecker (1973) defined seed priming as a pre-sowing treatment in which seeds are soaked in an osmotic solution that allows them to imbibe water and go through the first stage of germination, but does not permit radicle protrusion through the seed coat. The seeds then can be dried to their original moisture contents and stored or planted by conventional techniques. Priming has the ability to improve the mean performance of a seed lot and also reduces variation within a seed lot for a wide range of species.

Different stages of seed during priming: There are three main stages during germination (tri-phasic uptake of water).

- *Stage 1:* This stage is recognized with a rapid initial uptake of water that is usually completed in 6- 24h depending on the species. All seeds, even dead ones, take up water rapidly during this stage.
- *Stage 2*: This is the plateau phase of water uptake during which there is initiation of nucleic acid and protein synthesis in preparation for the emergence of the radicle. This stage may last two or three times as long as stage one.
- *Stage 3*: This is a stage characterized by the rapid uptake of water, cell expansion and the protrusion of radicle through the seed coat The stage 3 is relatively short and the differences in the times of germination between the seeds of a population are associated with differences in the duration of *stage 2*.

The 'good' seeds always germinate earlier than the 'poor' seeds. The seed priming enables 'good' seed to be held back, allowing 'poor' seeds in the lot to catch up the development. The hydration of the seeds can be regulated using osmotica (osmo-priming), salt (halo-priming) and inorganic or organic carriers (solid matrix priming).

Seed Storage

Seed is stored from harvest to the next planting. Seed which could not be sold in the current sowing season is stored as carry-over stocks and is also stored as an assurance against price fluctuations, production of poor quality seeds in a season or in case of famine and war. Germplasm and breeder seed stock is stored for longer duration. The objective of seed storage is to minimize the loss of viability and vigour of seeds. The entire storage period is divided into following stages:

- Storage on plants
- Storage before processing
- Storage after processing
- Storage in transit
- Retail storage

Storage on plants: After the seed attains the physiological maturity, it has to stay on the plant till harvest. Several factors such as water stress, high or low temperature, diseases and insect pests attack may affect seed quality by reducing viability and vigour.

Storage before processing: After harvest and seed extraction, the seed is taken for drying, cleaning and processing. During this stage, the seed has high moisture content. Seed deterioration during this period is rapid if proper drying of the seed is not done. Mechanical injury during processing could also be the reason for damaging seed quality.

Storage after processing: This is the stage in which the seed remains for a longer period in stores. Seed longevity in storage depends on several factors, the important ones are:

- Relative humidity (RH) of the atmosphere
- Storage temperature
- Seed moisture content
- Physical conditions of the seed
- Kind/variety of seed
- Initial seed quality
- Gases during storage
- Micro flora and insects

Out of these, the first three factors i.e. RH of the atmosphere, temperature and seed moisture content are very important.

In Transit, at retail store or farmer's farm : If the seed loses its viability by improper storage during these storage places, the advantages of quality seed are forfeited. Therefore, seed should be adequately stored at these places too.

Seed Storage Behaviour

Seeds are classified into two distinct groups according to their storage physiology. The first group is described as **orthodox**. In this group, seed can successfully be dried to moisture contents as low as 5% without any injury and are able to tolerate freezing temperatures. These are long-lived seeds. Seeds of onion, chillies, tomato and brinjal, exhibit orthodox type of storage behaviour, wherein seeds could be dried and stored at low temperature without any adverse effect on germinability and longevity. The second group is described as **recalcitrant**. These seeds cannot be dried to moisture content below 30% without injury and are unable to tolerate freezing. As a result, these seeds are difficult to store successfully. It includes many important tropical plantation crops, many tropical fruits, and a number of timber species of both the tropics and the temperate latitudes. The rate of deterioration of recalcitrant seeds is not easy to control because they are damaged by drying and since they can neither be dried nor be cooled to sub-zero temperatures because they would be killed by freezing injury resulting from ice formation. Most recalcitrant seeds can only be stored for a few weeks or months without loss of viability.

Factors Affecting Seed Storage

Moisture Content: The moisture content of seed during storage is the most influential factor affecting the longevity of seed. It is important to harvest mature, relatively dry seeds to reduce the moisture content of the seed after harvest. In addition to this direct effect, high moisture content causes damage to seed by processing machinery and handling.

Relative humidity and temperature during storage: Dry and cool conditions are the best for seed storage. The Harrington's simple **Thumb Rule** serves as a useful guide and measure of the effect of the moisture content and RH on seed storage:

- 1% decrease in moisture content nearly doubles the storability of the seed.
- 10% decrease in temperature nearly doubles the storability of the seed.

Pre-harvest effects : About 95% of the total dry weight of a seed is storage material, which will be used by the germinating seed and seedling until it can produce its own photosynthetic and nutrient absorbing system. It is seen that an immature seed, a seed with an imbalanced chemical composition or one mechanically damaged permitting early entry of micro-organisms will have short storage life. Seeds with high initial germination perform better in storage than the seeds having low germination. The most important pre-harvest factor affecting seed viability and storability is weather, especially excessive moisture, pre-harvest rains, temperature fluctuations, diseases and insect-pests.

Seed Maturity: Several studies indicated that there is relationship between storage life and seed maturity. The mature seeds have longer storage life than the immature seeds. Studies conducted on several crop seeds showed marked losses in viability of immature seeds during storage as compared to mature seeds. In the process of maturity, seed gains dry weight and is considered mature when it has attained maximum dry weight.

Mechanical damage: Damage to the seed reduces the storability by allowing the pathogens to enter the seeds through cracks in the seed coat. Broken or cracked seed coats also enhance embryo damage by chemical treatments used as disinfectants. Both fungi and chemical damage reduce the keeping quality of stored seeds.

Seed Vigour: The vigour of seeds at the time of storage is an important factor that affects the storage life of seed. The seed lots containing some seeds low in vigour will deteriorate early during storage than the seed lots containing all vigorous seeds.

Genetic effects: It is well known that seeds of some species survive longer than the other. The seeds having hard impermeable seed coat survive longer. Seeds of different varieties absorb and lose moisture at different rates under the same conditions. Size of seed, thickness and texture of the seed coat, and chemical composition of seed influence the moisture absorption. All these effects are genetic and are inherited.

Ventilation : Poor ventilation especially when the floor has no vapour barrier, results in air convection currents due to temperature differences within the storage. This causes moisture movement within the storage and RH of cooler air near the floor increases. Increased moisture content causes rapid deterioration of seed at the bottom of stacks. Ensure proper ventilation around seed, stack only on pallets leaving at least 10 cm open space for air movement, leave 25-30 cm open space between stacks and walls, and 50-60 cm above stacks.

Seed Quality Testing

The main aim of seed quality testing is to obtain accurate and reproducible results about the planting value of the seed. The planting value of the seed is dependent on genetic identity and purity, physical purity, germination, moisture and health status of seed lot. In seed testing laboratories, seed samples are tested for the determination of the seed quality in accordance with the standardized seed testing procedures provided in the National or International rules of seed testing. Seed quality reflects the overall value of the seed for intended purpose. Poor quality seed leads to loss of money and potential crop. Seed quality consists of physical purity, germination potential, genetic purity and seed health. Physical purity represents percentage by weight of pure seed. Germination testing is one of the most acceptable indices of seed quality. In addition, good quality seed should be of uniform size, having good luster and must conform to the prescribed standards for above mentioned parameters. To assess the quality of seed, testing is executed. For carrying different quality tests, seed sampling is the foremost step.

Seed sampling: For conducting various quality tests, a representative seed sample is drawn from seed lot, in which the same constituents are present as in the whole seed lot and in the same proportion. It is therefore essential that the sample must be drawn and submitted to the seed testing laboratory as prescribed in the Rules for Seed Testing.

A sample from the seed lot is obtained with the help of triers by taking small portions at random from different places in the lot. This is primary sample. These primary samples are then mixed and called composite sample. From composite sample, smaller samples are obtained by one or more stages to get submitted samples. At every stage, thorough mixing is very necessary. These samples are submitted to the seed testing laboratory for various tests, like germination test, moisture test, vigour and viability, verification of species and cultivar, seed health tests, etc. Different minimum sample sizes are prescribed by ISTA (International Seed Testing Association) for different crops.

Physical purity testing: Purity analysis is executed to determine the % composition by weight of the sample being tested and to identify the various species of seeds and inert particles constituting the sample. The working sample is separated into three component parts i.e. pure seed, other crop seeds, inert matter. The percentage of each part is determined by weight.

Genetic purity testing: Genetic purity refers to the trueness to the type of variety of the seed. It is determined on the basis of distinct morphological characters of the variety expressed at seed, seedling and plant level by comparing its submitted sample with authentic sample under identical environmental situation. This quality test is done on the basis of other distinguishable varieties (ODV) determination and growout tests.

The ODV Determination: This test is conducted using whole submitted sample. First the 'authentic seed sample' is examined under magnification for the expression of seed characters and to get familiarized with the seed characters of the stated variety. Then the submitted sample is weighed and placed on the work board and examined under illumination and magnification. The doubtful (ODV) seeds are separated out and counted. The results of the investigation are expressed as number of seeds belonging to the other distinguishable varieties found in the actual quantity examined. In addition, number per unit weight (e.g. per kg) may be calculated. The weight of the working sample for ODV determination should be equivalent to the weight of submitted sample of the species under consideration.

Grow-Out Test (GOT): Grow-out test is a measure for testing genetic purity of the seed lot on the basis of morphological characters and is based on inherited characteristics of plants produced by the seeds under test. It is pre-requisite for seed certification of foundation seed of all crops and hybrids of certain species like castor, muskmelon, brinjal, etc. The minimum population required for taking the observations is 400 plants. The affectivity of growout test proves the genuineness of a cultivar. Usually the cultivar differences are more distinct, if growth conditions are favourable. Crop should be so grown that the genetic differences express themselves as clearly as possible. The authentic control sample from the originating plant breeder/breeding institute is to be maintained by the testing station/agency. A minimum of two hundred plants from control sample are raised along with the test crop. The plots are examined through out the growing season with special emphasis on the period from flowering to ripening. The plants are examined carefully at each stage of growth and plants which are obviously of other cultivars are recorded and % of other cultivars, other species or aberrant types found are calculated.

Germination testing: Seed germination is the ability of seed to develop into a normal plant under favourable conditions. The moisture, temperature and oxygen are essential factors for seed germination. The ultimate objective of testing germination is to know the seed quality with respect to the planting value of seed. Testing under field conditions is normally unsatisfactory, as the results cannot be repeated with reliability. Laboratory methods have, therefore, been evoked in which the external conditions are controlled to give the most regular, rapid and complete germination for majority of samples of a particular species.

Four hundred seeds in replicates of 100 each are taken at random from the pure seed fraction and spaced uniformly and adequately apart on the moist substrate. Replicates may be divided into sub-replicates of 50 or 25 seeds depending on size of seeds and amount of space needed between them. In laboratory, this test can be conducted using paper or sand substrate. Each seedling is critically counted on the day prescribed for first and final count and following three categories are observed:

(i) *Normal seedlings*: These seedlings show the potential for continued development into satisfactory plants when grown in good quality soil and under favourable conditions of moisture, temperature and light. To be classified as normal, a seedling must conform to one of the following categories:

Intact seedlings: Seedlings with all their essential structure well developed, complete in proportion and healthy.

Seedling with slight defects: Seedling showing slight defects in their essential structures provided they show an otherwise satisfactory and balanced development comparable to that of intact seedlings of same test.

Seedlings with secondary infection: Normal seedlings affected by fungi or bacteria from sources other than the parent seed.

(ii) *Abnormal seedlings*: Abnormal seedlings do not show the potential to develop into a normal plant when grown in good quality soil and under favourable conditions of moisture, temperature and light. The following seedlings are classified as abnormal:

Damaged seedlings: Seedlings with any of the essential structures missing or badly and irreparably damaged due to external causes.

Deformed or unbalanced seedlings: Seedling with weak development or in which essential structures are deformed or out of proportion due to internal physiological causes.

Decayed seedlings: Seedlings with any of their essential structures, diseased or decayed as a result of external or internal seed borne diseases.

(iii) *Un-germinated seeds*: Seeds which have not germinated by the end of the test period. These are classified as follows:

Hard seeds: Seeds which remain hard at the end of the test period because they have not absorbed water.

Fresh un-germinated seeds: Seeds which are neither hard nor have germinated but remain clean and firm and apparently viable at the end of the test period.

Dead seeds: Seeds which at the end of the test period are neither hard nor fresh and not have produced any part of a seedling. They are collapsed and milky paste comes out when pressed at the end of the test.

Seed moisture determination: Moisture content of the seed is one of the most important factors influencing the seed germination, viability, vigour, seed health and seed storability. Seed moisture content is determined by several methods with different degrees of accuracy. There are two methods for moisture determination namely Air Oven Drying Method and Moisture Meter Method.

Air Oven Drying Method: In this method, seed moisture is removed by drying for a specified period and temperature. The samples for moisture determination weighing approximately 50-100 g should be packed separately in moisture-proof containers, such as sealed tins or thick polyethylene bags. Then moisture determination must be made in two replicates weighing 4-5 g each. The working sample should be obtained by spoon method. Mechanical dividers should not be used for this purpose. While drawing the working sample for moisture determination, care should be taken so that the sample is not exposed to the humidity, otherwise the seed will gain or lose moisture before the actual determinations. Depending upon the type of seed , some seeds are needed to be ground finely before the moisture content determination, some other seeds are coarsely ground (Legumes) whereas some seeds which are of very large size may be cut with knife. When the seed belongs to the species for which grinding is necessary and the moisture content is more than 17%, pre-drying before grinding is necessary. The seeds are dried at 130-133°C temperature for 1-4 h. After drying, the seeds are allowed to cool in the desiccator at least for one hour and then weighed. The moisture content as a percentage by weight is calculated as follows:

$$\text{Moisture content (\%)} = \frac{\text{Loss in weight}}{\text{Initial weight of seed}} \times 100$$

Seed Viability: Seed viability refers to the living conditions of seed and is an index of seed germination. To make quick estimate of the viability of seed samples in general and of those showing dormancy in particular, different seed viability tests are done. One of these tests is tetrazolium test. In this test, a colourless solution of 2,3,5, tri-phenyl tetrazolium chloride (TTC) or bromide is used as an indicator. All living cells which respire are capable of reducing this TTC into a red colored compound Formazan by hydrogen ion transfer reaction catalysed by the enzyme dehydrogenase. Formazan being red, stable and non-diffusible distinguishes the red-coloured living parts of seeds from the colorless dead parts. In addition to completely stained viable seeds and completely unstained non-viable seeds, partially stained seeds also occur. Varying proportions of necrotic tissue are found in different zones of these partially stained seeds. The position and size of the necrotic areas in the embryo /gametophyte tissue determine whether such seeds are classified as viable or non-viable.

Seed Vigour : Seed vigour is the sum total of all those seed properties which determine the potential level of activity and performance of the seed lot during germination and seedling emergence. Seed vigour is the degree of aliveness and from seed testing point of view is the sum total of all seed attributes that favour stand establishment under varying field conditions. So vigour test is not a single measure like seed germination and viability but a concept describing several characteristics associated with various aspects of performance of geminating seed or subsequent seedling emergence. The vigour test is done to assess the seed quality needed to supplement the germination and viability tests.

A number of vigour tests are proposed by different workers

A. Physical Tests

(i) *Seed Size*: One hundred or one thousand seeds are drawn randomly from the submitted sample and weighed in grams. The seed lot having more seed weight is considered more vigorous.

(ii) *Physical soundness*: Seed lot with uniformly shaped, size and good lustered seeds is more vigorous. Shriveled, undersized, underdeveloped, discolored and insect damaged seed lots are weak with low vigour.

B. Growth Test

This test is based on the principle that vigorous seeds grow at faster rate than less vigorous seeds, even under favourable conditions. The seedling growth rate can be measured by determining first count, speed of germination, seedling dry weight and seedling length.

(i) *First Count*: Two different samples of the seed are taken and kept for germination test in 4 replications of 100 each. The number of normal seedlings emerged on the first count day (as specified for each crop species) are counted. The samples with higher number of normal seedlings are more vigorous than the samples with lesser normal seedlings.

(ii) *Speed of Germination*: Two different samples of the seed are taken and kept for germination test in 4 replications of 100 each. The number of seedlings emerging daily is counted from day of planting the seed on the substratum till the time of germination is complete. Then the germination index (GI) is computed as follows:

$$GI = n/d$$

Where n = No. of seedlings emerging daily and d = day after planting

The seed lot with greater seed germination index is more vigorous.

(iii) *Seedling length and dry weight*: Again the two or more different seed lots are taken and kept for regular germination test. After prescribed time, five seedlings are selected, their length is measured and mean seedling length is calculated. The seed samples with more seedling length are more vigorous. Similarly for getting seedling dry weight, 20 seedlings are kept in an oven at 100°C for 24 h. The seed samples with greater seedling dry weight are more vigorous.

C. Stress Tests

Accelerated ageing test: Seeds are subjected to accelerated ageing by increasing seed moisture content and temperature for days to weeks followed by air drying for determining seed vigour.

Seeds are spread in a single layer on a wire mesh placed in a desiccator containing water for 100% RH. The desiccator is kept in an incubator at required temperature. The seeds are drawn periodically to test for germination. Seeds showing germination even after accelerated ageing are vigorous.

D. Bio-Chemical Test

Membrane Integrity Test: Weakening of cell membrane of poor vigour seeds causes leakage of water soluble compounds like sugars, amino-acids, electrolytes, etc., when immersed in water. Thus the measurement of electrical conductivity (EC) of the seed leachates gives an accurate estimation of seed vigour with reference to seed membrane permeability. The seed sample is surface sterilized with 0.1% mercuric chloride for 5-10 min and washed thoroughly with distilled water. These seeds are immersed in 100 ml distilled water at 25±1°C for 10-12 h. The water (leachate) is decanted off in a clean and dry beaker and conductance of the leachate is measured. Lower the EC value, greater is the seed vigour.

Seed Health Testing : The quality seed should be pure, homogenous and free from seed borne diseases. The presence of seed-borne pathogens in excess of minimum certification standards may lead to rejection of the seed lot and could also become a major cause for poor seed germination and poor field performance of crop. Thus, seed health is of paramount importance in development of quality seed. Therefore, it is essential to test the seed for its health status before sowing and/or trading.

Seed Health Testing Techniques

The entire submitted sample, or a proportion of it, depending on the test method may be used as a working sample. Normally the working sample should not be less than 400 pure seeds or an equivalent weight taken from the submitted sample. Various methods are available for detection of a particular seed-borne pathogen.

Visual observation of dry seeds: The seed sample is examined for the presence of seed coat discolourations, seed abnormalities, spores on the seed surface, fruiting structures, other infected plant parts mixed with the seed, etc. in dry condition by visual observation or under low power binocular microscope.

Washing Test: The method is quite simple, quick and is adopted for knowing the seed surface contamination due to smut spores, oospores, chlamydospores and spores of different genera of fungi-imperfecti like *Fusarium, Curvularia, Helminthosporium, Cephalosporium, Alternaria, Pyricularia, etc.* Seeds are soaked in a flask and shaken over a mechanical shaker for 5 to 10 minutes and the supernatant thus obtained is centrifuged at 2500-3000 rpm for 10-15 minutes. The sediment is examined under the compound microscope for the identification of fungi. This method however, does not reveal mycelial infections and the viability of spores.

Examination after incubation: The working sample is examined for presence of disease organisms, pests and physiological disturbance in or on the seed and seedlings after a specific period of incubation. The media commonly used are blotters, agar, etc. These tests are effective for detection of a large number of fungi. The identification of the pathogen is based on fungal morphology (sporulation, fungal structures, colony characteristics, etc.)

Agar plate method: The seeds are plated in Petri dishes on agar. Malt Extract agar or Potato dextrose agar is the most commonly used. In testing for storage fungi, agar media with high salt concentrations such as 18% NaCl is used. The seeds (10 seeds / Petri dish) are soaked for 5 minutes in 1% sodium hypochlorite The seeds are examined on 5th and 8th day of incubation. The fungi can be easily identified based on the colony characteristics. The method is good for flax fungi and *Ascochyta* spp. in legumes.

Growing on test: The plants are grown under controlled conditions to allow the development of symptoms that may appear due to seed-borne infections. This method may be used for post-entry quarantine and for the detection of certain systemic pathogens not detectable by the simple methods or for virus and bacteria. Here the seeds are sown in soil, sand or any other sterile inert medium under optimum controlled conditions.

Tests Based on Bioassays

The Phage–Plaque method: This method is designed for the detection of bacterial pathogens. The seeds are macerated followed by incubation for 24 h to multiply bacteria. Sample from this material is transferred to sterile flasks and standard suspension of phage particle is added. A small amount is then placed immediately and after 6-12 h on plates containing indicator bacterium.

Biochemical Tests

Serological methods: These tests are employed particularly for the detection of viral and bacterial pathogens. These tests are based on *in vitro* reactions between antigens and antibodies and the results may be seen as agglutinations or precipitations. The serological procedures can be used only for specified viruses, for which antiserums are available. The most common tests are precipitation test, gel diffusion test, agglutination test, enzyme linked immunosorbent assay (ELISA), immunosorbent electron microscopy (ISEM), immunosorbent dilution plating (ISDP), solid phase radioimmuno assay, etc.

Verification of Variety / Cultivars : The ability to characterize and identify plant varieties is essential for seed production and certification procedures. Thus there is a need for reliable method of varietal identification. Traditional methods usually involve the observations and recording of morphological characters or descriptions. Other alternative methods are also being tried to distinguish different cultivars using bio-chemical (molecular) markers.

Morphological Markers : Different cultivars are commonly identified on the basis of taxonomic differences in the seeds, seedlings and mature plants. For this, generally the grow out test is conducted on the basis of morphological characters of plants produced by the seeds under test.

Molecular markers: There are alternative tests which can distinguish varieties on the basis of stable biochemical properties of seeds or seedlings. Analysis of seed or seedling proteins and isozymes using electrophoresis are the most widely used. Electrophoresis is the migration of charged particles under the influence of an electric field. The rate of movement of these particles depends primarily upon:

- *The charge*: Molecules with higher charge migrate faster than those with lower charge
- *The size of the molecule*: Particles with smaller molecular weights migrate faster than those with higher weights.

In electrophoresis, gel acts as molecular sieve with desired pore size. Proteins with higher charge density migrate faster, resulting in differential rates of movement of proteins when a mixture of different proteins is subjected to an electric field. By altering the gel pore size (using polymers at different concentrations) and the charge on the protein molecule (by changing the pH of the system) a high degree of resolution can be achieved for separation of protein molecules in a mixture. There are different methods of performing electrophoresis such as:

- Starch gel electrophoresis
- Polyacrylamide gel electrophoresis (PAGE)
- Agarose gel electrophoresis

Hybrid Seed Production

During 2005-2006, India produced 113.5 million t of vegetables from 7.2 million ha with a productivity of 15.76 t/ha. However, according to the present estimate, India will need to produce 215 million t of vegetables by 2015 to meet the domestic as well as export market requirements. Since increase in land area under vegetable cultivation is a limiting factor due to rapid urbanization and decrease in land holding, hybrid technology is considered one of the available options for achieving the above target to fulfill the requirement of food and nutritional security of the country. High genetic variability and appreciable heterosis with respect to yield and other desirable characters are reported in most of the vegetable crops. Moreover, favourable genetic systems like varied sex forms, genetic and cytoplasmic male sterility, gynoecy, self-incompatibility, comparatively easy technique of emasculation and pollination and advantage of getting large number of seeds from a single cross can be successfully exploited for production of F1 hybrids in most of the vegetables.

Advantages of F_1 Hybrids of Vegetables

The hybrid varieties of vegetables have major advantages of high productivity, earliness and uniformity. A tomato hybrid can yield 80-90 t/ha as compared to an yield of 30-35 t/ha by the best open pollinated variety(s). Similarly, cabbage hybrid can yield 50-60 t/ha against 25-30 t/ha by the best open pollinated variety. Superior head quality (compactness) of hybrid cabbage is a matter of attraction to farmers of peninsular India, because open pollinated varieties produce light (puffy) head under warm and high temperature conditions. Similarly, long field holding capacity of F1 cabbage gives remunerative prices to the North Indian farmers during off-season. Uniformity of F_1 hybrids is a great advantage in highly cross-pollinated crop like onion where large variation occurs in shape, size and colour of the bulb even in improved open pollinated varieties. Higher degree of resistance to yellow vein mosaic disease in hybrid okra is preferred by the farmers. Thick skinned fruit with uniform colour, high T.S.S. and lycopene content of hybrid tomatoes are suitable for long distance transportation and are also preferred by the processors.

Technologies for Production of Hybrids and Hybrid Seed

- Hand emasculation and hand pollination
- Use of genic male sterility and gene-cytoplasmic male sterility

- Use of self incompatibility
- Use of gynoecious lines
- Use of monoecious lines
- Use of growth substances like etheral, gibberellic acid and silver nitrate.
- Use of graft technology
- Application of biotechnology

Hand emasculation and hand pollination : This method is used for production of hybrid seeds in tomato, eggplant and pepper. In India, many private companies are producing hybrid seeds of solanaceous crops and of okra by this method. Punjab Agricultural University, Ludhiana has developed three hybrids of tomato (TH 2312, TH 802 and TH 1) and two hybrids of brinjal (BH 1 and BH 2) using this method which are being commercially grown by the farmers.

Self-Incompatibility : Sporophytic system of self-incompatibility is being used in cabbage, cauliflower, Chinese cabbage and Japanese radish. In Japan, "Nagaokra Kohai No. 1" hybrid was developed using two SI inbred lines in 1951 by Takii Seed Co. It is the first hybrid of cabbage in the world to be developed by exploiting SI system. Now several hybrids are developed using this technique in most of the crucifers, which are available in several countries of the region.

Self-incompatible lines are maintained either by bud-pollination, CO_2 treatment, weak electric current or with the spray of 2-3% sodium chloride (NaCl).

Male Sterility : The gene-cytoplasmic male sterility system in onion was first of all discovered in 1942 for hybrid seed production. Genic male sterility is being used in chilli and muskmelon. Punjab Agricultural University, Ludhiana is the first public organization to develop a hybrid in muskmelon "Punjab Hybrid" in muskmelon by using a male sterile line MS-1 and Hara Madhu. Very large quantity of seed of this hybrid is being produced by NSC, Department of Horticulture Punjab, private seed companies and progressive growers. This hybrid was released at the national level and recommended for cultivation in Punjab, Haryana, Rajasthan, Western U.P. and Madhya Pradesh. Genic male sterility in chilli is used to develop an F_1 hybrid at Punjab Agricultural University, Ludhiana. Chilli hybrid "CH-1" developed by crossing a male sterile line MS-12 and pollen parent LLS has become very popular. It is taken up for processing purpose by Pepsi Foods and Nijjar Agro Ltd. Punjab Agricultural University has also developed another hybrid in chilli namely CH3 using male sterility.

Genic and functional male sterility present in tomato, can also be used with advantage. Genic male sterility in tomato is utilized commercially in Bulgaria. It is associated with anthocyaninless marker which helps in identification of male sterile plants and purity of hybrid seeds.

Gene-cytoplasmic male sterility is being used in onion, carrot and celery in Japan and in pepper, onion, carrot and radish in Republic of Korea. By using this mechanism, hybrids are being produced in these crops in Japan. In India, Arka Lalima, Arka Pitamber (yellow) and Arka Kirtiman hybrids were developed by IIHR utilizing gene-cytoplasmic male sterility in onion.

Gynoecious lines : Gynoecious lines in cucumber are maintained by inducing staminate flowers with the application of silver nitrate solution @ 300-500 ppm or silver thiosulphate @ 400-500 ppm at 2-4 true leaf stage. However in muskmelon the chemical sprays do not induce separate staminate flowers but induce anthers in the pistillate flowers.

Gynoecious lines are being used for the production of hybrids. Gynoecism is used for producing commercial hybrids in cucumber all over the world. Thus hybrids developed by utilizing gynoecious lines are totally female with parthenocarpic genes and do not require pollination for fruit set. Use of gynoecious line(s) has an advantage of avoiding the cumbersome process of roguing fertile/sterile plants from the rows of male sterile line MS-1 followed by hitherto. It has made the hybrid seed production easier, efficient and economical. Punjab Agricultural University has also developed one hybrid MH-25 in muskmelon using gynoecious lines W1998.

Monoecious lines : Monoecious lines are used as seed parents by practising hand pollination and pinching of male flowers for producing hybrid seeds. Monoecious sex expression in many cucurbits is exploited. In bottle gourd, Pusa Meghdoot and Pusa Manjari were developed and these were the first hybrids developed and released by public sector, IARI.

Use of Growth Substances : In some countries, etheral is being used on monoecious cucurbits for regulation of flower array and production of hybrid seeds. The technique is commercially feasible in cucumber, muskmelon, cucurbits, gourds etc.

Grafting : There are severe problems of soil borne diseases and pests. It is strenuous to incorporate resistance to such stresses in the varieties. Therefore, grafting technique taking wild or other species as rootstock is practised. Japan is growing 32% tomato, 50% eggplant, 93% watermelon and 7.2% cucumber as grafted plants. Tomatoes are grafted on *L. pimpinellifolium* or *L. esculentum,* eggplant on *S. torvum* or *L. integrifolium,* watermelon on bottlegourd, *Cucurbita* spp. on *Benincasa hispida* and cucumber on *Cucurbita ficifolia.*

Biotechnology : Mass propagation of parental lines especially self-incompatible lines, gynoecious lines, male sterile lines etc. *in vitro* is an important area for maintenance and multiplication of parental lines. Now techniques for *in vitro* propagation of tomato, crucifers, cucurbits and root crops are available.

Seed Borne Diseases

Seeds are the vehicles, victims and suspects of pathogens in general and seed borne pathogens in particular, which may be carried as contaminants or as infections. The spread of major diseases from one region to another in a country and also internationally had been through the seed. Presence of seed-borne pathogens in excess of minimum certification standards leads to rejection of the seed lot and is a main cause for poor seed germination and poor field performance of crop. The seed borne diseases may be caused by fungi, bacteria and viruses or animal pests, such as nematodes, eelworms and insects or physiological disorders due to deficiency of trace elements. Some seed borne diseases are designated as objectionable for specific crops. The seeds of such crops are certified on the basis of number of infected seeds. The infected seeds are identified on basis of symptoms and reported in percentage.

References

Agrawal, P.K. 1988. Seed industry in India. An appraisal. *Seed Tech. News*. 18(1):1-13

Agrawal, R.L. 1980. *Seed Technology*. Oxford and IBH Publishing Company, New Delhi.

Anonymous, 1988. Indian Minimum Seed Certification Standards. The Central seed Certification Board, New Delhi.

Arya, P.S.1997. Vegetable Growing in Hills. M.D. Publication Pvt. Ltd., New Delhi.

Bhanchandani, P.M., Pal, N. and Choudhary, B.1970. Seed to seed method of seed production in onion. *Prog. Hort.* 2: 15-23.

Choudhary, B. 1983. Vegetables. National Book Trust, India, New Delhi. 7th edition.

De Tempe, J. and Binnerts, J. 1979. Introduction to methods of seed health testing.

Fageria, M.S., Arya, P.S. and Singh, A. K. 1995. New trends in cabbage seed production. *Scientific Hort.* 4: 102-107.

George, R.A.T. 1985. Vegetable seed production, London and New Delhi.pp.318.

Gray, D.1983. Improving the quality of vegetable seeds. *Span*.26(1) : 4-6.

Hawthorn, L.R. and Pollard, L.H.1954. Vegetable and Flower Seed Production. The Blackistan Company, Inc. New York.

Heydecker, W. 1973. University of Nottingham, School of Agriculture. Rep. 1973/74.

Heydecker,W. and Coolbear, P. 1977.Seed treatment for improved performance survey, an attempted prognosis. *Seed Sci. & Technol.*, 5:353-425.

Ibid 1966. The Seed Act. Govt. of India, Manager Publications, G.O.I., New Delhi.

Ibid 1968. The Seed Rules. Govt. of India, Manager Publications, G.O.I., New Delhi.

ISTA 1976. International Rules for Seed Testing. International Seed Testing Association, Zurich.

ISTA 1985. International Rules for Seed Testing. *Seed Sci and Tech.*13(2) : 307-520.

Justica, O.L. 1961. The Testing of Seeds. Year Book Agri. (U.S. Dept. Agriculture).

Khan, A. A.1992. Pre-plant physiological seed conditioning. Horticulture Rev.,13:131-181.

Korla, B.N 1989. Hybrid Seed Production in Vegetable Crops. All India Workshop- cum-Seminar on Temperate Vegetable Crops. p. 78-81.

Mehta, Y.R., Douglas, J.E. and Singh, A. 1973. Historical development of seed improvement in India. *Seed Res.* 1 : 1-12.

Naik, I.B. and Sriniwas, K. 1992. Seed Production of vegetable crops II. Onion-a review. *Agril Rev.* 13 (2): 59-80.

Neergaard, P. 1977. Seed Pathology. The Macmillan Press Ltd., London, p.1187.Nieuwhof, M (1969). Cole Crops. Leonard Hill, London.

Rajan, S and Markose, Baby L. 2007.Propagation of Horticultural Crops.New India Publishing Agency, New Delhi,p270.

Roberts, E.H.1972. Viability of Seeds. Chapman and Hall, London.

Rowse, H.R.1992. Methods of priming seeds.UK patent No. 5119589.

Saluke, D.K., Desai, B.B. and Bhat, N.R. 1987. Vegetable and Flower Seed Production. Agricole Publishing Academy.

Scott, J. M. 1989.Seed coatings and treatments and their effects on plant establishment. Adv. Agron. 42.43-83.

Seshadri, V.S. 1990. Vegetable seed production in India-problems, prospects and policies. Paper presented in International Conference on Seed Science & Technology, New Delhi, Feb. 1990.

Shanmugam, C.1998. Indian Seed Act-Rules and regulations. *Seed tech. News.*28 (4) :p180.

Singh, H.1984. A new technique for seed separation. *J. Res. Punjab Agricultural University, Ludhiana,* 21(3) :463-64.

Singh, H. and Gill, K.S. 1987. Procedures for breeder and foundation seed production in vegetable crops. PAU, Ludhiana.

Singh, H.B., Thakur, M.R and Bhagchandani, P.M.1960. Vegetable seed production in Kullu Valley, II. Seed production of root vegetables *Indian J. Hort.* 17 : 38-47.

Thompson, J.R. 1979. An Introduction to seed Technology. Leonard Hill, London. p.157-67.

Town, P.A.1969. Vegetable Seed Standards. Bull. No. 38.

Tripathi, S.N. and Singh, R.K. 1969. A note on the methods of seed raising in cauliflower. *Indian J. Hort.* 26: 184-86.

Verma, S.R. and Singh, H. 1988. Development of an axial flow vegetable seed extraction machine. *J. Agri. Engg.* 25 (1) : 98-104.

Whitaker, T.W. and Davis, C.N. 1962. Cucurbits, Leonard Hill, London.

The Science of Horticulture : Vol. 1, 2011
© *K.V. Peter (ed.), pp. 269-288*
New India Publishing Agency, New Delhi (India)
E-mail : newindiapublishingagency@gmail.com
Web: www.bookfactoryindia.com

Chapter 11

Use of Plant Growth Promoting Rhizobacteria (PGPR) in Vegetable Crops

SADHAN KUMAR, P.G. AND NIRMALA DEVI, S.

Plant Growth Promoting Rhizobacteria (PGPR) are free-living bacteria having beneficial effect on plants by root colonization. They stimulate plant's growth. Intensive research work is carried out on use of various PGPR as plant growth promoters on various crops. There are reports which show that PGPR promoted growth and reproductive parameters of cereals, pulses, ornamentals, vegetable crops, plantation crops and even tree species. The important PGPR used in vegetable crops are *Azospirillum, Azotobacter, Pseudomonas fluorescens*, Phosphate Solubilising Bacteria (PSB) and *Bacillus subtilis*.

Mechanism of PGPR on Growth and Yield

Brown (1970) suggested that the stimulation of plant growth by inoculating seeds with phosphatic solubilizers might be more due to production of growth hormones rather than phosphate solubilization. The tested phosphate dissolving bacteria (*Bacillus polymyxa, B. pulvifaciens* and *Pseudomonas striata*) and fungi (*Aspergillus awamori, A.niger* and *Penicillum digitatum*) produced auxins and gibberellins to an appreciable extent.

Different mechanisms are suggested for solubilization of inorganic phosphorous by phosphate solubilising microorganisms. Production of organic acids in the micro-environment around the root is the most important cause of phosphate solubilization (Subba Rao, 1982).

Bhandari (1989) found that the nitric acids produced during oxidation of inorganic compounds of nitrogen and sulphur react with insoluble phosphate (rock phosphate)which brought about solubilization. Addition of rock phosphate coupled with inoculation of phosphate solubilizing microorganisms gave good response in many crops. These microorganisms solubilize not only insoluble soil phosphates and insoluble forms of phosphatic fertilizers like rock phosphate, bone meal, etc., but also increase the efficiency of soluble forms of phosphatic fertilizers applied to soil (Somani *et al.,* 1990).

Kloepper *et al.* (1991) found that some bacteria solubilize organic phosphate by secreting phosphate or inorganic phosphate from soil particles by releasing organic acids and this could make phosphorous as well as micronutrients more readily available for plant growth in soil.

Subba Rao (1995) gave conclusive evidence to show that the soil bacteria dissolved rock phosphate, bone meal, di-and tri-calcium phosphate, calcium phosphate and calcium carbonate. Most of the studies on phosphate solubilization were done first by isolating microorganisms from the soil and then studying the solubilization under *in vitro* conditions.

Subba Rao, (1997) found that the source of extra phosphate taken up by growing mycorrhizal and non-mycorrhizal onions in soils was from the pool of 'liable' soil phosphate. Enzymes extracted by polymorphic phospho microorganisms in intimate association with plant roots mineralize phosphorous for plant uptake.

The efficiency of nitrogen fixation by *Azospirillum brasilense* increased with the decreased carbon supply and low PO_2 (Okon *et al.,* 1976a). The optimum temperature for N_2 dependent growth by *Azospirillum* is between 32 and 40 ^{0}C which is similar to the optimum temperature for other nitrogen fixing bacteria from tropical environment (Day, 1977).

The enhancement in yield with *Azospirillum* is due to production of growth substances like IAA and GA which might have increased the availability and uptake of nutrients through plant roots (Deka *et al.,* 1996). In potato plantlets grown under *in vitro, Azospirillum* (P5JN) increased cytokinin content by inducing synthesis in the early stages of plant growth and development (Lazarovits and Nowak, 1997).

Azospirillum brasilense produced high quantities of IAA. This was the most likely explanation for increased seedling root length, root hairs, root branching and root surface areas (Kloepper, 2003).

Prabu *et al.* (2003) reported that phytohormones produced by the *Azospirillum* stimulated root growth and induced changes in root morphology, which in turn improved assimilation of nutrients and yield in okra.

Bashan and de-Bhasan (2004) reported that the gibberellins produced by *Azospirillum* contributed to increased root hair development. Application of gibberellin increased density of root hairs at levels similar to *Azospirillum* inoculation.

The significant increase in root biomass indicated particularly the role of growth hormones. The increased root surface might have resulted in a higher nutrient uptake and a better growth than N_2 fixation by the bacteria (Venkateswarulu and Rao, 1983).

Nelson and Maloney (1992) reported the competition for nutrients in biological control by *Pseudomonas fluorescens.* The *in vitro* antagonistic activity was based on competition and correlated with disease suppression. Anjaiah *et al.,* (1998) described about the antibiotics involved in disease suppression by *Pseudomonas fluorescens* as phenazine-1-carboxamide and anthranilate.

The mechanism of PGPR involved in growth promotion are due to increased production of auxin, gibberellin, cytokinin and ethylene (Kloepper and Schroth, 1981; Dubeikosky *et al.,* 1993; Glick, 1995). Handelsman and Stabb (1996) reported that the antibiosis is now often implicated as an important mechanism of biological control, resulting from the fact that it was an attractive mechanism to study which provides a highly effective mode of action.

In *Bacillus cereus* (strain UW85), production of kanosamine and zwittermicin A was suggested important for biocontrol activity (Milner *et al.,* 1998).

Competition between pathogenic and saprophytic microorganisms for organic materials released from roots can reduce growth and pathogenic activity of the pathogens. For this mode of action, the classical approach of comparing biocontrol activities of specific catabolic mutants with wild type strains is not simple, since anyone of a number of substrates could be utilized (Loper *et al.,* 1997).

Inoculation of lettuce plants with bacteria increased the cytokinin content of both shoots and roots. Accumulation of zeatin and its riboside was the highest in roots two days after inoculation, when their content was 10 times higher than in control. Changes in the content of other hormones (ABA and IAA) were observed at the end of experiments only. Accumulation of cytokinins in inoculated lettuce plants was associated with an increase in plant shoot and root weight of approximately 30% over 8 days. (Arkhipova *et al.,* 2005).

Effect of PGPR on growth and growth attributes

PGPR	Crop	Effect	Reported by
Azospirillum	Chilli	Increased plant height, number of primary, secondary and tertiary branches and days to 50% flowering	Amirthalingam (1988).
Pseudomonas strains	Tomato, cucumber, lettuce and potato	Increased shoot and fruit weight	Van Peer and Schippers (1988)
Azospirillum at 2.5 kg/ha	Okra	Increased plant height, shoot girth and fruit yield	Parvatam *et al.,* (1989).
50% recomm-ended dose of nitrogen + *Azospirillum* (2 kg/ha)	Okra	The highest root length and plant height	Sankaranarayanan *et al.,* (1995).
Azospirillum and *Azotobacter*	Knolkhol	Increase in leaf number / plant	Chalto *et al.* (1997)
Azospirillum	Okra	Increase in dry matter	Selvi *et al.* (1997)
Azospirillum and *Azotobacter*	Cabbage	Higher values for growth parameters like number of outer leaves and head volume	Verma *et al.* (1997)
Azospirillum and *Azotobacter*	Cauliflower	Increased chlorophyll content and leaf area	Bambal *et al.,* (1998)
Bacillus subtilis, Bacillus pumilis and *Curobacterium sp*	Tomato	Intensive plant growth promotion	Raupach and Kloepper (1998)
Azospirillum and phosphate solubilising bacteria (PSB)	Onion	Significant improvement in plant height, number of leaves, root length and neck girth	Thilakavaty and Ramaswamy (1999)

contd...

contd...

FYM + *Azospirillum* + phosphorus solubilizing bacteria (PSB).	Okra	Plant height, leaves / plant, nodes / plant, internodal length and fruits / plant were maximum	Naidu *et al.* (2000)
Azospirillum + phosphobacteria (2 kg/ha)	Brinjal	Favourably influenced the growth parameters and yield	Nanthakumar and Veeraragavathatham (2000).
Azospirillum + phosphate solubilising bacteria	Brinjal	Recorded the highest plant height, number of primary branches and number of leaves.	Anburani and Manivannan (2002)
Pseudomonas fluorescens + *Azospirillum* + *Glomus mosseae*	Tomato	Maximum root length, shoot and fruit dry weight	Muthuraju *et al.*, (2002).
Phosphate solubilising bacteria (PSB) + *Azospirillum*	Okra	Increase in the seedling length and root length.	Sajindranath *et al.*, (2002)
Azospirillum and *Azotobacter*	Cabbage	Increase in number and weight of non wrapper leaves / plant	Sharma (2002)
Bacillus subtilis strains SE34 and IN937b	Chilli, brinjal and okra	Enhancement in emergence rate, germination %, plant height, fresh weight and dry weight	Amruthesh *et al.*, (2003)
Azospirillum lipoferum and *Bacillus polymyxa*	Tomato	Increase in shoot and root fresh and dry weight	Badawy *et al.*, (2003)
Pseudomonas fluorescens	Watermelon	Better germination and plant vigour	Nallathambi *et al.*, (2003).
Azospirillum + phosphorus solubilizing bacteria (PSB)	Okra	Increased nodes / plant, number of leaves, leaf area, fresh and dry weight of plants	Prabu *et al.*, (2003).
Bacillus licheniformis	Tomato and pepper	Increased the height of plants and the leaf area	Lucas-Garcia *et al.*, (2004).

contd...

contd...

Azospirillum	Onion	Higher dry weight of shoot and bulb	Navale and Wani (2004)
Azospirillum and *Azotobacter*	Brinjal	Increased plant height (44.8 cm), leaves / plant (43) and shoot dry weight (30.8 g)	Wange and Kale (2004)
Azospirillum + *Azatobacter*	Lettuce	Increased number of leaves (44), average leaf area (200.6 cm^2), fresh shoot weight (125 g), dry matter weight (10.6 g) and root length (10.7 cm) / plant	Wange and Kale (2004)
Pseudomonas fluorescens + *Azospirillum*	Cauliflower	Maximum value of plant height, number of branches, canopy spared and dry matter production	Singh and Singh (2005)
Azospirillum, Pseudomonas fluorescens and *Azotobacter*	Onion	Increase in leaves / plant, plant height, shoot thickness, yield and bulb characteristics	Jha *et al.* (2006)

Effect of PGPR on Pest and Disease Occurrence

Raupach and Kloepper (1998) studied the effect of PGPR mediated disease suppression in cucumber against angular leaf spot (1996) and against a mixed infection of angular leaf spot and anthracnose (1997). The three-way mixture of PGPR strains (*Bacillus subtilis* GB03 plus *Bacillus pumilus* INR7 plus *Curtobacterium flaccumfaciens* ME1) as seed treatment showed intensive plant growth promotion and disease reduction to a level statistically equivalent to the synthetic elicitor Actigard applied as a spray.

In cucumber, it was found that bacterized plants suppressed the development of mite population by more than 40% on the leaves of susceptible cultivar Corona. The plants with PGPR were less injured in relation to the plants without bacteria. Infested leaves of cultivar Aramis showed a lower level of inhibition in photosynthesis compared to injured Corona leaves and the presence of bacteria had a positive effect on this process (Tomczyk *et al.*, 1999).

Zehnder *et al.* (1999) noticed that the tomato plants treated with PGPR demonstrated a reduction in the development of disease symptoms and often a reduction in the incidence of viral infection and increase in yield.

Effect of PGPR on yield and yield attributes

PGPR	Crop	Effect	Reported by
Azospirillum and *Azotobacter*	Cabbage	Increased the head volume, head weight and yield/ha	Verma *et al., (*1997)
Azospirillum + *Azotobacter* + 100% N	Cauliflower	Increased yield and early maturity of curds	Bambal *et al.* (1998)
Azospirillum	Okra	Increased yield	Ganeshe *et al.,* (1998)
Azospirillum 2kg/ha and phosphobacteria 2 kg/ha	Cucumber	Earliness, increased yield / vine and increased dry matter production	Nirmala and Vadivel (1999)
Azospirillum + PSB	Onion	Increased yield	Thilakavaty and Ramaswamy (1999)
Azospirillum	Okra	Increased yield and the highest benefit:cost ratio	Ganeshe *et al.* (2000)
Azospirillum +PSB.	Okra	Increased number of fruits /plant, weight of fruits /plant and fruit yield	Naidu *et al.* (2000)
Azospirillum	Okra	Increased yield	Asha Raj and Geetha Kumari (2001)
Azospirillum	Okra	Higher number of pods / plant and average fruit weight	Bahadur and Manohar (2001)
PSB and *Rhizobium*	French bean	Increased yield	Baronia (2001)
PSB	Brinjal	Increased the plant dry matter, N and P uptake and yield	Gaikwad and Wani (2001)
Azospirillum + Phosphate solubilising bacteria	Brinjal	Increased fruit set % and fruit yield	Anburani and Manivannan (2002)
Azospirillum	Tomato	Increased yield	Harikrishna *et al.,* 2002).
Azospirillum and *Azotobacter*	Okra	Enhanced dry matter content	Kamali *et al.* (2002)
Pseudomonas flourescens + *Azospirillum* + *Glomus mosseae*	Tomato	Increased yield	Muthuraju *et al.,* (2002)
Azospirillum and PSB.	Brinjal	No effect on fruit length, fruit girth, fruits / plant and fruit yield	Naidu *et al.* (2002)
Azospirillum	Tomato	Increased yield	Sengupta *et al.* (2002)

contd...

contd...

Azospirillum	Cabbage	Increased the number and weight of non-wrapper leaves, head length and width, weight of head and yield	Shalini *et al.* (2002)
Azospirillum + 60 kg N/ha	Cabbage	Higher head weight and yield	Sharma(2002).
Azospirillum	Cabbage	Increased yield	Sharma (2002)
Azospirillum brasilense+ *Bacillus polymyxa*.	Cabbage	Highest fruit girth, fruit weight, fruits / plant, and fruit yield	Devi *et al.* (2003)
Azospirillum + PSB	Okra	The highest benefit:cost ratio	Prabu *et al.*(, 2003)
20 tonnes PM + 5 kg *Azospirillum* + PSB.	Tomato	Increased fruit number	Raut *et al.* (2003)
Azospirillum + *Azotobacter*	Cabbage	Increased weight of head, diameter of head and number of folded leaves and yield	Singh and Khare(2003).
Azospirillum and *Azotobacter* with higher levels of fertilizer N (75 and 100%)	Okra	Higher yields	Swani *et al.* (2003)
75 kg N/ha + *Azotobacter* + *Azospirillum*	Okra and bitter gourd	Improved the shoot dry weight, fruit number, fruit weight and yield	Wange and Kale (2003)
PSB.	Cabbage	Increased head weight, head length, head diameter and number of outer leaves	Bahadur *et al.* (2004)
Bacillus licheniformis	Tomato	Increased number and diameter of fruits	Lucas-Garcia *et al.* (2004)
Azospirillum + *Glomus*	Onion	Higher yield	Navale and Wani (2004)
NPK + foliar MF + CCP + *Azospirillum*	Okra	Higher yields	Selvi *et al.* (2004)
Azospirillum + *Azotobacter*+ 150 kg N/ha	Cabbage	The highest average head diameter (15.9 cm), head length (11.9 cm) and head weight (380 g) as well as yield (11.5 t/ha)	Wange and Kale (2004)

contd...

contd...

Azospirillum + *Azotobacter* + 150 kg nitrogen/ha	Lettuce	Higher yields	Wange and Kale (2004)
Azospirillum + 100% N + P K	Onion	The Highest bulb yield, highest dry matter production	Mahanthesh *et al.*, (2005)
Azospirillum + 75% N + 100% PK or PSB + 75% P + 100% NK	Cauliflower	Increased curd diameter, curd height and curd yield	Narayanamma *et al.*,(2005)
Azospirillum + recommended dose of NPK @ 120:60:60 kg/ha	Cauliflower	Maximum weight of curd and curd yield	Singh and Singh (2005)
Azospirillum, *Pseudomonas fluorescens* and *Azotobacter* along with recommended dose of nitrogen and phosphorous.	Onion	Increase in both vertical and horizontal bulb diameter	Jha *et al.*, (2006)
Phosphate Solubilizing Bacteria	Tomato	Higher yield	Rabindra and Srivastava (2006)

Effect of PGPR on Quality attributes

PGPR	Crop	Effect	Reported by
Azospirillum	Knol khol	Increase in dry matter and vitamin C	Chalto *et al.*, (1997)
Azospirillum and Phosphate solubilising bacteria (PSB) + 45:45:35 kg NPK/ha.	Onion	Improvement of quality parameters like shape index, colour of bulb and pyruvic acid content	Thilakavathy and Ramaswamy (1999)
35 tonnes FYM + *Azospirillum* + Phosphorus Solubilizing Bacterium	Okra	Increase in crude protein	Naidu *et al.*, (2000)

contd...

contd...

Azospirillum	Okra	Increase in crude protein,reduction in crude fibre and higher ascorbic acid content	Asha Raj and Geetha Kumari (2001)
Azotobacter	Tomato	Increase in Total Soluble Solid (TSS)	Chaurasia *et al.*, (2001).
PSB	Tomato	No effect on total soluble solids	Chaurasia *et al.*, (2001)
Azospirillum	Tomato	Increase in total soluble solids and ascorbic acid	Sengupta *et al.*, (2002)
PSB	Broccoli	Higher carotenoid	Bahadur *et al.*, (2003)
Azospirillum	Tomato	Increase in Vitamin C content and shelf life. Reduced nitrate content	Raut *et al.*, (2003)
PSB	Tomato	Increase in TSS and ascorbic acid	Pardeep and Sharma (2004)
Azospirillum	Cauliflower	Higher vitamin C	Narayanamma *et al.*, (2005)
Azospirillum	Onion	Increase in total soluble solids	Jha *et al.*, (2006)
Rhizobium + PSB + press mud @ 5t/ha.	Peas	Higher vitamin C content, carbohydrates, protein, phosphorous and sulphur	Nirmala *et al.*, (2006)

The plant growth promoting rhizobacteria (PGPR) were tested for their ability to control Rhizoctonia rot and damping-off caused by *Rhizoctonia solani* AG4, *Pythium aphanidermatum* and *P. ultimum* on cucumber. Presence of the PGPR reduced the incidence of disease (Chen-XiaoBin *et al.*, 1999).

Bansal *et al.* (1999) revealed that the PGPR inoculation reduced root-knot infection in comparison with the check and carbofuran treated bottle gourd plants. Among all the bacteria, CO-70 followed by HT-54 was effective in promoting plant growth and reducing nematode infection.

PGPR treatments resulted in reduced Tomato Mosaic Virus incidence and disease severity and in some cases, a corresponding increase in fruit yield. The PGPR could become a component of an integrated program for management of this virus in tomato (Murphy *et al.*, 2000).

Tomczyk and Kiekiewicz (2001) reported that in cucumber and tomato, the spider mite-feeding caused only a small decrease of protein content in

Aramis (less susceptible), while presence of PGPR in the root system caused an increase of soluble proteins in the leaves. In infested leaves of Romatos (highly susceptible) without PGPR, an evident increase of amino acid content was found opposite to plants with PGPR.

PGPR control a wide range of pathogens like fungi, bacteria and viruses and they are known to control these pathogens by biocontrol mechanism which may be by decomposition or antagonism. However, the most studied phenomenon was the induction of systemic resistance by these bacteria in the host plant thereby containing the invading pathogens (Van Loon *et al.,* 1998)

Diseases of tomato caused by root-knot nematodes, *Fusarium, Phytophthora* and *Pythium* were not affected by PGPR treatments. PGPR formulation LS261 reduced number of root-knot nematode galls on pepper. Incidence of wilt symptoms on tomato was significantly lower in PGPR treated plots and the highest in the untreated plots. Yield of extra large tomato fruit and total yield increased with PGPR formulation LS256 (Kokalis-Burelle *et al.,* 2002).

Pseudomonas fluorescens strain PF1 induced systemic resistance against soil born pathogens of solanaceous vegetables like chilli and tomato (Ramamoorthy *et al.,* 2002).

In tomato, combination of three strains *viz., Pseudomonas fluorescens* strain CHA0, COP and COT was more effective for the suppression of tomato spotted wilt virus when compared to individual strain application or all possible two strains combination (Kandan, 2000).

Jetiyanon and Kloepper (2002) indicated that four mixtures of PGPR and one individual strain treatment significantly reduced severity of all four diseases (CMV of cucumber, bacterial wilt of tomato, anthracnose of long cayenne pepper and damping off of green kuang futsoi) when compared to the non bacterized control. Most mixtures of PGPR provided greater disease suppression than individual PGPR strains. These results suggest that the mixtures of PGPR can elicit induced systemic resistance to fungal, bacterial and viral diseases in tomato, long cayenne pepper, green kuang futsoi and cucumber.

Shojai *et al.* (2003) reported that the use of antagonists like *Bacillus subtilis,* against mildews and applications of milk and *Bacillus thuringiensis* (Bactospein) against trichoplusiani were effective. Results suggested possibility of the off-season cultivation of tomato and cucumber in the greenhouse through the proper implementation of a suitable IPM programme. Combination of PGPR strains *viz., Bacillus subtilis* GBO3 and *Bacillus amyloliquefaciens* IN937a with the carrier chitosan to the tomato leads to protection against CMV (Murphy *et al.,* 2000).

Sharma *et al.* (2003) reported that in okra the maximum reduction in number of galls was obtained with VAM, *Azotobacter* and *Azospirillum* alone. Combined application of phorate and microbes increased root-knot galling, indicating the probable deleterious effect of phorate on microbes or vice versa. The *Azotobacter* with phorate at 1 kg a.i/ha showed higher galling, while *Azospirillum* and VAM with phorate at 1 kg a.i/ha showed little difference.

Shylaja *et al.* (2003) revealed that the activity of peroxidase was better in the tolerant cultivar than in the susceptible. The susceptible cultivar seeds were treated with *Bacillus subtilis* and the peroxidase assay was repeated. Increase in the peroxidase activity was observed in tomato roots of the susceptible cultivar treated with *Bacillus subtilis* when compared with untreated. The 30 day old seedlings obtained from seeds of susceptible cultivar treated with *Bacillus subtilis* enzyme activity were significantly taller than the resistant cultivar not treated with bio agent, suggesting the induction of systemic resistance in the susceptible cultivar was due to root colonization by *Bacillus subtilis.*

Nallathambi *et al.* (2003) mentioned that the seed treatment of watermelon with bacterial isolates caused better seed germination when compared to untreated control. As for management of virus disease is concerned, none of the treatment could completely eliminate the disease. Seed treatments followed by foliar sprays of bacterial isolates *Pseudomonas fluorescens* (CIAH-111) recorded fewer incidences when compared with untreated plants.

Kavitha *et al.* (2003) reported that the treatments with the isolate *Bacillus subtilis* recorded the lowest incidence of damping off disease in chilli. An incidence of 15.66 % followed by 16.07 % in treatments with *Pseudomonas spp* and significant increase in plant growth were recorded because of treatments with these antagonistic bacteria. The combined bacterization of tomato seeds with *Pseudomonas fluorescens* and *Bacillus subtilis* performed better in reducing the damping off incidence and increased the root and shoot length more effectively than individual application of PGPR strains.

Ramesh and Korikanthimath (2003) reported that the *Pseudomonas fluorescens* and *Trichoderma* could be combined for the management of damping off in nursery as both possess different mode of action for pathogen suppression and growth promotion. For wilt management in brinjal, *Pseudomonas fluorescens* may be used as it induced systemic resistance in plants against the systemic pathogen.

Kloepper, (2003) reported that the *Bacilli* PGPR strains increased plant growth in the absence of pathogens and if the plants are then inoculated with foliar pathogens, induced resistance is manifested by reduced disease severity compared to controls.

Anith *et al.* (2004) observed that when the PGPR and Actigard applications were combined, Actigard plus *Pseudomonas putida* 89B61 or BioYield reduced bacterial wilt incidence when compared with the untreated control. Incorporation of S-H mixture into infested soil 2 weeks before transplanting reduced bacterial wilt incidence in one experiment. Combination of Actigard with the S-H mixture significantly reduced bacterial wilt incidence in tomato.

Siddiqui (2004) mentioned that the poultry manure with *Pseudomonas fluorescens* was the best combination for the management of *Meloidogyne incognita* on tomato but improved management of M. incognita can also be obtained if goat dung is used with *Pseudomonas fluorescence* or poultry manure with *Azospirillum brasilense.*

Sendhilvel *et al.* (2005) reported that the cowpea seed and the soil application of talc based formulation of *Pseudomonas fluorescens* (SVPF2) significantly reduced the root rot incidence.

Sally *et al.* (2005) reported that the seed treatment and soil application of *Trichoderma viride, T. harzianum, Pseudomonas fluorescens, Azotobacter chroococcum* reduced the damping off incidence considerably in brinjal, chilli and tomato. Application of these microbes also increased the germination(%) and promoted vigour of solanaceous vegetable seedlings as compared to control.

Several studies indicated that PGPR may stimulate production of biochemical compounds associated with host defense, massive accumulation of phytoalexins, phenolic compounds, increases in the activities of PR- proteins, defense enzymes and transcripts and enhanced lignification (Swamiyappan *et al.*, 2005).

Meena and Marimuthu (2005) reported that the potential of *Pseudomonas fluorescens* for the management of fruit of chillies was evaluated under green house and field conditions. Among the various strains of *Pseudomonas fluorescens* tested *in vitro*, the strain pf1 significantly inhibited the mycelial growth of *Colletotrichum capsici*. Combined application of talc-based powder formulation of *Pseudomonas fluorescens* to seed (10g/ kg seed) and foliage (1kg/ ha) effectively controlled the fruit rot disease. Early and high induction of various defense compounds was observed due to *Pseudomonas fluorescens* treatment.

References

Amirthalingam, S. 1988. Studies on the effect of *Azospirillum*, nitrogen and NAA on growth and yield of chilli. *S. Indian Hort.* 36: 218

Amruthesh, K.N., Niranjan Raj, S., Kiran, B., Shetty, H.S. and Reddy, M.S. 2003. Growth promotion by plant growth promoting rhizobacteria in some economically important crop plants. *Proceeding of 6th international plant growth promoting rhizobacteria* (PGPR) *workshop*, 5-10 Oct., Calicut, India, pp. 97-103

Anburani, A. and Manivannan, K. 2002. Effect of integrated nutrient management on growth in brinjal (*Abelmoschus esculentus*) cv. Annamalai. *S. Indian Hort.* 50(4/6): 377-386

Anith, K.N., Momol, M.T., Kloepper, J.W., Marois, J.J., Olson, S.M. and Jones, J.B. 2004. Efficacy of plant growth-promoting rhizobacteria, acibenzolar-S-methyl, and soil amendment for integrated management of bacterial wilt on tomato. *Plant Dis.* 88(6): 669-673

Anjaiah, V., Koedam, N., Nowak-Thompson, B., Loper, J.E., Hofte, M., Tamabong, J.T. and Cornelis, P. 1998. Involment of phenazines and anthranilate in the antagonism of *Pseudomonas aeruginose* PNA 1 and Tn5 derivatives towards *Fusarium* spp. and *Pythium* spp. *Mol. Plant Microbiol. Interaction.* 11: 847-854

Arkhipova, T., Veselov, S., Melentiev, A., Martynenko, E. and Kudoyarova, G. 2005. Ability of bacterium *Bacillus subtilis* to produce cytokinins and to influence the growth and endogenous hormone content of lettuce plants. *Plant Soil.* 272(1-2): 201-209

Asha Raj, K. and Geetha Kumari, V.L. 2001. Influence of organic manures and *Azospirillum* inoculation on bhendi (*Abelmoschus esculentus* L. Moench) yield, NPK uptake and nutrient status of soil. *J. Soil Biol. Ecol.* 21(1&2): 12-16

Badawy, F.H., El-Dsouky, M.M., Sadiek, H.S. and Abo-Baker, A.A. 2003. Response of tomato to inoculation with single and multi-strain inoculants of different bacterial species. *Assiut. J. Agric. Sci.* 34(5): 275-286

Bahadur, A. and Manohar, R.K 2001. Response of okra to biofertilizers. *Veg. Sci.* 28(2) 197-198

Bahadur, A., Singh, J. and Singh, K.P. 2004. Response of cabbage to organic manures and biofertilizers. *Indian J. Hort.* 61(3): 278-279

Bambal, A.S., Verma, R.M., Panchbhai, D.M., Mahorkar, V.K. and Khankhane, R.N. 1998. Effect of biofertilizers and nitrogen levels on growth and yield of cauliflower (*Brassica oleracea var. botrytis*). *Orissa J. Hort.* 26(2): 14-17

Baronia, A.K. 2001. Studies on minimizing expenditure on fertilizer through biofertilizers in French bean (*Phaseolus vulgaris* L.). *J. Soil Crops.* 11(1): 40-42

Bashan, Y. and de-Bashan, L.E. 2004. *Azospirillum* – plant relationships: environmental and physiological advances. *Canadian J. Microbiol.* 43: 103-121

Bhandari, S.C., Somani, L.L. and Gulati, I.J. 1989. *Biofertilizers*, Scintific Publishers, Jodhpur, 327p

Brown, M.E. 1970. Field evaluation of phosphate solubilizing bacteria for growth promotion. *J. Appl. Bact.* 35: 433-436

Chalto, M.A., Gandroo, M.Y. and Zargar, M.Y. 1997. Effect of *Azospirillum* and *Azotobacter* on growth, yield and quality of knolkhol (*Brassica oleracea* var. *gongylodes* L.). *Veg. Sci.* 24(1): 16-19

Chaurasia, S.N.S., Nirmal, D., Singh, K.P. and Kalloo, G. 2001. *Azotobacter* improves shelf life of tomato (*Lycopersicon esculentum*). *Indian J. Agric. Sci.* 71(12): 765-767

Chen-XiaoBin, X.B., Zhang, B.X., Lou, B.G. and Li, J.Y. 1999. Effect of plant growth-promoting rhizobacteria on disease control of cucumber seedlings. *J. Zhejiang Univ. Agric. Life Sci.* 25(6). 578-582

Day, J.M. and Dobereiner, J. 1977. Physiological aspects of N_2 fixation by *Spirillum* from *Digitaria* roots. *Soil Biol. Biochem.* 7: 107-112

Deka, B.C., Bora, G.C. and Shadeque, A. 1996. Effect of *Azospirillum* on growth and yield of chilli (*Capsicum annum*) cv. Pusa Jwala. *Haryana J. Hort. Sci.* 25(2): 44-46

Devi, H.J., Maity, T.K. and Paria, N.C. 2003. Effect of different sources of nitrogen on yield and economics of cabbage. *Environ. Ecol.* 21(4): 878-880

Dubeikovsky, A.N., Mordukhova, E.A., Kochethov, V.V., Polikarpova, F.Y. and Boronin, A.M. 1993. Growth promotion of black currant soft wood cuttings by recombinant strain *Pseudomonas fluorescens* BSP53a synthesizing an increased amount of indole-3-acetic acid. *Soil Biol. Biochem.* 25: 1277-1281

Gaikwad, R.M. and Wani, P.V. 2001. Response of brinjal (cv. Krishna) to phosphate solubilizing biofertilizers. *J. Maharashtra Agric. Univ.* 26(1): 29-32

Ganeshe, R.K., Pandey, R.P. and Rawat, A.K. 1998. Comparative performance of biofertilizers and chemical nitrogen on the yield of okra (*Abelmoschus esculentus* (L. Moench). *Hort. J.* 11(2): 53-58

Ganeshe, R.K., Pandey, R.P. and Rawat, A.K. 2000. Influence of biofertilizers and nitrogen on growth, yield, shelf life and economics of okra (*Abelmoschus esculentus* L. Moench). *JNKVV Res. J.* 32(1/2): 33-37

Glick, B.R. 1995. The enhancement of plant growth by free-living bacteria. *Canadian Dis.* 84: 306-308

Handelsman, J. and Stabb, E.V. 1996. Biocontrol of soil born plant pathogens. *Plant Cell.* 8: 1855-1869

Harikrishna, B.L., Channal, H.T., Hebsur, N.S., Dharmatti, P.R. and Sarangamath, P.A. 2002. Integrated nutrient management (INM) on availability of nutrients, uptake and yield of tomato. *Karnataka J. Agric. Sci.* 15(2): 275-278

Jetiyanon, K. and Kloepper, J.W. 2002. Mixtures of plant growth-promoting rhizobacteria for induction of systemic resistance against multiple plant diseases. *Biol. Control.* 24(3): 285-291

Jha, A.K., Pal, N., Saxena, A.K., Singh, D. and Jha, G.K. 2006. Co inoculation effect of VAM and PGPR on growth and yield of onion. *Indian J. Hort.* 63(1): 44-47

Kamali, I.A., Zargar, M.Y. and Chattoo, M.A. 2002. Effect of microbial inoculants, chemical nitrogen and their combination on brinjal (*Solanum melongena* L). *Veg. sci.* 29(1): 87-89

Kandan, 2000. Induction of systemic resistance against tomato spotted wilt virus (TSWV) in tomato by fluorescent *Pseudomonas* strains. M.Sc.(Ag) thesis, Tamil Nadu Agricultural University. Coimbatore. 117p

Kavitha, K., Nakkeeran, S., Chandrasekar, G., Fernando, W.G.D., Mathiyazhagan, S., Renukadevi, P. and Krishnamoorthy, A.S. 2003. Exploitation of *Pseudomonas chlororaphis* and *Bacillus subtilis* for the management of damping off of chilli (*Capsicum annuum* L.). *Proceeding of 6th international plant growth promoting rhizobacetria* (PGPR) *workshop,* 5-10 Oct., Calicut, India, pp.153-156

Kloepper, J.W. 2003. A review of mechanisms for plant growth promoting by PGPR. *Proceeding of 6th international plant growth promoting rhizobacetria* (PGPR) *workshop,* 5-10 oct., Calicut, India, 81-90

Kloepper, J.W. and Schroth, M.N. 1981. Development of a powder formulation of rhizobacteria for inoculation of potato seed pieces. *Phytopathology.* 71: 590-592

Kokalis-Burelle, N., Vavrina, C.S., Rosskopf, E.N. and Shelby, R.A. 2002. Field evaluation of plant growth-promoting Rhizobacteria amended transplant mixes and soil solarization for tomato and pepper production in Florida. *Pl. Soil.* 2002. 238(2): 257-266

Lazarovits, G. and Nowak, J. 1997. Rhizobacteria for improvement of plant growth and establishment. *Hort. Sci.* 32: 188-192

Loper, J.E., Nowak-Thomson, B., Whistler, C.A., Hagen, M.J., Corbell, N.A., Henkels, M.D. and Stockwell, V.O. 1997. Biological control mediated by antifungal metabolite production and resource competition: an overview. In: Ogoshi, A., Kobayashi, K., Homma, Y., Kodama, F., Kondo, N. and Akino, S. (eds.) *Plant growth promoting rhizobacteria – present status and future prospects.* Fac Agricultural Hokkaido University, Sapporo, pp. 73-79

Lucas-Garcia, J.A., Probanza, A., Ramos, B., Ruiz-Palomino, M. and Gutierrez Manero, F.J. 2004. Effect of inoculation of *Bacillus licheniformis* on tomato and pepper. *Agronomie.* 24(4): 169-176

Mahanthesh, B., Venkatesha, J., Radhakrishna, D., Umesha, K. and Veere Gowda, R. 2005. Studies on the impact of bio fertilizers with levels of NPK on onion (*Allium cepa* L.) cv. Bellary Red. *Proceedings of national symposium on biotechnological interventions for improvement of horticultural crops: Issues and Strategies.* 10-12 Jan, Kerala Agricultural University, Thrissur, Kerala, pp. 369-372

Meena, B. and Marimuthu, T. 2005. Induction of systemic resistance against die-back and fruit rot of chillies by *Pseudomonas fluorescens. Proceedings of national symposium on biotechnological interventions for improvement of horticultural crops: Issues and Strategies.* 10-12 Jan, Kerala Agricultural University, Thrissur, Kerala, pp. 359-361

Milner, J.L., Silo-Shu, L.A., Lee, J.C., He, H., Clardy, J. and Handelsman, J. 1998. Production of kanosamine by *Bacillus cereus* UW85. *Appl. Environ. Microbiol.* 62: 3061-3065

Murphy, J.F., Zehnder, G.W., Schuster, D.J., Sikora, E.J., Polston, J.E. and Kloepper, J.W. 2000. Plant growth-promoting rhizobacterial mediated protection in tomato against tomato mottle virus. *Plant Dis.* 84(7): 779-784

Muthuraju, R., Boby, V.U., Suvarna, V.C. and Jayasheela, N. 2002. Interactive effect of *Glomus mosseae, Pseudomonas fluorescens* and *Azospirillum brasilense* on growth and yield of tomato. *J. Soil Biol. Ecol.* 22(1&2): 8-15

Naidu, A.K., Kushwah, S.S. and Dwivedi, Y.C. 2000. Performance of organic manures, bio and chemical fertilizers and their combinations on microbial population of soil and growth and yield of okra. *JNKVV Res. J.* 33(1/2): 34-38

Naidu, A.K., Kushwah, S.S. and Dwivedi, Y.C. 2002. Influence of organic manures, chemical and biofertilizers on growth, yield and economics of brinjal. *S. Indian Hort.* 50(4/6): 370-376

Nallathambi, P., Umamaheswari, C., Joshi, H.K. and Dhandar, D.G. 2003. Evaluation of fluorescent *Pseudomonas* against virus diseases of Mateera (Watermelon). *Proceeding of 6th international plant growth promoting rhizobacteria* (PGPR) *workshop,* 5-10 Oct., Calicut, India, pp.181-184

Nanthakumar, S. and Veeraragavatham, D. 2000. Effect of integrated nutrient management on growth parameters and yield of brinjal (*Solanum melongena* L.) cv. PLR.1. *S. Indian Hort.* 48(1/6): 163-167

Narayanamma, M., Chiranjeevi, C.H., Reddy, I.P. and Riazuddin Ahmed, S. 2005. Integrated nutrient management in cauliflower (*Brassica oleracea* var. *Botrytis* L.). *Veg. Sci.* 32(1): 62-64

Navale, A.M. and Wani, P.V. 2004. Effect of *Glomus mosseae* and *Azospirillum lipoferum* inoculation on growth and yield of onion (*Allium cepa* L.) cv. B-780. *Veg. Sci.* 32(2): 186-189

Nelson, E.B. and Maloney, A.P. 1992. Molecular approaches for understanding biological control mechanisms in bacteria: studies of the interaction of *Enterobacter cloacae* with *Pythium ultimum. Canadian J. Plant Path.* 14: 106-114

Nirmala, D., Singh. R.K., Kumar, A. and Singh, J. 2006. Effect of inorganic and biofertilizers on biomass, quality and yield parameters of vegetable pea (*Pisum sativum* L.). *Int. J. Agric. Sci.* 2(2): 618-620

Nirmala, R. and Vadivel, E. 1999. Effect of combined application of organic manure and biofertilizers on growth and productivity of cucumber. *S. Indian Hort.* 47:(1) 252-254

Okon, Y., Albrecht, S.L. and Burris, R.H. 1976a. Factors affecting growth and nitrogen fixation of *Spirillum lipoferum. J. Bact.* 127: 1248-1254

Pardeep, K. and Sharma, S.K. 2004. Effect of phosphorus sources on cabbage-tomato cropping sequence at Solan. *Haryana J. Hort. Sci.* 33(3/4): 272-273

Parvatham, A., Vijayan, K.P. and Nazar, A. 1989. Effect of *Azospirillum* on growth and nutrient uptake of Pusa Sawani bhendi (*Abelmoschus esculentus* L. Moench). *S. Indian Hort.* 37: 227-229.

Prabu, T., Narwadkar, P.R. and Sajindranath, A.K. 2003. Economics of integrated nutrient management in okra. *J. Maharashtra Agric.Univ.* 27(3): 316-318.

Prabu, T., Narwadkar, P.R., Sajindranath, A.K. and Rafi, M. 2003. Effect of integrated nutrient management on growth and yield of okra (*Abelmoschus esculentus* L. Moench) cv. Parbhani Kranti. *Orissa J. Hort.* 31(1): 17-21.

Rabindra, K. and Srivastava, B.K. 2006. Residual effect of integrated nutrient management on growth, yield and yield attributes of tomato. *Indian J. Hort.* 63(1): 98-100.

Ramamoorthy, V., Raguchander, T. and Samiyappan, R. 2002. Induction of defense-related proteins in tomato roots treated with *Pseudomonas fluorescens* Pf1 and *Fusarium oxysporum* f.sp. *lycopersici. Plant Soil.* 239: 55-68.

Ramesh, R. and Korikanthimath, V.S. 2003. Use of rhizosphere bacteria and soil fungi for the management of damping off and wilt in brinjal. *Proceeding of 6th international plant growth promoting rhizobacteria* (PGPR) *workshop,* 5-10 Oct., Calicut, India, pp. 309-316.

Raupach, G.S. and Kloepper, J.W. 1998. Mixtures of plant growth-promoting rhizobacteria enhance biological control of multiple cucumber pathogens. *Phytopathology.* 88(11): 1158-1164.

Raut, R.L., Naidu, A.K., Jain, P.K. and Rajwade, V.B. 2003. Influence of organic and chemical sources of nutrients on the yield, fruit quality and storage of tomato in Madhya Pradesh, India. *JNKVV Res. J.* 37(1): 30-33.

Sajindranath, A.K., Narwadkar, P.R., Prabu, T. and Rathod, N.G. 2002. Effect of biofertilizers and growth regulators on germination in okra. *S. Indian Hort.* 50(4/6): 538-542.

Sally, K.M., Sadhankumar, P.G., Nirmala Devi, S. and Krishnan, S. 2005. Influence of beneficial microbes on damping off disease and seedling vigour in solanaceous vegetables. *Proceedings of national symposium on biotechnological interventions for improvement of horticultural crops: Issues and Strategies.* 10-12 Jan., Kerala Agricultural University, Thrissur, Kerala, pp. 391-392.

Selvi, D., Rani, P. and Perumal, R. 1997. Microfood with and without organics and biofertilizers on growth and development of bhendi. *Madras Agric. J.* 84(10): 625-626.

Selvi, D., Thiageshwari, S., Santhy, P. and Rajkannan, B. 2004. Effect of integrated nutrient management on yield and economics in okra in an Inceptisol. *J. Maharashtra Agric. Univ.* 29(2): 122-124.

Sendhilvel, V., Buvaneswari, D. Kanimozhi, S. Mathiyazhagan, S., Kavitha, K. and Raguchander, T. 2005. Manegement of cowpea root-rot caused by *Macrophomina phaseolina* (Tassi) Goid. using plant growth promoting rhizobacteria. *J. Biol. Control.* 19(1): 41-46.

Sengupta, S.K., Dwivedi, Y.C. and Kushwah, S.S. 2002. Response of tomato (*Lycopersicon esculentum* Mill.) to bioinoculants at different levels of nitrogen. *Veg. Sci.* 29(2): 186-188

Shalini, S.B., Channal, H.T., Hebsur, N.S., Dharmatti, P.R. and Sarangamath, P.A. 2002. Effect of integrated nitrogen management on yield of knolkhol and population of *Azospirillum* in vertisol. *Karnataka J. Agric. Sci.* 15(1): 151-153

Shankaranarayanan, R., Ahmed, S.H. and Alagesan, V. 1995. Effect of *Azospirillum* on improved varieties of bhendi. *S. Indian Hort.* 45: 52-53

Sharma, H.K., Pankaj, K. and Mishra, S.D. 2003. Effect of plant growth promoter microbes on the root-knot nematode Meloidogyne incognita on okra. *Curr. Nematology.*14(1/2): 57-60

Sharma, S.K. 2002. Effect of *Azospirillum, Azotobacter* and nitrogen on growth and yield of cabbage (*Brassica oleracea* var. *capitata*). *Indian J. Agrl. Sci.* 72(9): 555-557

Shojai, M., Ostovan, H., Zamanizadeh, H., Labbafi, Y., Nasrollahi, A., Ghasemzadeh, M. and Rajabi, M.Z. 2003. The management of cucumber and tomato intercrops and implementation of non-chemical control of pests and diseases for organic crop production in the greenhouse. *J. Agric. Sci. Islamic Azad Univ.* 9(2): 1-39

Shylaja, Rao, M.S., Dhananjay, N., Redddy, M.S. and Rahiman, A.B. 2003. Induction of systemic resistance in a susceptible cultivar of tomato by *Bacillus subtilis. Proceeding of 6th international plant growth promoting rhizobacteria* (PGPR) *workshop,* 5-10 Oct., Calicut, India, pp. 591-595

Siddiqui, Z.A. 2004. Effects of plant growth promoting bacteria and composed organic fertilizers on the reproduction of *Meloidogyne incognita* and tomato growth. *Bioresource Technol.* 95(2): 223-227

Singh, K. and Khare, R.K. 2003. Effect of *Azospirillum* and N on growth and yield of cabbage (*Brassica Oleraceae* Var. *Capitata*) under rainfed condition. *Proceedings of 6th international plant growth promoting rhizobacteria (PGPR) workshop,* Oct 5-10, Calicut, Kerala, India, pp. 109-110

Singh, V. and Singh, S.S. 2005. Effect of inorganic and biofertilizers on production of cauliflower (*Brassica oleracea* L var. *botrytis*). *Veg. Sci.* 32(2): 146-149

Somani, L.L., Bhandari, S.C., Saxena, S.N. and Vyas, K.K. 1990. *Biofertilizers.* Scientific Publishers, Jodhpur, 361p

Subba Rao, N.S. 1982. *Biofertilizers in agriculture.* Oxford and IBM Publishing Co. Ltd. Delhi. 278p

Subba Rao, N.S. 1997. *Soil microorganisms and plant growth.* Oxford and IBH Publishing Co. Pvt. LTD. New Delhi. 311p

Swamiyappan, R., Harish, S., Radjacommare, R. and Saravanakumar, D. 2005. Exploitation of plant growth promotiong rhizobacteria (PGPR) in horticultural crops. *Proceedings of national symposium on biotechnological interventions for improvement of horticultural crops: Issues and Strategies.* 10-12 Jan., Kerala Agricultural University, Thrissur, Kerala, pp. 30-41

Swani, A.K., Pattanayak, S.K., Jena, M.K. and Nayak, R.K. 2003. Effect of integrated use of bio-inoculants and fertilizer nitrogen on growth, yield and nitrogen economy of okra. *J. Indian Soc. Soil Sci.* 51(2): 145-150

Thilakavaty, S. and Ramaswamy, N. 1999. Effect of inorganic and biofertilizers on yield and quality parameters of multiplier onion (*Allium cepa* var. *aggregatum*). *Veg. Sci.* 26(1): 97-98

Tomczyk, A. 1999. *The use of plant growth promoting rhizobacteria (PGPR) to decrease the susceptibility of cucumber to spider mites.* International Organization for Biological and Integrated Control of Noxious Animals and Plants (OIBC/OILB), West Palaearctic Regional Section (WPRS/SROP). 22(1): 251-254

Tomczyk, A. and Kielkiewicz, M. 2001. Changes in content of proteins and free amino acids in the foliage of mite-infested glasshouse cucumber and tomato treated with plant growth-promoting rhizobacteria (PGPR). *J. Plant Prot. Res.* 41(1): 61-65

Van Loon, L.C., Bakker, P.A.H.M. and Pieterse, C.M.J. 1998. Systemic resistance induced by rhizosphere bacteria. *Annu. Rev. Phytopathology.* 36: 453-483

Van Peer, R. and Schippers, B. 1998. Plants growth response in bacterization with selected *Pseudomonas* spp. strains and rhizosphere microbial development in hydroponic cultures. *Canadian J. Microbiol.* 35: 456-463

Venkateswarlu, B. and Rao, A.V. 1983. Response of pearl millet to inoculation with different strains of *Azospirillum brasilense. Plant Siol.* 74: 379-386

Verma, T.S., Thakur, P.C. and Singh, A. 1997. Effect of biofertilizers on vegetable and seed yield of cabbage. *Veg. Sci.* 24(1): 1-3

Wange, S.S. and Kale, R.H. 2003. Response of okra and bittergourd to diazotrophs with graded levels of nitrogen. *J. Soil Crops.* 13(2): 271-274

Wange, S.S. and Kale, R.H. 2004. Effect of microbial inoculants with graded levels of inorganic nitrogen on broccoli and lettuce. *J. Soil Crops.* 14(1): 18-21

Zehnder, G.W., Yao-ChangBin, Murphy, J.F., Sikora, E.R., Kloepper, J.W., Schuster, J., Polston, J.E., Yao, C.B. and Agrawal, A.A. 1999. Microbe-induced resistance against pathogens and herbivores: evidence of effectiveness in agriculture. (ed.); Tuzun, S. (ed.); Bent, E. *Induced-plant-defenses-against-pathogens-and-herbivores: biochemistry, ecology and agriculture.* AD: Department of Entomology, Auburn University, Auburn, AL 36849, USA, pp. 335-355.

Horticulture Biotechnology

The Science of Horticulture : Vol. 1, 2011
© K.V. Peter (ed.), pp. 289-302
New India Publishing Agency, New Delhi (India)
E-mail : newindiapublishingagency@gmail.com
Web: www.bookfactoryindia.com

Chapter 12 Methodology for Gene Sequencing in Plants

JAYALAKSHMY, V.G.

The term DNA sequencing refers to methods for determining the order of the nucleotide bases, adenine, guanine, thymine and cytosine, in a molecule of DNA. In molecular terms ,the definitive understanding of a DNA molecule comes from determining its nucleotide sequence.The function of a gene can be deduced from its nucleotide sequence,for example by comparing the sequences of genes with known functions.

The first DNA sequences were obtained by academic researchers, using laborious methods based on 2 - dimensional chromatography in the early 1970s. Following the development of dye-based sequencing methods with automated analysis, DNA sequencing has become easier and orders of magnitude faster. Knowledge of DNA sequences of genes and other parts of the genome of organisms have become indispensable for basic research studying biological processes, as well as in applied field like diagnostic or forensic research. The advent of DNA sequencing has significantly accelerated biological research and discovery. The rapid speed of sequencing attained with modern DNA sequencing technology has been instrumental in the sequencing of the human genome, in the Human Genome Project. Related projects, often by scientific collaboration across continents, have generated the complete DNA sequences of many animal, plant and microbial genomes.

RNA sequencing, which was technically easier to perform than DNA sequencing, was one of the earliest forms of nucleotide sequencing. The major landmark of RNA sequencing is the sequence of the first complete gene and the complete genome of Bacteriophage MS2, identified and published by Walter Fiers and his coworkers at the University of Ghent (Ghent, Belgium), between 1972 and 1976.Prior to the development of rapid DNA sequencing methods in the early 1970s by Frederick Sanger at the University of Cambridge in England and Walter Gilbert and Allan Maxam at Harvard, a number of laborious methods were used. For instance, in 1973, Gilbert and Maxam reported the sequence of 24 basepairs using a method known as wandering-spot analysis.The chain-termination method developed by Sanger and coworkers in 1975 soon became the method of choice, owing to its relative ease and reliability.

I. Maxam and Gilbert's Chemical Degradation Method

In 1976-1977, Allan Maxam and Walter Gilbert developed a DNA sequencing method based on chemical modification of DNA and subsequent cleavage at specific bases. Although Maxam and Gilbert published their chemical sequencing method two years after the ground-breaking paper of Sanger and Coulson on plus-minus sequencing, Maxam-Gilbert sequencing rapidly became more popular, since purified DNA could be used directly, while the initial Sanger method required that each read start be cloned for production of single-stranded DNA. However, with the improvement of the chain-termination method, Maxam-Gilbert sequencing has fallen out of favour due to its technical complexity prohibiting its use in standard molecular biology kits, extensive use of hazardous chemicals and difficulties with scale-up.

The method requires radioactive labelling at one end and purification of the DNA fragment to be sequenced. Chemical treatment generates breaks at a small proportion of one or two of the four nucleotide bases in each of four reactions (G, A+G, C, C+T). Thus a series of labeled fragments is generated, from the radiolabelled end to the first 'cut' site in each molecule. The fragments in the four reactions are arranged side by side in gel electrophoresis for size separation. To visualize the fragments, the gel is exposed to X-ray film for autoradiography, yielding a series of dark bands each corresponding to a radio labeled DNA fragment, from which the sequence may be inferred. Also sometimes known as 'chemical sequencing', this method originated in the study of DNA-protein interactions (footprinting), nucleic acid structure and epigenetic modifications to DNA, and within these it still has important applications.This chemical cleavage method uses double-stranded DNA samples and so does not require cloning of DNA into an M13 phage vector to produce single-stranded DNA as is the case with the Sanger-Coulson method. It involves modification of the bases in DNA followed by chemical base-specific cleavage.

This techniques involve the following steps:

Stages

1. Double-stranded DNA to be sequenced is labelled by attaching a radioactive phosphorus (^{32}P) group to the 5' end. Polynucleotide kinase enzyme and ^{32}P-dATP are used here.

2. Using dimethyl sulphoxide and heating to 90°C, the two strands of the DNA are separated and purified (*e.g.* using gel electrophoresis and the principle that one of the strands is likely to be heavier than the other due to the fact that it contains more purine nucleotides (A and G) than pyrimidines (C and T) which are lighter.

3. The mixture is divided into 4 samples, each treated with a different reagent having the property of destroying either only G, or only C, or A and G or T and C. This part of the process involves alteration of bases (*e.g.* dimethylsulphate methylates guanine) followed by removal of altered bases. Lastly, piperidine is used for cleavage of the strand at the points where bases are missing.

Base specificity	Chemicals used for base alteration	Chemicals used for altered base removal	Chemicals used for strand cleavage
G	Dimethylsulphate	Piperidine	Piperidine
A+G	Acid	Acid	Piperidine
C+T	Hydrazine	Piperidine	Piperidine
C	Hydrazine + alkali	Piperidine	Piperidine
A>C	Alkali	Piperidine	Piperidine

4. If reactions are arranged to give only one, or a few, cleavages per DNA molecule, a nested set of end-labelled DNA fragments of different lengths is produced.

5. The samples are run together on a sequencing gel which separates the fragments by electrophoresis depending on their size. DNA bands in the gel are visualized by autoradiography (^{32}P-labelled 5' end fogs photographic film). and the sequence is determined from positions of bands in four lanes.

II. Sanger's Dideoxy Nucleotide Synthetic Method

Freidrich Sanger had initially developed a method for DNA sequencing, which utilized DNA polymerase to extend DNA chain length. This was termed 'Plus-minus method'. Subsequently he developed a more powerful method, utilizing single stranded DNA as template for DNA synthesis in which 2', 3' dideoxy nucleotides were incorporated leading to termination of DNA synthesis. These dideoxy nucleotides are used as triphosphates (ddNTP) and can be incorporated in a growing chain, but they terminate synthesis, since they cannot form a phosphodiester bond with next incoming deoxynucleotide triphosphate (dNTP). A mixture of different length fragments is produced depending where the interruptions occurred. As with the Maxam-Gilbert method, the mixture of fragments is run on a gel and the sequence read off.

Following steps are involved in Sanger's dideoxy method for DNA sequencing.

1. Sample DNA to be sequenced is spliced into M13 vector DNA. Infected *E. coli* host cells release phage particles containing single-stranded (ss) recombinant DNA including the sample sequence. DNA is extracted from phage for sequencing.

2. A short oligonucleotide primer (usually chemically synthesized and sometimes labelled) is added to the ss recombinant DNA. The primer anneals at a position which will act as the starting point for synthesis of the complementary strand. (The DNA polymerase to be used requires a primer to initiate complementary strand synthesis.)

3. DNA polymerase (*e.g.* Klenow fragment of DNA polymerase I or something similar) is then added in the presence of:

   a) The 4 normal nucleotides: d-ATP, d-CTP, d-GTP and d-TTP (one or more of which are labelled with ^{32}P).

   b) A low concentration of 4 analogues of the normal nucleotides are prepared in separate incubation mixes. The analogues are dideoxynucleotides (ddNTP) which are identical to normal nucleotides except that the hydroxyl groups (OH) in the sugar ring are replaced with hydrogens (H). Just like the normal substrates, the analogues (dd-ATP, dd-CTP, dd-GTP and dd-TTP) have A, C, G or T bases attached. The DNA polymerase can use the analogues as substrates and cannot in fact distinguish them from the normal nucleotides.

4. Complementary strand synthesis occurs away from the primer (and away from the 5' end). However, when an analogue becomes incorporated into the new strand, chain termination occurs and further

synthesis ceases because the 3' end of the analogue lacks the necessary hydroxyl group to allow further chain extension, unlike the normal nucleotide. Therefore, in each of the 4 separate incubation mixes an assortment of partially synthesized radioactively labelled double stranded DNA molecules will be found. The DNA fragments will vary in length depending at which point the analogue becomes incorporated. Since incorporation is random, the population of molecules in a mix should represent every position of a particular base.

5. Each of the 4 mixes is run together on a sequencing gel which separates the fragments by electrophoresis depending on their size. The gel contains urea which causes denaturation of the double-stranded DNA. The process is carried out at high voltage to generate heat which prevents the strands re-associating. This is done for all the four reaction mixtures on adjoining lanes in the gel.

DNA bands in the gel are visualized by autoradiography (^{32}P-label fogs photographic film). The gel is used for autoradiography so that the position of different bands in each lane can be visualized.The DNA sequence is read directly from the gel in a similar way to a Maxam-Gilbert sequencing gel. A dark band in a lane indicates a DNA fragment that is the result of chain termination after incorporation of a dideoxynucleotide (ddATP, ddGTP, ddCTP, or ddTTP). The relative positions of the different bands among the four lanes are then used to read (from bottom to top) the DNA sequence.

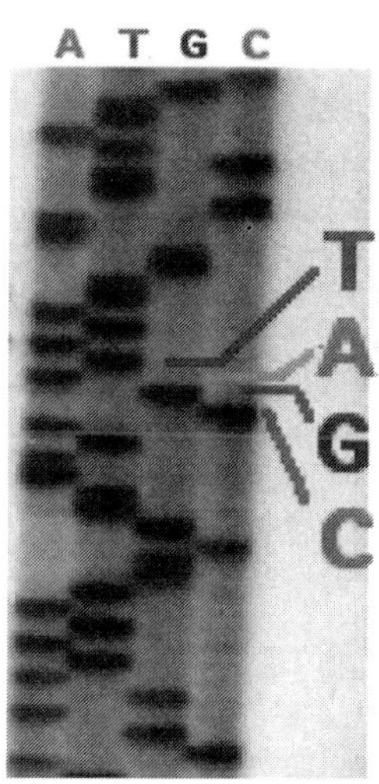

Part of a radioactively labelled sequencing gel

Technical variations of chain-termination sequencing include tagging with nucleotides containing radioactive phosphorus for radiolabelling, or using a primer labeled at the 5′ end with a fluorescent dye. Dye-primer sequencing facilitates reading in an optical system for faster and more economical analysis and automation. The later development by Leroy Hood and coworkers of fluorescently labeled ddNTPs and primers set the stage for automated, high-throughput DNA sequencing

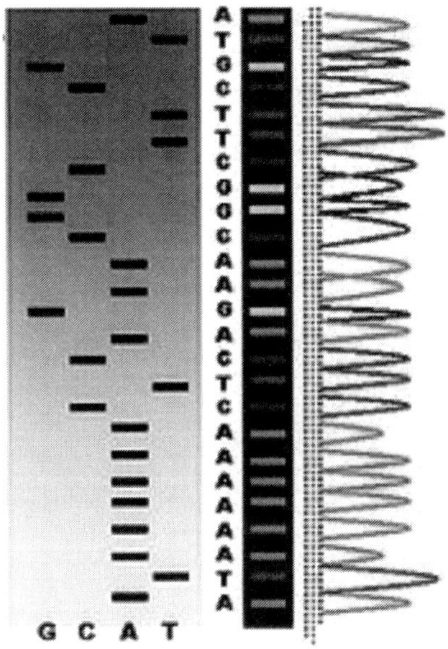

Sequence ladder by radioactive sequencing compared to fluorescent peaks

III. Automatic DNA Sequencers

A variant of the above dideoxy method was later developed ,which allowed the production of automatic sequencers (eg.Perkin Elmer's ABI prism 310 or 3700 Genetic analyzer ; Amersham Pharmacia Biotech's MegaBACE 1000 and 4000). In this new approach different flurescent dyes are tagged either to the oligonucleotide primer (dye primers) in each of the four reaction tubes (blue for A,red for C etc.) or to each of the four ddNTPs (dye terminators) used in a single reaction tube ; when four tubes are used, they are pooled. After the PCR reaction is over, the reaction mixture is subjected to separation of synthesized fragments through electrophoresis..

Dye-terminator sequencing utilizes labelling of the chain terminator ddNTPs, which permits sequencing in a single reaction, rather than four reactions as in the labelled-primer method. In dye-terminator sequencing, each of the four dideoxynucleotide chain terminators is labelled with fluorescent dyes, each of which with different wavelengths of fluorescence and emission. Owing to its greater expediency and speed, dye-terminator sequencing is now the mainstay in automated sequencing. Its limitations include dye effects due to differences in the incorporation of the dye-labelled chain terminators into the DNA fragment, resulting in unequal peak heights and shapes in the electronic DNA sequence trace chromatogram after capillary electrophoresis. This problem has been addressed with the use of modified DNA polymerase enzyme systems and dyes that minimize incorporation variability, as well as methods for eliminating "dye blobs". The dye-terminator sequencing method, along with automated high-throughput DNA sequence analyzers, is now being used for the vast majority of sequencing projects

Depending upon the electrophoretic system used, whether slab gel electrophoresis or capillary electrophoresis, following two types of automatic sequencing systems have been designed

1.Slab Gel Sequencing Systems

These systems make use of ultra thin (75mm) slab gels and involve running of atleast 96 lanes per gel, the runs being completed twice daily. In these systems, automation in sample loading of sequencing gels has also been achieved, by using a plexiglass block having wells that are same distance apart as the comb teeth cut in a porous membrane, used as a comb for drawing samples by capillary action. Each well in plexiglass block is filled with a sample (PCR dideoxy reaction mixture), so that when the porous membrance comb is lowered into the sample wells in the plexiglass block, the samples are drawn up automatically into the comb teeth by capillary action. Using this approach of employing porous combs, automated loading of upto192, 384 or 480 samples per gel has been achieved. The porous comb with the samples is placed between the glass plates of the gel apparatus above the flat surface of the polymerized gel and the samples are driven from the comb into the gel by electrophoresis.

2.Capillary Gel Sequencing Systems

In these sytems slabgel electrophoresis is replaced by capillary gel electrophoresis to analyse DNA samples.In these systems ,instead of scanning DNA as it migrates through 96 lanes each in a series of poured slabgels, DNA is sent through a set of 96 capillary tubes filled with a polymerized gel. As the DNA fragments pass through these capillary tubes ,DNA fragments are scanned.In the original models of the above old slab gel machines, gels must be poured and reagents frequently loaded interrupting the sequencing. In capillary gel sequencing systems, the robot moves the DNA samples and reagents through the tubes continuously, requiring attention only once a day.

The system produces a steady flow of data, each representing one of the four DNA bases.

IV. Direct DNA Sequencing Using PCR

Polymerase Chain Reaction has also been used for sequencing the amplified DNA product. This method also called ligation mediated PCR (LMPCR). This is more fast and reliable and can utilize either the whole genomic DNA or cloned fragments for sequencing a particular DNA segment. It involved 2 steps.

i) Generation of sequencing templates (double/single stranded) using PCR

ii) Sequencing of PCR products either with the thermplabile DNA polymerase or with the thermostable Taq DNA polymerase. Thus, the DNA sequencing method using PCR eliminates the need of cloning the DNA in single stranded DNA phage vector ie, M13.

Although double stranded DNA product of PCR can be utilized for sequencing, this may cause difficulty due to reassociation after denaturation thus preventing the sequencing primer from annealing to its complementary sequence to allow extension. To reduce this problem, either a variant of the standard method for sequencing double stranded DNA is employed or single stranded DNA templates are produced following asymmetric PCR.

A number of thermolabile DNA polymerases are used for sequencing of *in vitro* amplified DNA. Alternatively, Taq DNA polymerase (thermostable enzyme used for PCR, can also be utilized for sequencing reactions. In either case Sanger's synthetic method involving dideoxy nucleotide (ddNTP) for chain termination , is used for sequencing.However Maxam and Gilbert's chemical degradation method can also be used.

V. DNA Sequencing Through Transcription

DNA sequencing could be successfully achieved using RNA polymerase enzyme for transcription, instead of using 'sequenace' or 'thermo sequenace' used in Sanger's dideoxy nucleotide method of chain termination. This method of sequencing described as 'transcriptional sequencing' makes use of 3′ d NTPs for terminating transcription in the same manner as ddNTPs are used for terminating DNA synthesis in Sanger's method (Whenever 3′dNTP is available at 3′ position , no OH or PO4 group will be available for the phosphodiester bond formation).The 3′ dNTPs can be labeled with fluorescent (rhodamine) dyes , so that four dNTPS with four colour dyes (TMR-3′dUTP, ROX-3′dCTP, R6G-3′dATP, R110-3′dGTP) can be used along with sufficient quantities of all the four NTPs (ribonucleoside triphosphates).This will alow the RNA synthesis to proceed normally with the help of RNAP, till occasionally a 3′dNTP (dye labeled) is incorporated. This will yield RNA chains of different lengths, each having at its terminus, a specific dNTP, which will be identified by its colour in an automatic sequencer.

This method is believed to be superior, firstly because incorporation of NTP during transcription does not require a primer, thus making it faster (240 bases/sec) than by Taq polymerase (60 bases/sec.). Secondly because a large amount of product can be obtained from small template (600 hundred molecules of RNA from one molecule of DNA) and thirdly, because direct PCR sequencing can be done, without the elimination of primers, and 2′ dNTPs. The template in this method will have to be prepared by PCR using primers, to which phage promoter sequences are appended, which facilitate binding of RNAP for transcription. The residual primers and 2′-dNTPs need not be removed, because they will not interfere with RNA synthesis during transcription.

VI. DNA Sequencing by Hybridization Using Microarrays on DNA Chips

The main objectives of several genome projects is the large scale sequence analysis of genomes and its application to individual case studies in medicine and research. Sequencing by hybridization for detection of single nucleotide polymorphisms (SNPs) using DNA microchips was developed to meet this objective. A sequencing microchip is developed either by immobilizing or by parallel synthesis of a complete set of 8-mers (ie. $4^8 = 65536$) on a solid support. The dimension of the microchip can be further reduced by using printing technology and robotics. The technique involves the hybridization of a short fragment of unknown sequence with a set of short oligonucleotides, particularly 8 mers of known sequences. Identification and analysis of a set of overlapping oligomers that form perfect duplexes with the DNA of interest permits reconstruction of the target DNA sequence.For example, if we wish to determine the sequence of a 17 base target DNA fragment , and hybridize it with complementary ,overlapping 8-mers to form perfect duplexes , the overlap of seven or fewer adjacent 8-mers showing perfect match enables reconstruction of the complementary DNA sequences of the fragment.

Oligonucleotides shorter or longer than the 8-mers can also be used for sequencing, but while the use of shorter oligos leads to decrease in stability of the duplexes formed, the use of longer oligos leads to loss of discrimination between perfect base pair matches and mismatches. Therefore, the sequencing microchip is mainly composed of 8 - mers. The hybridization is generally carried out at lower temperature, so that discrimination between perfect and imperfect duplexes can be achieved.

The limitation of the above technique is that the fragment to be sequenced should be small in size, although it should still be longer than n-1 where n is the length of the olligonucleotides immobilized on microchip. Analysis of long DNA fragments may be facilitated by fragmentation of the target DNA into smaller pieces prior to hybridization experiments. However, the method cannot be applied to repeat DNAs such as microsatellites or minisatellites.

VII. DNA Sequencing by DE – MALDI – TOF Mass Spectrometry

Matrix Assisted Laser Desorption/ionisation mass spectrometry has the potential to rapidly acquire DNA sequence information. It is used successfully for conformation of the sequences of short synthetic oligonucleotides by mass analysis of products resulting either from i) synthesis failures (as in sanger's method) or due to ii) time -dependent exonuclease digestion. In both cases different products (oligonucleotides) will differ by single nucleotides. The mass of each fragment can be accurately determined and the fragments can be arranged by differences of single nucleotides that can be identified by mass differences. This can be applied only to short oligonucleotides, since the resolution and sensitivity of MALDI-TOF go down with increasing size.

For the purpose of DNA sequencing, MALDI-TOF mass spectrometry can also be used in combination with Sanger's termination reactions. In this approach ,mass spectra of the four Sanger's sequencing reactions are overlaid and each sized products connected to one of the four bases. The advantage of this approach is that MALDI-TOF MS is quick,requiring only minutes per sample and no gels,radioactivity or fluorescent labeling is required. Although in theory this method works, in practice, problems are faced due to the presence of buffer salts and other components in the reaction mixture.

VIII. Large-Scale Sequencing Strategies

Current methods can directly sequence only relatively short (300-1000 nucleotides long) DNA fragments in a single reaction. The main obstacle to sequencing DNA fragments above this size limit is insufficient power of separation for resolving large DNA fragments that differ in length by only one nucleotide.

Genomic DNA is fragmented into random pieces and cloned as a bacterial library. DNA from individual bacterial clones is sequenced and the sequence is assembled by using overlapping DNA regions.

Large-scale sequencing aims at sequencing very long DNA pieces, such as whole chromosomes. Common approaches consist of cutting (with restriction enzymes) or shearing (with mechanical forces) large DNA fragments into shorter DNA fragments. The fragmented DNA is cloned into a DNA vector, and amplified in *Escherichia coli*. Short DNA fragments purified from individual bacterial colonies are individually sequenced and assembled electronically into one long, contiguous sequence. This method does not require any pre-existing information about the sequence of the DNA and is referred to as *de novo* sequencing. Gaps in the assembled sequence may be filled by primer walking. The different strategies have different tradeoffs in speed and accuracy; *shotgun methods* are often used for sequencing large genomes, but its assembly is complex and difficult, particularly with sequence repeats often causing gaps in genome assembly.

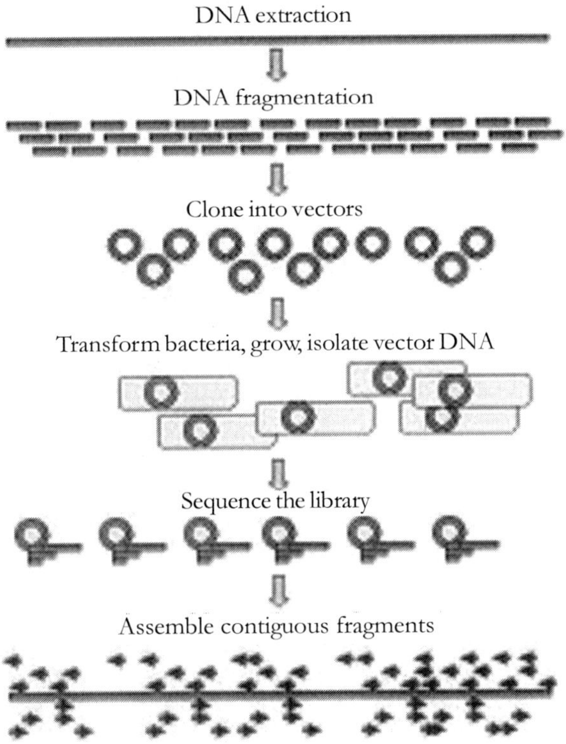

IX. New Sequencing Methods

1. High-Throughput Sequencing

The high demand for low-cost sequencing has driven the development of high-throughput sequencing technologies that parallelize the sequencing process, producing thousands or millions of sequences at once. High-throughput sequencing technologies are intended to lower the cost of DNA sequencing beyond what is possible with standard dye-terminator methods.

2. *In vitro* Clonal Amplification

Molecular detection methods are not sensitive enough for single molecule sequencing, so most approaches use an *in vitro* cloning step to amplify individual DNA molecules. Emulsion PCR isolates individual DNA molecules along with primer-coated beads in aqueous droplets within an oil phase. Polymerase chain reaction (PCR) then coats each bead with clonal copies of the DNA molecule followed by immobilization for later sequencing. Emulsion PCR is used in the methods by Marguilis *et al.* (commercialized by 454 Life Sciences) and Shendure and Porreca *et al.* (also known as "polony sequencing") and SOLiD sequencing, (developed by Agencourt, now Applied Biosystems). Another method for *in vitro* clonal amplification is *bridge PCR*, where fragments are amplified upon primers attached to a solid surface. The single-molecule method developed by Stephen Quake's laboratory (later commercialized by Helicos) skips this amplification step, directly fixing DNA molecules to a surface.

3. *Parallelized Sequencing*

DNA molecules are physically bound to a surface, and sequenced in parallel.*Sequencing by synthesis*, like dye-termination electrophoretic sequencing, uses a DNA polymerase to determine the base sequence. Reversible terminator methods (used by Illumina and Helicos) use reversible versions of dye-terminators, adding one nucleotide at a time, detect fluorescence at each position in real time, by repeated removal of the blocking group to allow polymerization of another nucleotide. Pyrosequencing (used by 454) also uses DNA polymerization, adding one nucleotide species at a time and detecting and quantifying the number of nucleotides added to a given location through the light emitted by the release of attached pyrophosphates.

4. *Sequencing by Ligation*

This enzymatic sequencing method uses a DNA ligase to determine the target sequence. Used in the polony method and in the SOLiD technology, it uses a pool of all possible oligonucleotides of a fixed length, labeled according to the sequenced position. Oligonucleotides are annealed and ligated; the preferential ligation by DNA ligase for matching sequences results in a signal informative of the nucleotide at that position.

5. *Microfluidic Sanger Sequencing*

In microfluidic Sanger sequencing, the entire thermocycling amplification of DNA fragments as well as their separation by electrophoresis is done on a single chip (approximately 100 cm in diameter) thus reducing the reagent usage as well as cost. In some instances, researchers have shown that they can increase the through-put of conventional sequencing through the use of microchips. Research will still need to be done in order to make this use of technology effective.

References

Braslavsky I, Hebert B, Kartalov E, and Quake SR 2003. "Sequence information can be obtained from single DNA molecules". *Proc. Natl Acad. Sci. U.S.A.* 100 (7): 3960-4..

Edwards J.R., Ruparel H. and Ju.J. 2005. "Mass-spectrometry DNA sequencing". *Mutation Research* 573 (1-2): 3-12.

Fiers W, Contreras R, Duerinck F, *et al* 1976. "Complete nucleotide sequence of bacteriophage MS2 RNA: primary and secondary structure of the replicase gene". *Nature* 260 (5551): 500-7.

Gilbert W and Maxam A 1973. "The nucleotide sequence of the lac operator". *Proc. Natl. Acad Sci. U.S.A.* 70 (12): 3581^.

Gupta. P.K.2005.Biotechnology and genomics, Rastogi publications, Gangothri Shivaji road, Meerut-250 002.

Hall N 2007. "Advanced sequencing technologies and their wider impact in microbiology". *J. Exp. Biol* 210 (Pt 9): 1518-25..

Hanna GJ, Johnson VA, and Kuritzkes DR, *et al* 2000. "Comparison of sequencing by hybridization and cycle sequencing for genotyping of human immunodeficiency virus type 1 reverse transcriptase". *J. Clin. Microbiol.* 38 (7): 2715-21.

Ju J, Ruan C, Fuller CW, Glazer AN, and Mathies RA 1995. "Fluorescence energy transfer dye-labeled primers for DNA sequencing and analysis". *Proc. Natl. Acad. Sci. U.S.A. 92* (10): 4347-51.

Margulies M, Egholm M, and Altman WE, *et al* 2005. "Genome sequencing in microfabricated high-density picolitre reactors". *Nature* 437 (7057): 376-80.

Maxam AM and Gilbert W 1977. "A new method for sequencing DNA". *Proc. Natl. Acad. Sci. U.S.A.* 74 (2): 560.

Min Jou W, Haegeman G, Ysebaert M, and Fiers W 1972. "Nucleotide sequence of the gene coding for the bacteriophage MS2 coat protein". *Nature* 237 (5350): 82-8.

Ronaghi.M, Karamohamed S., Pettersson B., Uhlen M. and Nyren P. 1996. "Real-time DNA sequencing using detection of pyrophosphate release". *Analytical Biochemistry* 242: 84-9.

Sanger F and Coulson AR 1975. "A rapid method for determining sequences in DNA by primed synthesis with DNA polymerase". *J. Mol Biol.* 94 (3): 441-8.

Sanger F, Nicklen S and Coulson AR 1977. "DNA sequencing with chain-terminating inhibitors". *Proc. Natl. Acad. Sci. U.S.A.* 74 (12): 5463-7.

Shendure J, Porreca GJ, and Reppas NB, *et al* 2005. "Accurate multiplex polony sequencing of an evolved bacterial genome". *Science* 309 (5741): 1728-32.

Smith LM, Fung S, Hunkapiller MW, Hunkapiller TJ and Hood LE 1985. "The synthesis of oligonucleotides containing an aliphatic amino group at the 5' terminus: synthesis of fluorescent DNA primers for use in DNA sequence analysis". *Nucleic Acids Res.* 13 (7): 2399-412.

Smith LM, Sanders JZ, and Kaiser RJ, *et al* 1986. "Fluorescence detection in automated DNA sequence analysis". *Nature* 321 (6071): 674-9.

The Science of Horticulture : Vol. 1, 2011
© *K.V. Peter (ed.), pp. 303-324*
New India Publishing Agency, New Delhi (India)
E-mail : newindiapublishingagency@gmail.com
Web: www.bookfactoryindia.com

Chapter 13 Molecular Tools for Characterisation of Horticultural Crops

BEHERA, T.K. AND SABINA ISLAM

Use of morphological markers has continued till today for characterization of several crops including horticultural crops with the Mendel's principles of inheritance by following visible traits in the progenies of sexual crosses. Genetic markers play an essential role in the study of variability and diversity, in the construction of linkage maps, and in the diagnosis of individuals or lines carrying certain linked genes. Within this context, the limitations of morphological markers became quickly apparent. They tend to be restricted to relatively a few traits, display a low degree of polymorphism, are often environmentally variable in their manifestation, and can depend on the expression of several unlinked genes. Furthermore, some may affect plant viability or seed set, distorting gene frequencies in the progeny. Morphological markers were largely supplemented by biochemical markers, particularly isoenzymes that could be easily scored by electrophoresis (Ganapathy and Scandalios 1973; Tanksley 1983). The limitations of isoenzmyes as markers, in particular both the limited number of polymorphic enzymes that can be conveniently stained and the environmental effects on expression pattern, were apparent already twenty years ago (Tanksley 1983). The shortcomings drove the development of markers based on DNA polymorphisms. These marker types generate "fingerprints," distinctive patterns of DNA fragments resolved by electrophoresis and detected by staining or labeling. A molecular marker is a nucleotide sequence corresponding to a particular physical location in the genome. Its sequence needs to be polymorphic enough between plant accessions to allow its pattern of inheritance to be easily followed.

Ideal Genetic Markers (highly dependent on application and species involved)

- No detrimental effect on phenotype
- Co-dominant in expression
- Single copy
- Economic to use
- Highly polymorphic
- Easily assayed
- Multi-functional
- Highly available (Un-restricted use)
- Genome-specific in nature (especially when working with polyploids)
- Can be multiplexed
- Ability to be automated

Markers are basically characterised as (1) Morphological (also 'classical' or 'visible') markers: Visually characterized phenotypic characters like flower colour, seed shape, growth habits or pigmentation. (2) Biochemical markers, which include allelic variants of enzymes called isozymes. Isozyme markers are differences in enzymes which are detected by electrophoresis and specific staining. (3) DNA (or molecular) markers, which reveal sites of variation in DNA. Among these, DNA markers are the most widely used markers because they are abundant; arise from different classes of DNA mutations like substitution mutations (point mutations), rearrangements (insertions or deletions), errors in replication of tandemly repeated DNA. They are selectively neutral because they are usually located in non-coding regions of DNA and are practically unlimited in number and not affected by environmental factors and/or the developmental stage of the plant. They are used in the construction of linkage maps, assessing the level of genetic diversity within germplasm and cultivar identity.

The development of RFLPs (restriction-fragment length polymorphisms) began in the world with the development of human genetics (Botstein *et al.* 1980). This method exploits variation in the occurrence of restriction sites in genomic sequences hybridizing to a cloned probe. Originally, RFLP analysis required Southern blotting and hybridization making the method fairly slow and laborious. This method is still used to generate "anchor" markers, used by many researchers to make consensus recombination maps, though it is often implemented with the polymerase chain reaction (PCR) to generate the polymorphic fragments. The advent of PCR made possible many other marker methods. These are two broad categories: methods which detect specific, cloned, and sequenced targets in the genome; methods which use conserved or general primers which amplify from many anonymous sites throughout the genome. Of the later category, the first to be developed was Randomly Amplified Polymorphic DNA (RAPDs; Williams *et al.* 1990). The RAPDs are indeed rapid, and need only short primers of random sequence, but suffer from low polymorphism information content (PIC), poor correlation with other marker data and problems in reproducibility due to low annealing temperatures in the reactions. Among the methods targeting known sites, an important one to emerge around 1990 was the detection of simple sequence repeats (SSRs) or microsatellites in plants (Zhao and Kochert 1993). These require sequence data for each locus, slowing development in new species. On the other hand, the Amplified Fragment Length Polymorphism method (AFLP), introduced in the mid 1990s, is an anonymous marker method. It detects restriction sites by amplifying a subset of all the sites for a given enzyme pair in the genome by PCR between ligated adapters (Vos *et al.* 1995). To some extent, both AFLP and RFLP detect single nucleotide polymorphisms (SNPs) at restriction sites. Today, perhaps the marker systems receiving the greatest attention concern the detection of SNPs. These are generally not random, but are found in genes of interest. All of these methods, RFLP, RAPDs, microsatellites, AFLP and SNPs, detect variations at one or only a few nucleotide positions.

DNA markers have already played a major role in the genetic characterization and improvement of many crop species. They have also contributed to and greatly expanded our abilities to assess biodiversity, reconstruct accurate phylogenetic relationships, and understand the structure, evolution and interaction of plant and microbial populations. Molecular marker systems reveal variation in genomic DNA sequence. The first generation of molecular markers, RFLP, were based on DNA-DNA hybridisation and were slow and expensive. The invention of the polymerase chain reaction (PCR) to amplify short segments of DNA gave rise to a second generation of faster and less expensive PCR-based markers.

DNA markers may be broadly divided into three classes based on the method of their detection: 1) hybridization-based and 2) polymerase chain reaction (PCR)-based. Restriction fragment length polymorphisms (RFLPs) were the first generation of hybridization-based markers with substantial impact in agricultural biotechnology. These were subsequently followed by amplification-based technologies derived from the polymerase chain reaction (PCR). Notably, the use of arbitrary oligonucleotide primers in the amplification reaction facilitated the study of previously uncharacterized genomes. The more common of these PCR-based DNA techniques are Amplified Fragment Length Polymorphism (AFLP), Simple Sequence Repeat (SSR), Inter-Simple Sequence Repeat (ISSR) and Randomly Amplified Polymorphic DNA (RAPD). Other techniques have also been developed and are widely used in genetic analysis.

1. Salient Features of DNA Markers

1.1 Hybridization-Based

RFLP: It creates unlimited number of loci, co-dominant, and many detection systems. It can be converted to SCARs, robust in usage and it detects in related genomes. It does not require sequence information. But it is labour intensive and requires large quantity of DNA. It has low levels of polymorphism, long exposure times and needs considerable skill.

1.2 PCR Based

1.2.1. Randomly Amplified Polymorphic DNA (RAPD): Results are obtained quickly, fairly cheap. No sequence information is required. Relatively small DNA quantities are required, high genomic abundance, good polymorphism, can be automated, highly sensitive to laboratory changes, low reproducibility, cannot be used across populations, often see multiple loci, dominant.

1.2.2. Simple Sequence Repeat (SSR): It is fast, highly polymorphic and robust. It can be automated and requires very small DNA. It is co-dominant, multi-allelic and multiplexing is possible. It does not require radioactivity. It requires high developmental and start-up costs, usually single loci even in polyploids.

1.2.3. Inter-Simple Sequence Repeat (ISSR): Highly polymorphic, robust in usage. It can be automated, usually dominant.

1.2.4. Amplified Fragment Length Polymorphism (AFLP): Small DNA quantities are required, no sequence information required. It can be automated, evaluated of up to 100 loci, dominant.

1.2.5. Single nucleotide polymorphism (SNP): Robust in usage, polymorphism is identifiable. Different detection methods are available, suitable for high throughput. It can be automated but requires very high development costs and sequence information.

1.2.6. Sequence characterized amplified region (SCAR)/ Cleavage amplified polymorphism (CAP): These are co-dominant and require small quantities of DNA, highly reliable, usually single locus, very labour intensive

1.2.7. Microarray: Single base changes, highly abundant and polymorphic, co-dominant. Small DNA quantities are required, highly reliable, usually single locus, suitable for high throughput, no gel system. It can be automated and it has high development and start-up costs.

New developments in genomic research gave access to an enormous amount of sequence information as well as new insights on the function and interaction of genes and the evolution of functional domains, chromosomes and genomes. In this context, functional and comparative genomics can help in comparative genetic mapping and linkage analysis of useful horticultural traits.

The present chapter is restricted to the use of molecular markers for characterization in terms of genetic diversity analysis and DNA fingerprinting, phylogenetic studies and hybridity test with suitable examples in different horticultural crops (vegetables, fruits and ornamental crops).

2. Characterization, Fingerprinting and Genetic Diversity

Genetic fingerprinting is the unambiguous identification of an individual (based on the presence or absence of alleles at different markers) or a population (based on frequencies of alleles of the markers). This is an absolute measure and does not change depending on other individuals or populations under study. Genetic diversity is a study undertaken to classify an individual or population compared to other individuals or populations. This is a relative measure, as the distance between any pair of entries in the study is greater or lesser depending on all pairwise comparisons that can be made in the study.

Fingerprinting and diversity studies may be used for many different reasons, including:

1. Proprietary information - Cultivar ID, protection of breeders rights
2. Determining relationships for breeding programs -choosing parents (Increasing/decreasing genetic diversity(GD); heterosis)
3. Choosing parents for mapping studies – maximise polymorphisms
4. Determining relationships in germplasm repositories - avoiding redundancies; identify "missing" diversity; construction of core/base collection
5. Associating molecular marker data with phenotypic traits of interest

6. Monitoring changes in populations over time - due to selection (artificial or natural) in breeding programs or genebanks

Fingerprinting and diversity studies are both done using molecular markers. These are segments of chromosomes which do not necessarily encode any traits and are not affected by the environment but which are inherited in a Mendelian fashion. Matching alleles in two individuals are assumed to be Identical By Descent (IBD). Some segments of the chromosome change faster than others (i.e. coding vs. non coding DNA, RFLPs vs. SSRs) so use fast changing markers for closely related individuals and slow changing markers for less related individuals (different species). Different marker types therefore have different usefulness in fingerprinting individuals and populations, and other considerations must be remembered when choosing a marker for fingerprinting studies as well. A good marker for fingerprinting studies will be cheap to run, or gives a lot of information per run; very repeatable between assays, people, machines, and laboratories; experience very low error rate and easy and unambiguous to score; and contain many alleles but not many rare alleles (high information content).

The laboratory methodology, fingerprinting and diversity studies require the following steps:

- Isolation of DNA.
- Digestion, hybridization, and/or amplification of DNA into specific fragments.
- The sizing /or separation of DNA fragment combinations or patterns into a set of individual DNA fingerprints.
- The comparison of DNA fingerprints from different individuals
- Calculation of similarity (or dissimilarity) coefficients for all pairs of entries in the genetic study
- Grouping entries based on similarity or creation of a dendrogram or graph to visualize the differences

2.1 Vegetable Crops

Cultivated tomato (*Solanum lycopersicum* L.) germplasm shows limited genetic variation. Many DNA marker systems are used for genetic diversity studies in wild and cultivated tomatoes, but their usefulness for characterizing phenotypic differences among very closely related cultivars remain uncertain. Garcia-Martinez *et al.* (2006) used 19 selected simple sequence repeat (SSR) markers and 7 amplified fragment length polymorphism (AFLP) primer combinations to characterize 48 cultivars of tomato, mainly traditional cultivars from the south-east of Spain. A better grouping of the 'Muchamiel' cultivars was observed with SSR markers, whereas the grouping of cultivars of 'De la pera' type was the best achieved with AFLPs. However, both types of markers adequately grouped cultivars of the main types, confirming utility of SSR and AFLP markers for identification of traditional cultivars of tomato.

The concept of genetic base broadening was introduced to mitigate this lack of diversity in modern crops to utilize more effectively the rich genetic resources available in wild relatives, old varieties, and landraces (Saavedra *et al.* 2001). The richness of available gene pools in wild relatives, makes these plants an interesting case for study of genetic base broadening programmes to assure better responses in future breeding generations to biotic and abiotic stresses. Villand *et al.* (1998) used 41 random amplified polymorphic DNA (RAPD) primers, generated 98 polymorphic RAPD markers and same were used to estimate relationships among 96 accessions collected from a wide geographic range. Differences in RAPD marker frequencies indicated uniqueness of accessions from the New and Old World collections. In addition, accessions from Ecuador, Peru and Chile had a larger magnitude of marker diversity than Old World accessions. Wang *et al.* (2007) characterised 11 tomato cultivars and one wild relative with the sequence related amplified polymorphism (SRAP) marker system. They observed 10 out of 15 primer combinations used could detect 55 polymorphic loci. An average of 3.7 cultivars could be distinguished with each primer combination, and accessions used in the study could be distinguished with the combination of ME8/EM8-1 and ME6-1/EM14-1. Cluster analysis (UPGMA) showed the cultivars which share similar horticultural traits clustered at a lower genetic distance.

Gaikward *et al.* (2008) studied genetic relatedness of 38 Indian bitter gourd cultigens (commercial varieties and cultivated landraces originating from different agroecological zones) by amplified fragment length polymorphism (AFLP) analysis. Six primer combinations yielded a total of 519 bands of which 404 (77.8%) were polymorphic among the cultigens examined. Genetic similarities among cultigens ranged 0.44 to 0.88, indicating that the bitter gourd cultigens examined were genetically diverse. The relatively high marker indices (6.2 to 19.4), polymorphic information content of the markers used (0.20 to 0.25), and multiplex ratios (28.9 to 77.4) collectively indicate that the AFLP markers used are discriminatory in bitter gourd and that the analysis of broad-based cultigens described provides valuable baseline information for advancing initial breeding strategies for this crop species. The studies using RAPD (Dey *et al.* 2006) and ISSR (Singh *et al.* 2007) markers were not able to uniquely discriminate all the bitter gourd accessions examined, whereas AFLP analysis was discriminatory and allowed for a more complete dissection of unique differences among accessions of bitter gourd within and between collection sites (Behera *et al.* 2008). Pavlikaki *et al.* (2004) applied RAPD marker traditional cucumber cultivars namely Knossos, Kalivia, Kos2, Ikaria, Samos, Crete1, Crete2 and Crete3. Fourteen of the 20 primers used showed consistently reproducible and simple amplification products. Assaying the variation of the accessions, yielded 42 RAPD markers, 33 of which were polymorphic.

Liu *et al.* (2008) used 3 molecular marker systems, RAPD, ISSR and SRAP for identification and genetic diversity analysis of 35 elite late-bolting radish cultivars. These were detected by 35 RAPD primers, 22 ISSR primers and 17 SRAP primer combinations, the proportions of polymorphic bands were 85.44%, 85.2% and 85.41%, respectively, and the mean genetic similarity coefficients between pairs of genotypes were 0.781, 0.787 and 0.764, respectively. In another study, Zhao *et al.* (2007) analyzed seven radish cultivars by using the optimized SRAP and AFLP marker system. All cultivars can be distinguished clearly with SRAP primer combinations and 14 AFLP primer combinations. Total of 155 bands were observed with 11 SRAP primer combinations, of which 84 were polymorphic bands. Wang *et al.* (2006) revealed that RAPD marker was the most polymorphic (93.4%) among the three marker types. An average of 23.25 polymorphic fragments amplified with each AFLP primer combination was obtained with a relatively low polymorphic ratio. The dendrogram based on RAPD and SRAP data showed high accordance with the classification according to the morphological traits. The genetic diversity of radish germplasm was highly associated with the color of fresh root skin and their origin.

Random amplified polymorphic DNA (RAPD) was used(Madhou *et al.* 2005) to estimate the degree of genetic diversity between and within 2 populations of wild radish (WARR 5 and WARR 6) and found to exhibit multiple herbicide resistance compared with a susceptible population (WARR 7). It is believed that weed species with high degrees of genetic variation show potential for developing resistance to herbicides. Of the 13 RAPD primers screened, 9 primers generated 97 polymorphic bands concomitant with a high level of polymorphism (82%) between the wild radish populations, characteristics of an outbreeding species. Analysis of molecular variance (AMOVA) showed a markedly higher proportion of diversity within populations (87%) as opposed to between populations (13%). Only one marker, OPC19-8, was found unique to the WARR 7 population but absent in WARR 5 and in most individuals of the WARR 6 populations. This marker may potentially be correlated with herbicide susceptibility. The two resistant wild radish populations were found closely related (0.7% dissimilar) to each other, whereas the susceptible population was genetically dissimilar to them by 2.3%.

Daucus carota L. accessions and related wild subspecies were examined with AFLP and ISSR markers which revealed extensive variation within *D. carota*. Although cultivated carrot and wild *D. carota* subspecies can cross freely, cultivated and wild carrots clustered separately, supporting the possibility that human selection for desirable horticultural traits has artificially reduced gene flow between cultivated and wild forms. Our analyses support the likelihood that North American *D. carota* populations arose due to introduction of weedy materials rather than escape of cultivated forms (Bradeen *et al.* 2002).

Molecular characterization can quantify and partition variation among genebank holding. The assessment of genetic identity, relatedness, and structure among entries and collections becomes a priority to characterize and conserve crop genetic resources. Phippen *et al.* (1997) tested 14 phenotypically uniform accessions of *Brassica oleracea* var. *capitata* (cabbage) with nine RAPD primers. These amplifications generated 110 fragments, of which 80 were polymorphic ranging in size from 370 to 1720 bp and those were sufficient to distinguish between all 14 accessions. This proposed grouping would concurrently save approximately 70% for each cycle of regeneration (approximately 20-25 years at the most) which alternatively could then be used for other priorities in *B. oleracea* conservation and use. Molecular markers provide gene bank curators with additional sources of information to better plan and organize collection holdings and use finite financial support in a more effective manner. Thirteen SSRs were used (Tonguc and Griffiths, 2004) to assess genetic similarity between 54 *B. oleracea* cultivars, belonging to 3 groups (cabbage, cauliflower and broccoli). The varietal group with the narrowest genetic variation in the study was cauliflower (*B. oleracea* var. *botrytis*) followed by broccoli (*B. oleracea* var. *italica*) and cabbage (*B. oleracea* var. *capitata*) groups.

2.2 Fruits Crops

Karihaloo *et al.* (2003) analysed genetic diversity of Indian mango cultivars using RAPD markers. The similarity between pairs of cultivars ranged from 0.318 to 0.75 with a mean of 0.565. Majority of the cultivars from northern and eastern regions of India clustered together and were separate from southern and western cultivars. Analysis of molecular variance revealed that 94.7% of the genetic diversity in mango existed within regions. They further observed that, differences among regions were significant; northern and eastern regions formed one zone and western and southern regions formed another zone of mango diversity in India. Ravishankar *et al.* (2004) studied genetic relatedness among polyembryonic and monoembryonic mango cultivars each grown in the west coast of southern India using random amplified polymorphic DNA (RAPD) markers. Dendrogram analysis of RAPD grouped the cultivars into 2, based on embryo types and the grouping indicated different genetic base of monoembryonic and polyembryonic types of Indian mango cultivars. Souza and Lima (2004) studied genetic variability in mango genotypes by RAPD markers. Bootstrap consistency test, indicated three diverse groups ("Mallika" and "Amrapali", "Van Dyke", "Palmer" and "Keitt", and "Irwin" and "Glenn").

Srivastava *et al.* (2007) assessed genetic diversity of mango germplasm using three different PCR based marker system methods {(RAPD, ISSR, and Directed Amplification of Minisatellite DNA (DAMD)}. A total of 158, 69 and 59 distinct DNA fragments by RAPD, ISSR and DAMD, respectively were amplified and approximately 85%, 64% and 90% were polymorphic, respectively and DAMD was judged to be the best method to analyse mango germplasm. Pandit *et al.* (2007) analysed genetic diversity of mango cultivars using ISSR markers. Twelve different, cultivar-specific bands were detected for six cultivars. Non-Indian mango cultivars were found genetically diverged from Indian mango germ pool and no separation was observed between North Indian and South Indian cultivars.

Kim *et al.* (2000) generated SCAR markers for identification of *Pyrus pyrifolia* cultivars. The positive identifications of SCAR markers for *Pyrus pyrifolia* cultivars were obtained with LCH332, LCH332-1.4, LCH350, LCH351, LCH384, and LCH387 specific primer pairs. Lee *et al.* (2004) generated RAPD, SCAR and conserved 18S rDNA markers for classifying and identifying cultivars of *Pyrus pyrifolia* (Japanese pear) and *P. communis* (European pear). Amplification involving sequence-specific primer pairs LCH346UP and LCH346DOWN resulted in a loss of polymorphism. Compared with *P. pyrifolia* cultivars, *Pyrus communis* cultivars lacked the restriction enzyme site of KpnI and had one less RsaI site. RAPD markers were used to characterize Portuguese *Vitis vinifera* cultivars, representative of the "Douro" and "Vinhos Verdes" regions. All the studied cultivars were distinguished at a similarity level between 0.62 and 0.90 (Leal *et al.* 2004).

Rootstock identification after grafting is very difficult following traditional characterization methods due to the absence of observable phenotypic traits. Wunsch *et al.* (2004) employed molecular marker tools (SSR) for characterization of genotypes, comprising of several species and interspecific hybrids, useful as rootstocks for sweet cherry. All the genotypes studied except one pair of siblings and one pair of accessions collected in the same location were unequivocally distinguished with the combination of the 62 alleles obtained with the 9 polymorphic SSRs.

Liu *et al.* (2006) reported the use of molecular markers for core collection of pomelo germplasm. They adopted stepwise reduction method based on SSR and AFLP markers to screen the core collection of pomelo germplasm. After the evaluation of the number of polymorphic loci, percentage of polymorphic loci, number of alleles, number of effective alleles, Nei's gene diversity index and Shannon's information index, 25 samples were chosen as core collection. A collection of sour orange (*Citrus aurantium* L.) accessions coming from Sicily and other countries was investigated by two polymerase chain reaction (PCR)-based DNA marker technologies. Inter-simple sequence repeat (ISSR) primers and randomly amplified polymorphic DNA (RAPD) primers were used to identify and to evaluate the genetic variability and relationship of accessions. The ISSR and RAPD analysis of sour orange accessions found a high grade of genetic diversity in foreign accessions, while a low variability was detected in local accessions (Siragusa *et al.* 2006).

Ji *et al.* (2007) analysed genetic diversity of 'A' genome among banana cultivars using SSR markers. All microsatellite loci were polymorphic among cultivars. The averaged pair wise genetic distances among cultivars were estimated from 0.00 to 34.27% with an average of 12.45%, which revealed a narrow range of variability among cultivars, except for five cultivars. Goulao and Oliveira (2001) employed SSR and ISSR markers for molecular characterisation of apple cultivars. The similarity coefficient between cultivars ranged from 0.20 to 0.87 for SSR analysis and from 0.71 to 0.92 for ISSR methodology.

2.3 Ornamental Crops

Characterization of commercial cultivars of chrysanthemum through random amplified polymorphic DNA (RAPD) analysis revealed that varieties could be distinguished from each other and the level of similarity between cultivars was not very high (Martin *et al.* 2002). The RAPD technique was extended for diversity estimation of chrysanthemum, carnation, orchid and other floricultural crops. Qin *et al.* (2002) reported RAPD analysis for diversity estimation of genetic resource of chrysanthemum. High polymorphism was observed across cultivars. They also found that the cultivars with similar petal type had high similarity coefficient. Molecular characterization of carnation cultivars using RAPD revealed extensive genetic variation among cultivars and they were clustered into 10 groups based on similarity (Su *et al.* 2004). Genetic variation in chrysanthemum (*Chrysanthemum morifolium*) genetic resource representing standard, spray and no pinch-no stake groups were examined using random amplified polymorphic DNA. RAPD fragment(s) unique to a particular cultivar was observed (Sehrawat *et al.* 2008).

Miao *et al.* (2007) employed ISSR analysis data for establishing genetic relationship of chrysanthemum [*Dendranthema* × *grandiflora* (Ramat.) Kitamura] cultivars at molecular level. The similarity coefficient between cultivars ranged 0.294 to 0.802. The cultivars were divided into 6 groups and the genetic relationship of the cultivars was related to their petal types. Amplified fragment length polymorphism generated data were used for genetic diversity studies in chrysanthemum cultivars. Cluster analyses showed that there were close genetic relationships among cultivars of the same petal type but the flower colour had no clear pertinence with the cluster results (Wu *et al.* 2007). Chiu *et al.* (2008) evaluated genetic variation in the genetic resource of chrysanthemum using inter simple sequence repeat (ISSR) DNA markers. The average genetic similarity in summer and autumn cultivars were 0.43 and 0.32, respectively. Cluster analysis revealed that there were 3 major groups in summer chrysanthemums and 2 major groups in autumn cultivars. Liu *et al.* (2008) resorted to AFLP analysis to study the genetic diversity in chrysanthemum genetic resource. Cluster analysis classified the 26 cultivars into 5 groups based on pedigree.

The genetic variation in desert rose (*Adenium obesum*) clones and hybrids from diverse geographical region was studied using ISSR. The poly AC motif was the most abundant and gave the highest rate of polymorphism. Genetic similarity ranged nil to 88%. The Arab clones recorded the highest genetic similarity and clustered into one group (Chou and Tsai 2005). Gan *et al.* (2006) evaluated genetic diversity among *Cymbidium* accessions using RAPD markers. The genetic similarity (GS) among accessions ranged 0.503 to 0.765, with a mean of 0.598. They were able to distinguish all accessions by RAPD markers.

Zhang *et al.* (2001) explored AFLP technique for cultivar identification in modern rose (*Rosa hybrida* L.). They studied modern rose cultivars originated from French rose breeders. High level of AFLP polymorphism was observed among cultivars. They concluded that AFLP technique can effectively be used for cultivar identification. Pirseyedi *et al.* (2008) reported use of AFLP markers for diversity estimation of Iranian Damask rose (*Rosa damascena* Mill.). The banding pattern showed extreme variability among the studied genotypes. Tejaswini *et al.* (2008) studied genetic relationship in carnation using RAPD and ISSR markers. Polymorphic information content (PIC) generated by RAPD was higher than that realized from ISSR indicating superiority of RAPD for discriminating genotypes and for analysing diversity.

RAPD analysis was applied to study the genetic relationship of cultivars of *Gladiolus hybridus*. A high level of polymorphism was noticed in the banding pattern of different cultivars and could be used for identification of different cultivars. The results revealed that the germplasm resource of *G. hybridus* cut flower cultivars had a narrow genetic base at molecular level (Wang *et al.* 2008).

3. Phylogenetic Relationship

In biology, phylogenetic is the study of evolutionary relatedness among various groups of organisms (e.g., species, populations), discovered through molecular sequencing data and morphological data matrices.

3.1 Vegetable Crops

Kochieva *et al.* (2004) used AFLP, RAPD and ISSR marker systems to determine genetic variation and phylogenetic relationships within the genus *Capsicum*. In total, 1921 bands were used in the genome analysis of 61 accessions representing 11 species of the *Capsicum*. The genetic data confirmed the recognition of *C. frutescens* and *C. chinense* as separate species with high bootstrap values. This study further supported the recognition of *C. praetermissum* as a separate species inside the *C. baccatum* complex. Molecular markers revealed high genetic similarities of *C. pubescens* with *C. eximium* and *C. cardenasii*. *C. tovarii* is found genetically closer to *C. baccatum* (based on AFLP and RAPD data) or *C. praetermissum* (based on ISSR data). Molecular data supported the close relationships of *C. galapagoense* with *C. frutescens* (RAPD analysis) and *C. annuum* (ISSR analysis) and this species can therefore be considered a member of the *C. annuum* complex. Bosland and Baral (2004) studied the taxonomic relationship between *C. frutescens* and *C. chinense* using morphological, molecular (RAPD) marker and sexual compatibility studies. All three approaches provided evidence that *C. frutescens* and *C. chinense* are distinct species. Diagnostic morphological differences, RAPD marker-based cluster analysis and hybrid analyses were used to delimit *C. frutescens* and *C. chinense*. In many cases, two morphological characters, calyx constriction and flower position could separate accessions of *C. frutescens* from *C. chinense*. The RAPD-based analysis clearly separated accessions of *C. frutescens* and *C. chinense* into two distinct groups. The average genetic similarity within *C. frutescens* and *C. chinense* accessions was 0.85 and 0.8, respectively, whereas the average genetic similarity between *C. frutescens* and *C. chinense* accessions was only 0.38. The progenies from *C. frutescens* and *C. chinense* hybridization had reduced fertility. Based on these evidences, *C. frutescens* and *C. chinense* represent two morphologically diagnosable, phylogenetically distinct, and reproductively isolated species (Baral and Bosland, 2004).

3.2 Fruit Crops

Phylogenetic relationships among *Mangifera* species were analysed by comparing amplified fragment length polymorphism (AFLP) markers. The common mango (*M. indica*) was closely related to banana mango (*M. sylvatica*), *M. laurina*, and *M. oblongifolia. M. macrocarpa, M. foetida*, and *M. odorata*. Two species *M. gedebe* and *M. griffithii* were placed in distant positions among the *Mangifera* species tested, and they are distantly related to *M. indica* (Eiadthong *et al.* 2000).

Gulsen and Roose (2004) employed inter-simple sequence repeats (ISSR) and simple sequence repeats (SSR) or microsatellite markers to measure genetic diversity and phylogenetic relationship among lemon accessions, related taxa, and three proposed ancestral species, *Citrus maxima, C. medica*, and *C. reticulata*. All lemons, rough lemons and sweet lemons, including some other suspected hybrids were clustered with citrons. Most lemons had nearly identical marker phenotypes, suggesting that they originated from a single clonal parent *via* a series of mutations. Citrons contributed the largest part of the lemon genome and a major part of the genomes of rough lemons, sweet lemons and sweet limes.

Korbin *et al.* (2002) employed RAPD and ISSR-PCR for strawberry (*Fragaria* × *ananassa*), apple (*Malus domestica*) and *Ribes* species germplasm characterization. Twenty primers for RAPD (among 60 tested) and seven for ISSR (among 10 tested) were chosen as creating polymorphic DNA bands and differentiating the investigated genotypes. In many cases, both RAPD- and ISSR-based genetic similarity confirmed relatedness connected with biological origin and with the place where the cultivar was developed.

Monte *et al.* (2000) studied genetic relationships among *Pyrus* species and cultivars using AFLP and RAPD markers. Both the techniques revealed a similar level of molecular polymorphisms and proved their reliability to assess genetic relationships among pear cultivars. *P. pyrifolia* cultivars exhibit a closer genetic relatedness, clustering apart from *P. communis* cultivars. Most of the Portuguese cultivars tend to cluster together, indicating to constitute a relatively independent genetic pool.

3.3 Ornamental Crops

The relationship of commercially available carnation cultivars and wild species of carnation was carried out through RAPD analysis. Twenty five out of 30 cultivars were identified by 1-6 RAPD markers and 13 wild species were identified by 2-19 markers. The cluster analysis classified the genotypes into eight major groups (Lee *et al.* 2002). Deak *et al.* (2005) studied genetic variation in rose genotypes from Hungary and the Carpathian Basin based on AFLP markers. The main groups of the AFLP dendrogram were in agreement with the commonly recognized major sections of the genus.

Chung *et al.* (2006) utilized RAPD analysis data for studying the genetic relationship and differentiation of *Paphiopedilum* and *Phragmipedium*. The similarity values ranged 0.629 to 0.882. The orchids examined were separated into two major subgroups. *Paphiopedilum* and *Phragmipedium* were successfully differentiated by RAPD and the results were in good agreement with morphological based classifications. RAPD technique was employed in determining the genetic relationship among different orchid species. Species of three morphologically distinct groups of Orchids belonging to the subfamily Epidendroideae and tribe Dendroebieae and Cymbideae were studied. The polymorphic patterns showed different degrees of genetic relationship and grouped all the species into 5 groups (Chakrabarti, 2007). Zhu *et al.* (2007) studied genetic diversity and relationship of cultivars of hybrid *Cymbidium*, and wild species (*C. aloifolium* and *C. floribundum*) using AFLP molecular markers. Genetic similarity among cultivars ranged from 0.3399 to 0.8223, with an average of 0.5783. The cultivars were divided into 4 groups and the grouping was consistent with the morphological classification based on the type of inflorescence, flower size or flower colour. All the cultivars had their characteristic bands and identified based on the AFLP markers.

4. Hybridity Test

A fast and accurate genetic purity test of F_1 hybrid plants is essential for seed production and accelerating advanced breeding generations in breeding programmes. DNA technology has great potential for enhancing purity assessment of hybrids.

4.1 Vegetable Crops

The genetic purity of 3 F_1 chilli hybrids (*C. annuum* 'Bangchang' × *C. chinense* 'CM021', *C. chinense* 'CM021' × *C. chinense* 'CM022' and *C. annuum* '83-168' × *C. annuum* 'KKU-cluster') was determined using 2 molecular marker techniques, RAPD and ISSR. RAPD analysis successfully detected all 3 F_1 hybridity, while ISSR detected only 2. This was due to the RAPD marker system producing a greater number of markers than the ISSR system. (Mongkolporn, *et al.* 2004).

The cabbage hybrid cultivar 'Xinxia 50' and its parents were analyzed with 83 RAPD primers, 49 ISSR primers and 56 SSR primers. Two RAPD primers (NAURP2068 and NAURP2079), two ISSR primers (NAUISR1034 and NAUISR1062) and one SSR primer (NAUSSR1011), which produced male and female parent-specific markers simultaneously in the hybrid, can be applied successfully for testing the genetic purity of this hybrid. The genotype of hybrid 'Xinxia 50' was determined using these five primers. Of the 228 F_1 hybrid seeds tested, 16 appeared to be false hybrids. Eight out of the 16 putative false hybrids exhibited similar band patterns to the female parent with all five identified primers, suggesting that they could be derived from selfing of the female parent plant.

4.2 Fruit Crops

Hajos and Hajdu (2003) utilized RAPD and microsatellite for characterization of two Hungarian wine grape (*Vitis vinifera*) hybrids and their parents. Andrade *et al.* (2005) used RAPD markers for identification of zygotic and nucellar seedlings in Citrus. The origin of 16.6% polyembryonic and 83.3% monoembryonic seeds of *C. reshni* was determined as sexual. All the seedlings of Robinson were zygotic. They suggested that RAPD markers allowed the efficient identification of zygotic and nucellar seedlings of *C. reshni* and Robinson.

4.3 Ornamental Crops

Zhang *et al.* (1997) employed RAPD technique for testing genetic purity in petunia and cyclamen seed. Bulked samples of petunia and cyclamen hybrid cultivars produced consistent RAPD marker profiles. Evaluation of individual seeds from a single cyclamen hybrid produced polymorphic banding patterns that were attributed to genetic variability present in the female and male inbred parents. They concluded that RAPD markers can be used to quickly assess the genetic purity of selected cultivars of these two flower seed crops. Roh and Ikeda (2003) studied genetic diversity of *Dendranthema pacificum* (Nakai) using RAPD technique. The RAPD analysis indicated that *D. pacificum* accession 1-7 collected in the trade in the United States could be a hybrid between *D. pacificum* and other species with ray florets.

Antonova *et al.* (2006) developed SSR marker system for identification and certification of sunflower hybrids. Molecular genetic passports were made on the basis of the nine loci. The average number of alleles per locus was 2.2. Dominant type of inheritance was found in two loci and co-dominant in seven. They inferred that the seven co-dominant loci are suitable for definition of hybrid vigour.

Molecular markers are being used extensively to investigate the genetic basis of horticultural traits and to facilitate the transfer and accumulation of desirable traits between breeding lines. A number of techniques have been particularly useful for genetic analysis. For example, collections of RFLP probes have been very versatile and important for the generation of genetic maps, construction of physical maps, establishment of syntenic relationships between genomes and marker assisted breeding. Numerous examples of specific genes that have been identified as tightly linked to RFLP markers are available for the improvement of specific agronomic traits in almost all major horticultural crops. Markers have also been used to identify the genes underlying quantitative variation for height, maturity, disease resistance and yield in virtually all major horticultural crops. In particular, the PCR-based techniques are useful in the assessment of biodiversity, the study of plant and pathogen populations and their interactions; and identification of plant varieties and cultivars. Amplified DNA techniques have produced sequence-tagged sites that serve as landmarks for genetic and physical mapping. It is envisioned that emerging oligonucleotide-based technologies derived from the use of hybridization arrays, the so-called DNA chips and oligonucleotide arrays, will become important in future genomic studies. However, these are still under development, are proprietary, or require the use of expensive equipment, and are therefore not yet suitable nor cost-effective for adequate transfer to developing countries. Clearly, the initial transfer of technology has only involved a selected group of techniques that are well established and/or seem to have a broad application (e.g., RFLP, SSR, ISSR, AFLP, RAPD, and SNPs). However, techniques are continuously changing and evolving, so technology transfer needs to keep pace with current developments in genomics.

References

Andrade RM, Villegas MA, Gutierrez EMA, Carrillo CG and Garcia VA 2005. Polyembryony and RAPD markers for identification of zygotic and nucellar seedlings in Citrus. *Agrociencia Montecillo* 39: 371-383.

Antonova TS, Guchetl SZ, Tchelustnikova, TA and Ramasanova, SA 2006. Development of marker system for identification and certification of sunflower lines and hybrids on the basis of SSR-analysis. *Helia* 29: 63-72.

Baral JB and Bosland PW 2004. Unraveling the species dilemma in *Capsicum frutescens* and *C. chinense* (Solanaceae): a multiple evidence approach using morphology, molecular analysis, and sexual compatibility. *Journal of the American Society for Horticultural Science*. 129: 826-832

Behera TK, Gaikward AB, Singh AK and Staub JE 2008. Relative efficiency of DNA markers (RAPD, ISSR and AFLP) in detecting genetic diversity of bitter gourd (*Momordica charantia* L.). *Journal of the Science of Food and Agriculture*. 88: 733-737.

Bosland PW and Baral JB 2004. The taxonomic answer to the species dilemma in *Capsicum chinense* and *C. frutescens*: a key issue for Capsicum improvement and conservation genetics. Proceedings of the XIIth EUCARPIA meeting on genetics and breeding of Capsicum and eggplant, Noordwijkerhout, Netherlands,17-19-May,2004, pp 32-37.

Botstein D, White RL, Skolnick M, and Davis RW 1980. Construction of a genetic linkage map in man using restriction fragment length polymorphisms. *Am J Hum Genet* 32:314–331.

Bradeen JM, Bach IC, Briard M, Clerc V-le, Grzebelus D, Senalik DA and Simon PW 2002. Molecular diversity analysis of cultivated carrot (*Daucus carota* L.) and wild *Daucus* populations reveals a genetically nonstructured composition. *Journal of the American Society for Horticultural Science* 127: 383-391.

Chakrabarti S 2007. Application of RAPD technique to study the genetic relationship of orchids. Recent trends in horticultural biotechnology, Vol-II-ICAR national symposium on biotechnological interventions for improvement of horticultural crops: issues and strategies, Vellanikkara, Kerala, India, 10-12 January 2005. pp 565-569.

Chiu YS, Ong PL and Tsai CT 2008. Evaluation of genetic variation of chrysanthemum (*Dendranthema × morifolium* Ramat.) varieties with inter simple sequence repeat DNA markers. *Journal of the Taiwan Society for Horticultural Science* 54: 1-9.

Chou MC and Tsai CT 2005. Evaluation of genetic variation of desert rose (*Adenium*) cultivars with inter simple sequence repeat (ISSR) markers. *Journal of the Chinese Society for Horticultural Science* 51:175-188.

Chung SY, Choi SH, Kim MJ, Yoon KE, Lee GP, Lee JS and Ryu KH 2006. Genetic relationship and differentiation of *Paphiopedilum* and *Phragmipedium* based on RAPD analysis. *Scientia Horticulture* 109: 153-159.

Deak T, Facsar G and Bisztray GD 2005. Genetic polymorphism of Rosa genotypes native to Hungary. *Acta Horticulture* 690: 57-62.

Dey SS, Singh AK, Chandel D and Behera TK 2006. Genetic diversity of bitter gourd (*Momordica charantia* L.) genotypes revealed by RAPD markers and agronomic traits. *Scientia Horticulture* 109: 21-28.

Eiadthong W, Yonemori K, Kanzaki S, Sugiura A, Utsunomiya N and Subhadrabandhu S 2000. Amplified fragment length polymorphism analysis for studying genetic relationships among *Mangifera* species in Thailand. *Journal of the American Society for Horticultural Science* 125: 160-164.

Gaikwad AB, BeheraTK, Singh AK, Devanshi Chandel, Karihaloo JL, Staub JE 2008 Amplified fragment length polymorphism analysis provides strategies for improvement of bitter gourd (*Momordica charantia* L.). *HortScience* 43: 127-133.

Gan N, Tan XH, Chen QB, Wei Y and Zheng YL 2006. Genetic diversity in Cymbidium based on RAPD markers and PCR-RFLP analyses of organellar DNAs. *Acta Horticulturae Sinica*. 33(2): 349-355.

Ganapathy PS and Scandalios JG 1973. Malate Dehydrogenase Isozymes in Haploid and Diploid *Datura* Species: Their use as markers in somatic cell genetics. *J Heredity* 64:186-188.

Garcia-Martinez S, Andreani L, Garcia Gusano M, Geuna F and Ruiz JJ 2006. Evaluation of amplified fragment length polymorphism and simple sequence repeats for tomato germplasm fingerprinting: utility for grouping closely related traditional cultivars. Genome 49: 648-656

Goulao L and Oliveira CM 2001. Molecular characterisation of cultivars of apple (*Malus* × *domestica* Borkh.) using microsatellite (SSR and ISSR) markers. *Euphytica* 122: 81-89.

Gulsen O and Roose ML 2001. Determination of genetic diversity and phylogenetic relations to citrus ancestors in lemons by DNA markers. *Bahce* 30: 53-63.

Hajos NM and Hajdu E 2003. Isozyme and DNA fingerprinting characterization of two Hungarian wine grape hybrids and their parents. *Acta Horticulturae* 603: 211-216.

Ji XR, Chen YY, Wang JY, Zheng LS, Wei SX, Xie ZS and Wu YT 2007. Analysis of genetic diversity of A genome among banana cultivars using SSR markers. *Journal of Fruit Science* 24: 783-787.

Karihaloo JL, Dwivedi YK, Sunil A and Gaikwad AB 2003. Analysis of genetic diversity of Indian mango cultivars using RAPD markers. *Journal of Horticultural Science and Biotechnology* 78: 285-289.

Kim CS, Lee GP, Han DH, Ryu, KH and Lee CH 2000. SCARs markers derived from RAPD for cultivar identification in *Pyrus pyrifolia*. *Journal of the Korean Society for Horticultural Science* 41(2): 125-128.

Kochieva EZ, Ryzhova NN, van-Dooijeweert W, Boukema IW and Arens P 2004. Assessment of genetic relationships in the genus Capsicum using different DNA marker systems. Proceedings of the XIIth EUCARPIA meeting on genetics and breeding of Capsicum and eggplant, Noordwijkerhout, Netherlands,17-19May, 2004, pp 44-50.

Korbin M, Kuras A and Zurawicz E 2002. Fruit plant germplasm characterisation using molecular markers generated in RAPD and ISSR-PCR. *Cellular and Molecular Biology Letters* 7: 785-794.

Leal F, Castro I, Guedes PH, Pinto CO, Ortiz JM and Martin JP 2004. Characterization of Portuguese grapevine cultivars using random amplified polymorphic DNA markers. *Acta Horticulture* 652: 401-405.

Lee SY, Yae BW and Kim KS 2002. Identification and analysis of genetic similarity in wild species and cultivars of carnation using RAPD. *Journal of the Korean Society for Horticultural Science* 43: 738-742.

Lee GP, Lee CH and Kim CS 2004. Molecular markers derived from RAPD, SCAR, and the conserved 18S rDNA sequences for classification and identification in *Pyrus pyrifolia* and *P. communis*. *Theoretical and Applied Genetics* 108: 1487-1491.

Liu Y, Sun ZH, Liu DC, Wu B and Zhou Q 2006. Screening the core collection of pomelo germplasm based on molecular marker. *Journal of Fruit Science* 23: 339-345.

Liu M, Zhang SH, Liang HY and Zhen ZX 2008. AFLP analysis on the genetic diversity of some chrysanthemum species. *Journal of Agricultural University of Hebei*. 31(1): 48-52.

Liu LW, Zhao LP, Gong YQ, Wang MX, Chen LM, Yang JL, Wang Y, Yu FM and Wang LZ 2008. DNA fingerprinting and genetic diversity analysis of late-bolting radish cultivars with RAPD, ISSR and SRAP markers. *Scientia Horticulturae* 116: 240-247.

Madhou P, Wells A, Pang ECK and Stevenson TW 2005.Genetic variation in populations of Western Australian wild radish. *Australian Journal of Agricultural Research* 56: 1079-1087.

Martin C, Uberhuaga E and Perez C 2002. Application of RAPD markers in the characterization of Chrysanthemum varieties and the assessment of somaclonal variation. *Euphytica* 127: 247-253.

Miao HB, Chen FD and Zhao HB 2007. Genetic relationship of 85 chrysanthemum [*Dendranthema* × *grandiflora* (Ramat.) Kitamura] cultivars revealed by ISSR analysis. *Acta Horticulturae Sinica* 34: 1243-1248.

Mongkolporn O, Dokmaihom Y, Kanchana-Udomkan C, and Pakdeevaraporn P 2004. Genetic purity test of F1 hybrid Capsicum using molecular analysis. *Journal of Horticultural Science and Biotechnology*. 79: 449-451.

Monte CL, Cabrita L, Oliveira C and Leitao J 2000. Assessment of genetic relationships among *Pyrus* species and cultivars using AFLP and RAPD markers. *Genetic Resources and Crop Evolution* 47: 257-265.

Pandit SS, Sirsha M, Giri AP, Pujari KH, Patil BP, Jambhale ND and Gupta VS 2007. Genetic diversity analysis of mango cultivars using inter simple sequence repeat markers. *Current Science* 93: 1135-1140.

Pavlikaki H, Navarro CP, and Fanourakis N 2004 Genetic relationships of different Greek landraces of cucumber (*Cucumis sativus*) as assessed by RAPDs. Progress in cucurbit genetics and breeding research Proceedings of Cucurbitaceae, the 8th EUCARPIA Meeting on Cucurbit Genetics and Breeding, Olomouc, Czech Republic, 12-17 July, 2004 pp 101-106.

Phippen WB, Kresovich S, Candelas F and McFerson JR 1997. Molecular characterization can quantify and partition variation among genebank holdings: a case study with phenotypically similar accessions of *Brassica oleracea* var. capitata L. (cabbage) 'Golden Acre'. *Theoretical and Applied Genetics* 94:227-234.

Pirseyedi SM, Mardi M, Davazdahemami S, Kermani MJ and Mohammadi SA 2005. Analysis of the genetic diversity 12 Iranian Damask rose (*Rosa damascena* Mill.) genotypes using amplified fragment length polymorphism markers. *Iranian Journal of Biotechnology* 3: 225-230.

Qin HL, You J and Gao JP 2002. RAPD analysis of 18 Chrysanthemum cultivars. *Acta Horticulturae Sinica* 29: 488-490.

Ravishankar KV, Chandrashekara P, Sreedhara SA, Dinesh MR, Lalitha A and Saiprasad GVS 2004. Diverse genetic bases of Indian polyembryonic and monoembryonic mango (*Mangifera indica* L) cultivars. *Current Science* 87: 870-871.

Roh MS and Ikeda H 2003. Genetic diversity of *Dendranthema pacificum* (Nakai) Kitam. native to Japan. *Acta Horticulturae* 620: 239-244.

Saavedra G, Spoor W and Harrier L 2001. Molecular markers and genetic base broadening in *Lycopersicon spp. Acta Horticulturae* 546: 503-507.

Sehrawat SK, Kumar R, Dahiya DS, Boora KS and Yadav R 2003. DNA fingerprinting of chrysanthemum cultivars using RAPDs. *Acta Horticulturae* 624: 479-485.

Singh AK, Behera TK, Chandel D, Sharma P and Singh NK 2007. Assessing genetic relationships among bitter gourd (*Momordica charantia* L.) accessions using inter simple sequence repeat (ISSR) markers. *Journal of Horticultural Science and Biotechnology* 82: 217–222.

Siragusa M, Pasquale F de, Abbate L and Tusa N 2006. Identification of sour orange accessions and evaluation of their genetic variability by molecular marker analyses. *HortScience* 41: 84-89.

Souza VAB de and Lima PS da C 2004. Genetic variability in mango genotypes detected by RAPD markers. *Acta Horticulturae* 645: 303-310.

Srivastava AP, Ramesh C, Sangeeta S, Shailendra R, Ranade SA and Vivek P 2007. A PCR-based assessment of genetic diversity, and parentage analysis among commercial mango cultivars and hybrids. *Journal of Horticultural Science and Biotechnology* 82: 951-959.

Staub JE and Ivandic V 2000. Genetic assessment of the United States national cucumber collection. *Acta Horticulturae* 510: 113-121.

Su YB, Lin C, Mao J, Mo XJ and Yang M 2004. RAPD identification of different carnation cultivars. *Acta Horticulturae Sinica* 31: 109-111.

Tanksley SD 1983. Molecular markers in plant breeding. *Plant Mol Biol Rep* 1:3-8.

Tejaswini, Paramesh H, Sreedhara SA and Lalitha A 2008. Molecular markers for working out genetic relationship among genotypes of carnation (*Dianthus caryophyllus* L.). *Indian Journal of Genetics and Plant Breeding* 68: 93-95.

Tonguc M and Griffiths PD 2004. Genetic relationships of Brassica vegetables determined using database derived simple sequence repeats. *Euphytica* 137: 193-201.

Villand J, Skroch PW, Lai T, Hanson P, Kuo CG, and Nienhuis J 1998. Genetic variation among tomato accessions from primary and secondary centers of diversity. *Crop-Science* 38: 1339-1347.

Vos P, Hogers R, Bleeker M, Reijans M, Van de Lee T, Hornes M, Frijters A, Pot J, Peleman J, Kuiper M and Zabeau M 1995. AFLP: a new technique for DNA fingerprinting. *Nucl. Acids Res* 23: 4407-4414.

Wunsch A, Gella R and Hormaza JI 2004. Molecular characterization of rootstocks for sweet cherry (*Prunus avium* L.). *Acta Horticulturae* 658: 599-602.

Wang JG, Che DD, Liu SK and Yang CP 2008. RAPD analysis of twenty six general species in *Gladiolus hybridus* Hort. *Bulletin of Botanical Research* 28: 321-324.

Wang Y, Gong YQ, Zhao TM, Liu G, Yu FM, Ye HL and Liu LW 2007. Optimization of SRAP-PCR system and cultivar molecular identification in tomato (*Lycopersicon esculentum* L.). *Journal of Nanjing Agricultural University* 30: 23-29.

Williams JGK, Kubelik AR, Livak KJ, Rafalski JA and Tingey S 1990 DNA polymorphisms amplified by arbitrary primers are useful as genetic markers. *Nucleic Acids Res* 18: 6531-6535.

Wu ZS, Li HL, Liu JH, Zuo ZR and Tian RC 2007. Analyses of genetic diversity among 65 Chrysanthemum cultivars based on AFLP. *Journal of Nanjing Forestry University Natural Sciences Edition* 31: 67-70.

Yang YJ, Zhang KC, Lin K and Jiang Q 2007. RAPD markers linked to the freestone gene of peach fruit. *Journal of Fruit Science* 24: 585-588.

Zhang D, Besse C, Cao MQ and Gandelin MH 2001. Evaluation of AFLPS for variety identification in modern rose (*Rosa hybrida* L.). *Acta Horticulturae* 546: 351-357.

Zhang J, McDonald MB and Sweeney PM 1997. Testing for genetic purity *Petunia* and *Cyclamen* seed using random amplified polymorphic DNA markers. *J Amer Soc Hort Sci* 32: 246–247.

Zhao LP, Liu LW, Gong YQ, Wang MX, Yu FM, and Wang LZ 2007. Cultivar fingerprinting in radish (*Raphanus sativus*) with SRAP and AFLP markers. *Bulletin of Botanical Research* 27: 687-693.

Zhao X and Kochert G 1993 Phylogenetic distribution and genetic mapping of a (GC)n microsatellite from rice (*Oryza sativa* L.). *Plant Mol Bio* 21: 607-614.

Zhu GF, Li DM and Guo ZF 2007. Genetic diversity and relationship of hybrid Cymbidium based on AFLP marker. *Acta Horticulturae Sinica* 34: 417-424.

The Science of Horticulture : Vol. 1, 2011
© K.V. Peter (ed.), pp. 325-346
New India Publishing Agency, New Delhi (India)
E-mail : newindiapublishingagency@gmail.com
Web: www.bookfactoryindia.com

Chapter 14 Transgenic Ornamentals with Reduced Ethylene Production and Perception

SHIGERU SATOH, YUSUKE KOSUGI AND
TAKAKO NARUMI

Ethylene is a primary plant hormone involved in the senescence of plant tissues including flowers and leaves. In carnation flowers, inhibition of the synthesis or action of ethylene delays the onset of senescence symptoms and lengthens the vase life of flowers. Leaves of cut chrysanthemum of some cultivars frequently become yellow, sometimes prior to the onset of flower senescence, in response to exogenous ethylene. Generation of transgenic plants with suppressed production or action of ethylene is an excellent way to lengthen the longevity of these ornamental plants. This paper reviews our studies on generation and characterization of carnations with reduced ethylene production and chrysanthemum with reduced sensitivity to ethylene.

Ethylene is a primary plant hormone involved in the senescence of cut flowers of many ornamentals like carnation, sweetpea and delphinium. It is produced in a large amount at a later stage of flower senescence. The increased ethylene production accelerates withering and dropping of petals. Inhibition of the synthesis or action of ethylene delays the onset of senescence symptoms and extends the vase life of flowers.

In cut flowers susceptible to ethylene, preservatives containing silver thiosulphate anionic complex ion (STS) (Veen, 1979) are currently being used to lengthen their vase life. However, STS is harmful to the environment, and may be banned in the near future. Thus, transgenic ornamentals with suppressed production or action of ethylene are excellent alternatives for the preservation of flower longevity. So far, transformation with transgenes related to the biosynthesis and action of ethylene was successfully used to down-regulate ethylene production or responsiveness to ethylene in petunia (Wilkinson *et al.* 1997), torenia (Aida *et al.* 1998) and carnation (Savin *et al.* 1995, Bovy *et al.* 1999). Cut flowers of these transgenic plants have a prolonged vase life compared with those of non-transgenic plants.

Carnations are used as ornamental plants in the form of potted plants as well as in the form of cut flowers. The preservatives described above cannot be used for potted carnation plants, since the chemicals are usually administered to cut flowers through vascular transport by immersing the cut stem end in solutions containing the preservative. Therefore, the generation of transgenic plants is a promising way to both lengthen the display time of potted carnation plants and do so the vase life of cut flowers of ordinary carnation plants.

Cut chrysanthemum is one of the most popular cut flowers and is of economic importance in the floriculture industry. Leaves of cut chrysanthemum frequently become yellow spontaneously, sometimes prior to the onset of flower senescence (Doi *et al.* 2003, 2004, Ferrante *et al.* 2003). Moreover, exposure to ethylene during storage, transportation or display of the cut flowers accelerates leaf yellowing although the sensitivity to ethylene differs with the cultivar (Reyes-Arribas *et al.* 2001, Doi *et al.* 2003). Leaf yellowing is a characteristic of leaf senescence; it makes cut flowers unsightly, lowers their quality, value and shortens vase life. Therefore, transgenic chrysanthemum with reduced ethylene sensitivity is expected to have very green leaves, and a greatly extended vase life.

In this chapter, we review our recent studies on generation and characterization of carnations with reduced ethylene production and chrysanthemum with reduced sensitivity to ethylene.

Transgenic Carnations with Reduced Ethylene Production

1. Senescence and Ethylene Production of Carnation Flowers

Cut carnation flowers produce a large amount of ethylene several days after full opening of the flower (Borochov and Woodson 1989, Abeles *et al.* 1992, Reid and Wu 1992). The produced ethylene causes in-rolling of petals and wilting of the whole flower. Therefore the prevention of synthesis or action of ethylene inhibits the onset of senescence symptoms and extends the vase life of the flower.

Ethylene is synthesized in plant tissues through the following pathway: L-methionine ® *S*-adenosyl-L-methionine ® 1-aminocyclopropane-1-carboxylate (ACC) ® ethylene. ACC synthase and ACC oxidase catalyze the last two reactions (Yang and Hoffman 1984, Kende 1993). Previously, three genes encoding ACC synthase (*DC-ACS1, DC-ACS2* and *DC-ACS3*) and one gene encoding ACC oxidase (*DC-ACO1*) identified in carnation (Wang and Woodson 1991, Park *et al.* 1992, Henskens *et al.* 1994, Jones and Woodson 1999). Recently, Harada *et al.* (2009) identified another ACC synthase gene, *DcACS1b*, which is very similar to *DcACS1a* (renamed from *DC-ACS1*), and Norikoshi *et al.* (2008) that for another ACC oxidase [tentatively *DcACO2(DC-ACO2)*]. The expression of these genes is regulated in a tissue-specific manner in relation to flower senescence. *DC-ACS1* and *DC-ACO1* genes play a crucial role in autocatalytic and massive ethylene production in the petals of carnation flowers undergoing senescence.

So far, transgenic carnation plants with suppressed production or action of ethylene have been generated; e.g., the line transformed with ACC oxidase (*DC-ACO1*) cDNA in antisense orientation (Savin *et al.* 1995) and the line transformed with an *Arabidopsis thaliana etr1-1* allele capable of rendering ethylene insensitivity (Bovy *et al.* 1999). Cut flowers of these transgenic lines have a prolonged vase life compared with those of the non-transgenic plants.

In a series of our studies, several lines of transgenic carnation were generated by transforming with *DC-ACO1* cDNA in sense orientation (Kosugi *et al.* 2002) and *DC-ACS1* cDNA in sense orientation (Iwazaki *et al.* 2004). In addition, carnation plants, which are used as potted plants, were transformed with *DC-ACO1* cDNA in sense and antisense orientation (Kinouchi *et al.* 2006).

2. Transformation Vectors

The transformation vectors harboring carnation ACC oxidase cDNA (*DC-ACO1, s/a ACO* transgenes) in sense or antisense orientation were constructed using pMLH2113-GUS (Mitsuhara *et al.* 1996) and pIG121-Hm (Ohta *et al.* 1990) (Fig. 1). The transformation vector pMLH2113/*sACO* was prepared by replacing a *GUS* gene with the coding sequence of *DC-ACO1* cDNA in sense orientation in pMLH2113-GUS (Kosugi *et al.* 2000). The transformation vectors, pIG121/*aACS* and pIG121/*sACS*, were prepared by replacing the *GUS* gene with *DC-ACS1* cDNA in sense (*sACS* transgene) or antisense (*aACS* transgene) orientation in pIG121-Hm (Iwazaki *et al.* 2004). Similary, the transformation vectors, pIG121/*aACO* and pIG121/*sACO*, were prepared using *DC-ACO1* cDNA instead of *DC-ACS1* cDNA (Fig. 1). The transformation vectors have an *NPT II* gene and an *HPT* gene, which enable selection of the transformants in the presence of Kanamycin or Hygromycin, respectively.

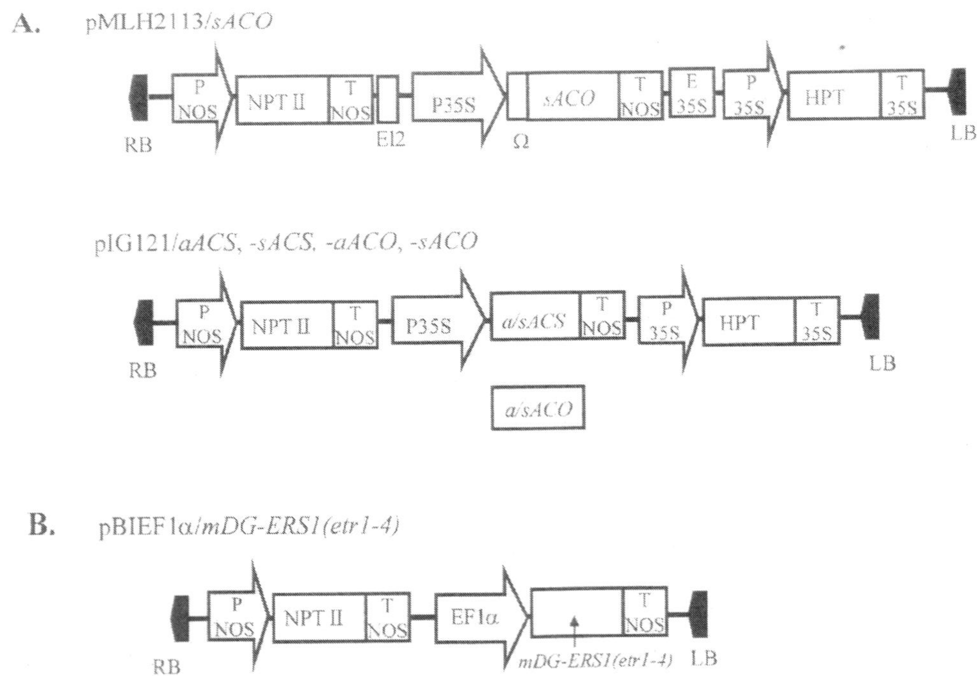

Figure 1. Schematic representation of transformation vectors. **A**. pMLH2113/*sACO*, pIG121/*aACO, -sACO*, *-aACS* and *-sACS* which harbor carnation *DC-ACO1* cDNA or *DC-ACS1* cDNA in sense or antisense orientation. **B**. pBIEF1a/*mDG-ERS1(etr1-4)* which harbor a chrysanthemum ethylene receptor *DG-ERS1* cDNA with one-nucleotide substitution corresponding to that present in *Arabidopsis etr1-4*. RB and LB, right and left border of T-DNA, respectively; NPT II, neomycin phosphotransferase II gene; P35S, cauliflower mosaic virus 35S (CaMV 35S) promoter; T35S, CaMV 35S terminator; PNOS, the promoter of nopalin synthase gene; TNOS, the terminator of nopalin synthase gene; E35S, CaMV 35S enhancer; El2, the upstream sequence (|419 ' |90) of CaMV 35S promoter x 2; W, 5'-untranslated sequence of tobacco mosaic virus; HPT, Hygromycin phosphotransferase gene; *sACS* and *aACS*, *DC-ACS1* cDNA in sense and antisense orientation, respectively; EF1a, the promoter of tobacco elongation factor 1a gene.

3. Generation and Cultivation of Transgenic Carnation Plants

Kosugi *et al.* (2000, 2002) transformed a carnation cultivar 'Nora' with *DC-ACO1* cDNA in sense orientation (*sACO* transgene) and generated a transgenic carnation line (sACO-1).

Transformation by *Agrobacterium*-mediated gene transfer and regeneration of carnation plants were performed by the method of Firoozabady *et al.* (1995) with modifications. Briefly, leaf explants from shoot culture were mixed with *Agrobacterium* in a nutrient solution [BD medium: MS medium supplemented with B5 vitamins, 3% sucrose, 0.059% MES, 0.5 mg l^{-1} BA and 0.5 mg l^{-1} 2,4-D] for 15 min, and cocultivated on another medium supplemented with 100 mM acetosyringone for a week at 24 °C in the dark. For selection of transformants, Hygromycin at 20 mg l^{-1} was used as a selection agent, since a preliminary observation showed that Kanamycin up to 100 mg l^{-1} allowed frequent escape formation. Survived explants were subcultured repeatedly on solid IT medium [MS medium supplemented with B5 vitamins, 3% sucrose, 0.059% MES, 0.5 mg l^{-1} indolebutyric acid, 0.5 mg l^{-1} thidiazuron and 0.25% Gerlite] until they formed normal shoots with distinct internodes. The shoots were rooted and transplanted to vermiculite in pots, then to soil. The plants were grown in a containment greenhouse until they flowered. Similarly, Iwazaki *et al.* (2004) generated 2 lines of 'Nora' carnation transformed with *DC-ACS1* cDNA in sense orientation (*sACS* transgene) and 3 lines transformed with *DC-ACS1* cDNA in antisense orientation (*aACS* transgene), respectively.

'Lillipot' carnation, which is used as potted plants, were transformed with *aACO and ACO* transgenes (Fig. 1) by the method described above, but the following modifications were done to improve transformation efficiency. Namely, acetosyringone at 100 mM was added to the medium for shoot culture prior to leaf explant preparation and pre-culture of *Agrobacterium* in addition to media for infection and coculture, and water was used instead of BD medium for infection and coculture.

4. Selection of Transformants and Transformation Efficiency

An example is shown below with the results of transformation of 'Lillipot' carnation. The leaf explants (shoot explants) of 'Lillipot' carnation were transformed with transformation vectors harboring *aACO or sACO* transgenes and subjected to selection with Hygromycin for one month, and subsequent re-differentiation of leaflets for one month. The leaflets were grown further for 3 months to form shoot clusters. The integration of *HPT* gene and *aACO or sACO* transgenes was confirmed by PCR-analysis (Fig. 2) using genomic DNAs extracted from leaflet segments of shoot clusters. Table 1 summarizes the results of transformation experiments in 'Lillipot' carnation. The transformation efficiency was 1.5-7.0 % for each transgene. As a whole, out of 600 lines from each leaf explant, 30 lines contained the *HPT* gene and of those lines 27 contained *aACO or sACO* transgenes (Fig. 2), giving an average transformation efficiency of 4.5%.

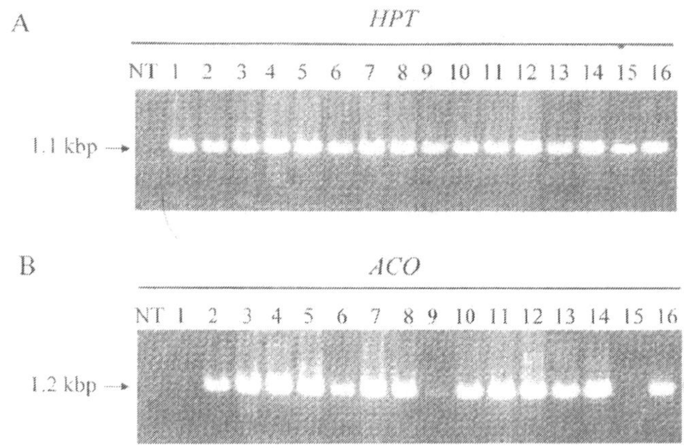

Figure 2. PCR analysis of integrated *HPT* gene and *sACO* transgene in 'Lillipot' transgenic lines. Genomic DNAs extracted from lines transformed with pMLH2113/*sACO* were analyzed by PCR with combinations of appropriate upstream and downstream primers for *HPT* gene, and those for *sACO* transgene. The arrows show 1.1 kbp PCR amplificates for *HPT* gene and 1.2 kbp PCR amplificates for *sACO* transgene. NT, non-transformed 'Lillipot'; Nos 1-16, respective lines transformed with pMLH2113/*sACO*.

5. Confirmation of Integration of Transgenes by Southern Blot Analysis

Integration of respective transgenes was confirmed by Southern blot analysis. Preliminary experiments showed that the sACO-1 line harbored at least 5 copies of the *sACO* transgene (Satoh *et al.* 2000). Among 5 transgenic lines transformed with *sACS* and *a ACS* transgenes (Iwazaki *et al.* 2004), the sACS-1 line harboring *sACS* transgene had the severest reduction in ethylene production in flowers (see below). Southern blot analysis revealed that the sACS-1 line had two copies of the *sACS* transgene (Fig. 3). Fig. 4 shows Southern blot analysis of four lines of 'Lillipot' carnation transformed with pMLH2113/*sACO*. This analysis indicated the integration of multiple copies of *sACO* transgenes into the lines.

6. Suppressed Senescence and Ethylene Production of Transgenic Carnation Flowers

The appearance of cut flowers during the senescence period (10 days) in the sACO-1 line significantly differed from that in non-transformed 'Nora' carnation (Fig. 5). Flowers of the sACO-1 line remained turgid without petal in-rolling until about day 10, but started to show desiccation and discoloration in the rim of petals on day 11 or later. By contrast, flowers of the non-transformed 'Nora' remained turgid until day 5, showed in-rolling of petals on day 6, and completely wilted on day 9. Petal in-rolling at the onset of flower wilting is a well-known characteristic of ethylene-dependent senescence of carnation flowers. The lack of petal in-rolling and prolonged vase-life in flowers of the sACO-1 line suggests suppressed ethylene production in the flowers. Actually, flowers of non-transformed 'Nora' showed a climacteric rise in ethylene production attaining a maximal rate on day 5, whereas flowers of the sACO-1 line produced a negligible amount of ethylene during the senescence period (Fig. 6).

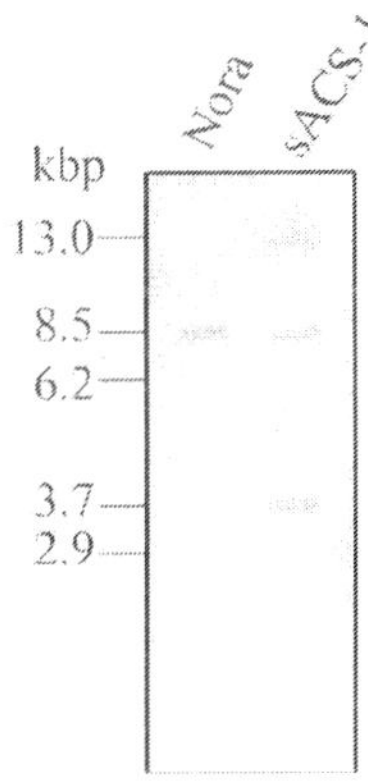

Figure 3. Southern blot analysis of *sACS* transgenes in the sACS-1 line derived from 'Nora' carnation. The *Hin*d III digests of genomic DNAs from non-transformed 'Nora' and the sACS-1 line were separated on agarose-gel and hybridized with a DC-ACS1 probe (560 bp).

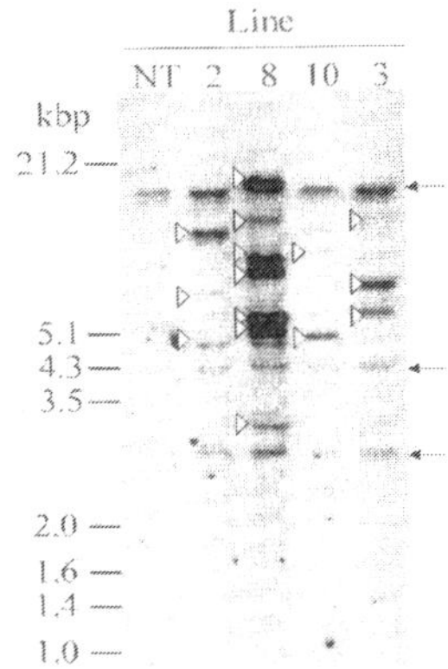

Figure 4. Southern blot analysis of 'Lillipot' lines transformed with pMLH2113/*sACO*. Four lines were randomly selected from 14 lines (Table 1) for *Hin*d III-digestion, respectively. The digests were separated on agarose-gel and hybridized with a 986-bp DC-ACO1 OR probe. Open arrowheads in the panel show extra bands that were absent in the digests of the non-transformed 'Lillipot' genomic DNA. NT, non-transformed 'Lillipot'.

The suppressed ethylene production in the sACO-1 line was accompanied by no accumulation of mRNA for the *DC-ACO1* gene in the gynoecia and mRNAs for the *DC-ACO1* and *DC-ACS1* genes in the petals (Fig. 7). Previously, it was shown that in naturally senescing carnation flowers, ethylene is first produced in the gynoecium; and the evolved ethylene acts on petals, as a diffusible signal, to induce the expression of *DC-ACO1* and *DC-ACS1* genes resulting in autocatalytic ethylene production in the petals (Shibuya *et al.* 2000). The results indicate that the *sACO* transgene inhibits the expression of *DC-ACO1*, probably by cosuppression, in the gynoecium, which then results in no ethylene production in the whole flower (gynoecium plus petals) of the sACO-1 line.

Figure 5. Senescence of carnation flowers of the non-transformed 'Nora' (**top**) and the sACO-1 transgenic line (**bottom**). Flowers of both lines were harvested at the full-opening stage (day 0), and then left in a glass vessel with their cut stem end in distilled water. Photographs were taken 0, 5, 6, 7, 8, 9 and 10 days after the start of the experiment.

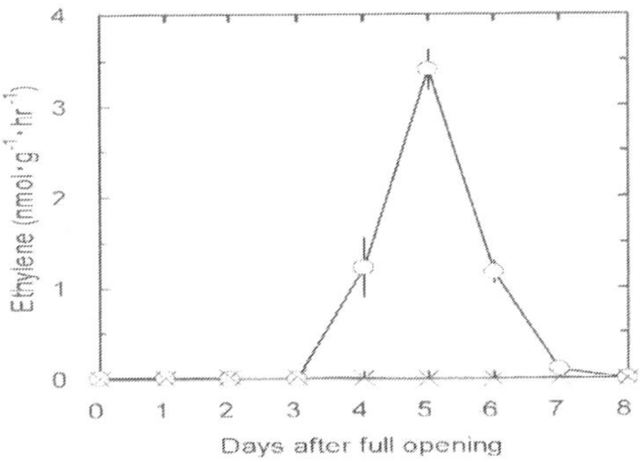

Figure 6. Ethylene production from cut carnation flowers during senescence in the non-transformed 'Nora' (control) and the sACO-1 lines. Five flowers each of both lines were harvested at the full-opening stage (day 0) and their ethylene production measured daily. Data are shown by the mean ± SE of 5 flowers. Z, 'Nora' control; ×, sACO-1.

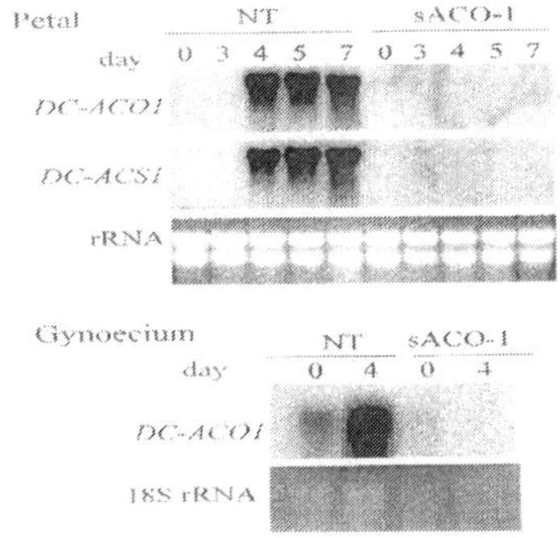

Figure 7. RNA gel blot analysis of mRNAs for *DC-ACO1* and *DC-ACS1* in petals and gynoecium of the non-transformed 'Nora' control (NT) and the sACO-1 lines during natural senescence. Petals were isolated from cut flowers 0, 3, 4, 5 and 7 days, but gynoecia 0 and 4 days after full opening of flowers. Total RNAs isolated from the petals were separated on an agarose gel and hybridized to DIG-labeled *DC-ACO1* and *DC-ACS1* probes, but in the case of total RNAs isolated from the gynoecium a ^{32}P-labeled DC-ACO1 probe was used .

Furthermore, cut flowers of 2 lines transformed with *sACS* transgene and 3 lines with *aACS* transgene (Iwazaki *et al.* 2004) showed suppressed ethylene production during natural senescence as compared with flowers of non-transformed 'Nora' carnation. Among 5 transgenic lines, the sACS-1 line harboring *sACS* transgene had the severest reduction in ethylene production in flowers, and the flowers lost their vase-life by drying and discoloring in the rim of petals, which are typical to ethylene-independent senescence of carnation flowers.

In 5 'Lillipot' lines transformed with *sACO* and *aACO* transgenes, cut flowers of the pMLH-sACO-2 line and -12 lines had vase-lives of 7.6 ± 0.4 and 6.3 ± 0.5 days, respectively, whereas that of the NT control line was 2.8 ± 0.2 days. Other transgenic lines had vase-lives insignificant from that of the non-transformed (NT) control. Flowers of the pMLH-sACO-2 and -12 lines eventually deteriorated with desiccation and discoloration in the rim of petals. Flowers of the NT control showed a climacteric rise in ethylene production, attaining a maximal rate (2.90 nmol^{-1} g^{-1} h^{-1}) on day 3. Flowers of the pMLH-sACO-2 and -12 lines produced very small amounts of ethylene during senescence period of 8 days and their maximum ethylene production rates were around 10% that of the NT control. With pMLH-sACO-2 and -12 lines, *DC-ACS1* and *DC-ACO1* transcripts were not detected in their petals during senescence of 6 days, but they were accumulated abundantly in the petals of the NT control flowers.

7. Analysis of Ethylene Production and Wilting in Carnation Petals Using the Transgenic sACO-1 Line

Ethylene, evolved from the gynoecium, induces autocatalytic ethylene production and wilting in carnation petals. The autocatalytic ethylene production is due to expression of *DC-ACS1* and *DC-ACO1*. Petal wilting in flower senescence is caused by decomposition of cell constituents by hydrolytic enzymes such as proteinase, lipase and nuclease. Cysteine proteinase (CP ase) is probably one of the enzymes responsible for hydrolytic degradation of cell components leading to cell death during senescence of petals (Jones *et al.* 1995, Panavas *et al.* 1998, 1999). A CPase gene (*DC-CP1*) is up-regulated during natural and exogenous ethylene-induced senescence of carnation petals. Previously, it was thought that the ethylene production and wilting were regulated in concert and could not be separated since the two events occur simultaneously.

Kosugi *et al.* (2000) examined the response of petals of transgenic carnation line sACO-1 to exogenous ethylene with regard to the induction of petal in-rolling, ethylene production and some genes involved in these events. As described above, cut flower of the sACO-1 line produced only a trace amount of ethylene during the senescence period, and had a vase life about twice longer than flower of the control 'Nora' carnation. Interestingly, the treatment with ethylene of petals detached from the sACO-1 flowers caused accumulation of transcripts for *DC-CP1* and petal wilting, but it did not induce accumulation of transcripts for *DC-ACO1* and *DC-ACS1* and ethylene production (Fig. 8). These findings indicated that the exogenous ethylene-induced expression of *DC-ACS1* and *DC-ACO1* genes, leading to autocatalytic ethylene production, and that of *DC-CP1* genes, leading to petal wilting, were regulated separately in the petals of the sACO-1 flowers. Based on these finding, Satoh *et al.* (2000) suggested the presence of two subset of ethylene responses leading to the expression of respective genes on carnation petals (Fig. 9).

Transgenic Chrysanthemum with Reduced Ethylene Sensitivity

1. Leaf Senescence and an Ethylene Receptor Gene of Chrysanthemum

As described in Introduction, it is possible that leaf yellowing is prevented and the vase life greatly extended in transgenic chrysanthemum plants with reduced ethylene sensitivity. Previously, a mutated ethylene receptor gene (*Arabidopsis etr1-1* gene) was successfully used to reduce sensitivity to ethylene in petunia (Wilkinson *et al.* 1997) and carnation (Bovy *et al.* 1999). However, in many cases transgenes of heterologous origin do not work well in chrysanthemum plants, probably because of gene silencing, raising doubt on the use of heterologous transgenes like the *etr1-1* gene in chrysanthemum. To solve this problem, it may be helpful to use ethylene receptor genes obtained from chrysanthemum itself to down-regulate ethylene responsiveness in transgenic chrysanthemum plants, because the native gene might be less affected by the gene-silencing mechanism. Narumi *et al.* (2005a) obtained a cDNA encoding ethylene receptor (DG-ERS1) from chrysanthemum [*Chrysanthemum morifolium*, formerly *Dendranthema grandiflorum* (Ramat.) Kitamura]. The cDNA (2427 bp) contained an open reading frame of 1920 bp coding for 640 amino acids. The predicted DG-ERS1 protein had an amino-terminal ethylene sensor domain and a histidine kinase domain, but lacked a receiver domain. Then, *DG-ERS1* mRNA levels in flower and leaf tissues of 'Seiko-no-makoto' chrysanthemum, which is an ethylene-sensitive cultivar, was examined. Real-time PCR analysis revealed that *DG-ERS1* mRNA was present in a large amount in ligulate corollas (hereafter, petals for convenience) and mature leaves when the flowers were fully open, and the amount decreased with time or in response to a 12-h ethylene treatment. Ethylene receptors have been shown to act as negative regulators of ethylene responses in plants (Hua and Meyerowitz 1998, Ciardi *et al.* 2000, Tieman *et al.* 2000). This indicates that the level of ethylene receptors is inversely related to the sensitivity to ethylene, and the decrease in the level of *DG-ERS1* mRNA in flower and leaf tissues of 'Seiko-no-makoto' chrysanthemum may increase the sensitivity to ethylene and, hence, accelerate the senescence of flowers and leaves. These findings suggest the involvement of DG-ERS1 in the perception of ethylene in cut chrysanthemum plants, especially in those of 'Seiko-no-makoto' cultivar.

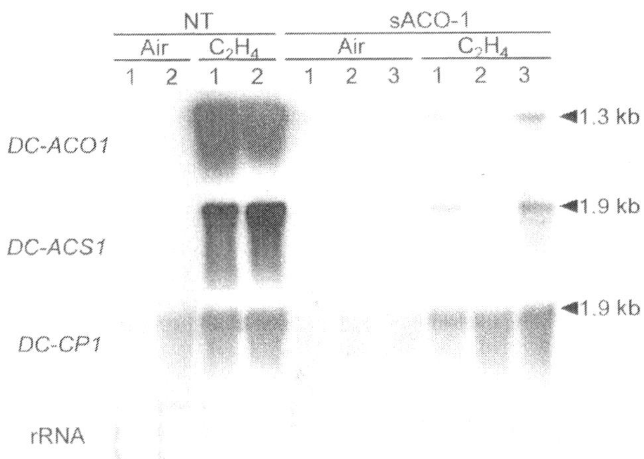

Figure 8. Accumulation of transcripts for *DC-ACO1*, *DC-ACS1* and *DC-CP1* in petals of control (NT) and sACO-1 transgenic lines after treatment with ethylene at 10 ml l^{-1} for 18 h. Analysis was performed with petals obtained from two individual flowers (1 and 2) for the control line and three individual flowers (1, 2 and 3) for the transgenic line.

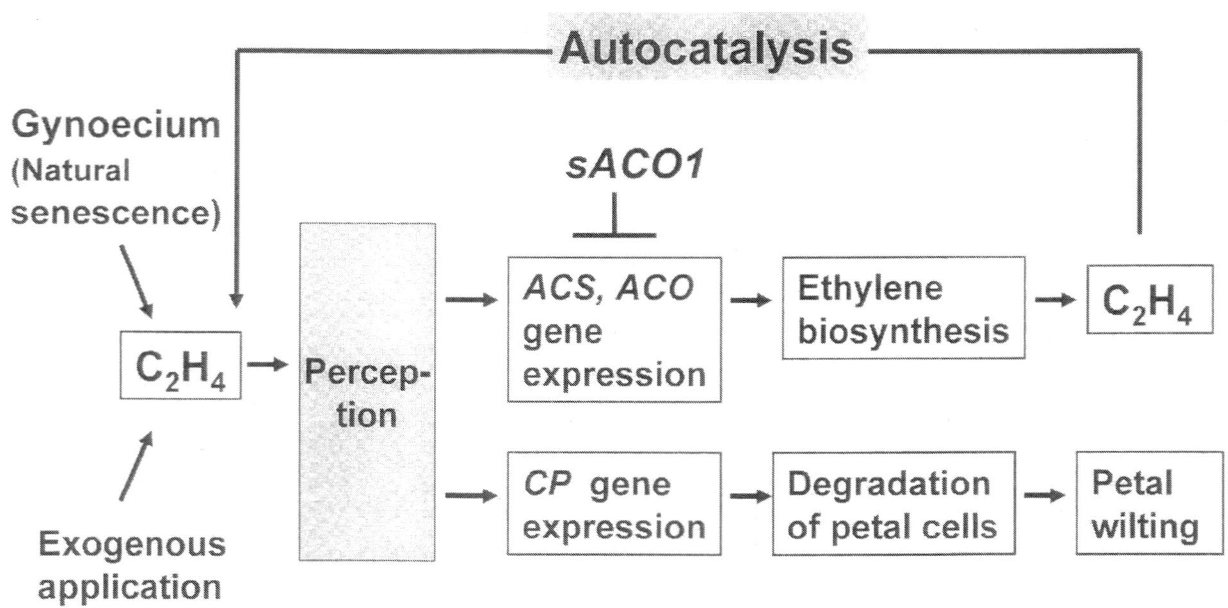

Figure 9. Two subsets of ethylene responses in carnation petals. The figure depicts two putative ethylene signaling pathways responsible for autocatalytic ethylene production and wilting during petal senescence in carnation flowers.

2. Transformation Vectors

Narumi *et al.* (2005b) produced mutated ethylene receptors genes (*mDG-ERS1*s) from the *DG-ERS1* cDNA by introducing one-nucleotide substitutions corresponding to those present in *Arabidopsis etr1-1, etr1-2, etr1-3, etr1-4,* and tomato *Nr,* and introduced each of the *mDG-ERS1*s into chrysanthemum plants. However, only the results with the mutated ethylene receptor gene *mDG-ERS1(etr1-4)* are shown hereafter, since it was revealed that this construct tended to be more effective in conferring reduced ethylene sensitivity in chrysanthemum than the others.

A one-nucleotide substitution was introduced into *DG-ERS1* cDNA with site-directed mutagenesis by PCR using appropriate primers; the mutation replaced A with T at the position 187 bp (when +1 is A of the ATG starting codon of *DG-ERS1* cDNA), resulting in the conversion of Ile-63 in DG-ERS1 to Phe-63 in the resultant mutated protein. This one-nucleotide substitution was the same as that present in the nucleotide sequence of mutated *Arabidopsis etr1-4* gene (Chang *et al.* 1993).

A transformation vector pBIEF1a/*mDG-ERS1(etr1-4)* (Fig. 1B) was constructed from pBIEF1a by replacing the *GUS* gene with the *mDG-ERS1(etr1-4)* cDNA. In turn, the pBIEF1a vector was constructed from pBI101 (Clonetech) by inserting the promoter of a tobacco elongation factor 1a (EF1a) gene. The construct in the transformation vector was named *EF1a/mDG-ERS1(etr1-4)*. The transformation vector contained the *NPT II* gene for the selection of transformants.

3. Generation and Cultivation of Transgenic Chrysanthemum Plants

The transformation vector was introduced into *Agrobacterium tumefaciens* EHA105 (Hood *et al.* 1993), and to 'Sei-Marine' chrysanthemum plants according to the procedure described previously (Aida *et al.* 2004). In brief, about one month after *Agrobacterium* inoculation, putative transformants (calli) were formed on the infected leaf explants. About two months after inoculation, adventitious buds emerged from the calli. The adventitious buds were detached, transplanted to the elongation medium and grown to normal plantlets. As a whole, a total of 62 independent regenerated lines were obtained from 1600 explants transformed with the transformation vector pBIEF1a/*mDG-ERS1(etr1-4)*. Table 2 summarizes the results of transformation experiments.

4. Selection of Transformants and Transformation Efficiency

Forty-five regenerated lines out of 62 were Paromomycin (an antibiotic related to Kanamycin)- resistant, i.e. putative transgenic lines. Finally, it was shown that 42 out of 45 lines contained the transgene *EF1a/mDG-ERS1 (etr1-4)*, with a 2.6% transformation efficiency (Fig. 10).

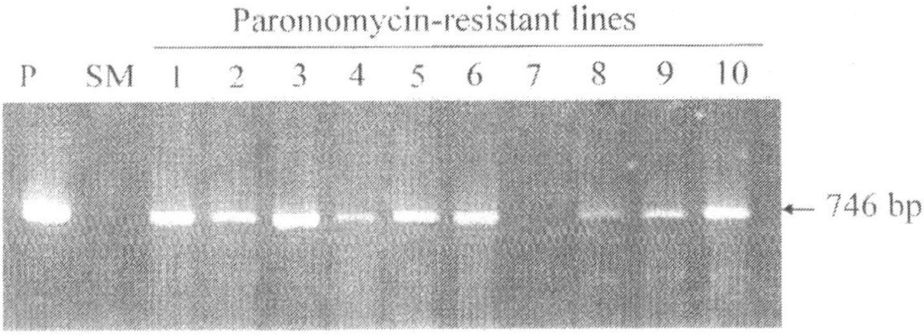

Figure 10. PCR analysis of integrated *FF1a/mDG-ERS1 (etr1-4)* transgene in non-transformed and transformed lines of 'Sei-Marine' chrysanthemum. Genomic DNAs extracted from Paromomycin-resistant lines of FF1a/mDG-ERS1(etr1-4) transformant were analyzed by PCR with a combination of appropriate upstream and downstream primers, whose complementary sequences located in the EF1a promoter and *mDG-ERS1(etr1-4)*, respectively. PCR analyses were conducted on all of the transformed lines, but only the results for ten randomly-selected lines are shown. In the figure, lines were randomly numbered for convenience. The arrow shows the PCR amplificate of 746 bp. P, pBIEF1a::*mDG-ERS1(etr1-4)* plasmid. SM, non-transformed 'Sei-Marine'.

5. Confirmation of Integration of the Transgene by Southern Blot Analysis

The integration of the transgene was confirmed by PCR and Southern blot analyses. PCR analysis gave amplification corresponding to a partial sequence of the *EF1a/mDG-ERS1(etr1-4)* transgene (Fig. 10). Southern blot analysis was conducted with *Hin*d III- and *Xba* I-digests of genomic DNAs obtained from the lines and the 300-bp long DG-ERS1 DNA probe. In Fig. 11, the results with 'Sei-Marine' and representative three transformed lines are shown. Ornamental chrysanthemums are usually hexaploids, and are known to be generated from diploid wild chrysanthemums by cross breeding. In the previous study, Southern blot analysis of genomic DNAs extracted from several cultivars of hexaploid chrysanthemums revealed the presence of multiple *DG-ERS1* genes and its homologous genes in the hexaploid chrysanthemum (Narumi *et al.*, 2005a). Six distinct bands hybridized to the DG-ERS1 DNA probe were detected in the *Hin*d III- and *Xba* I-digests of the genomic DNA of 'Sei-Marine' (Fig. 11). Irrespective of sensitivity to ethylene (as described in the next section), all the lines had one or two (and three in the line not shown in Fig. 11) extra bands in the *Hin*d III-digests of genomic DNAs, which were absent in 'Sei-Marine' genomic DNA (Fig. 11A). Moreover, with *Xba* I-digests of genomic DNAs, the DNA fragment of 1514-bp long, which corresponded to the partial sequence of the *EF1a/ mDG-ERS1(etr1-4)* transgene, was detected in all of the lines of the mDG-ERS1 (etr1-4) transformant (Fig. 11B), but not in non-transformed 'Sei-Marine'.

6. Reduced Ethylene Sensitivity in in vitro *Plantlets of the Transgenic Lines*

Sensitivity to ethylene was examined using the Paromomycin-resistant lines of the putative transformant. Four *in vitro* plantlets, 7-9 cm high with 8-

10 leaves, from each of 40 independent transformed lines were treated with ethylene at 10 ml l^{-1} for 6 days. The plantlets of respective lines showed various degrees of leaf yellowing in response to *in-vitro* exposure to ethylene, and were classified into three groups according to the magnitude of leaf yellowing (Fig. 12). Out of 40 lines, 25 lines showed ethylene sensitivity similar to that of the non-transformed 'Sei-Marine' (severe leaf yellowing), 8 lines reduced ethylene sensitivity (yellowing in one or two of the lowermost leaves of the plants), and 7 lines markedly reduced ethylene sensitivity (no leaf yellowing).

Narumi (unpublished results) detected the transcript of the mutated ethylene receptor gene *mDG-ERS1(etr1-4)* in leaves of all the transgenic lines, but did not find any relationship between the amount of *mDG-ERS1(etr1-4)* transcript and the magnitude of reduction in sensitivity to ethylene. The expression of the integrated *mDG-ERS1(etr1-4)* gene might confer reduced ethylene sensitivity through a mechanism similar to that responsible for ethylene insensitivity of the *Arabidopsis etr1-1* mutant (Chang *et al.* 1993).

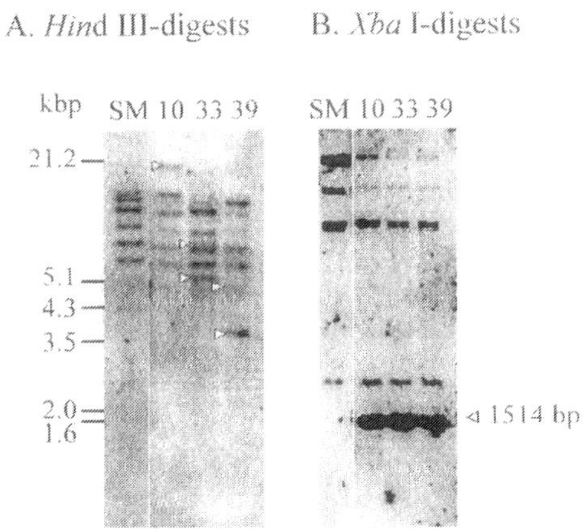

Figure 11. Southern blot analysis of lines of the mDG-ERS1(etr1-4) transformant of 'Sei-Marine' chrysanthemum. Three lines were randomly selected from the mDG-ERS1(etr1-4) transformant. The *Hin*d III- or *Xba* I-digests of genomic DNAs from the three lines were separated on agarose-gel and hybridized with a 300-bp DG-ERS1 DNA probe for *DG-ERS1* cDNA. Open arrowheads in the panel of *Hin*d III-digests (**A**) show extra bands that were not obtained with the non-transformed 'Sei-Marine' genomic DNA. The 1514-bp *Xba* I-fragment is indicated in the panel of *Xba* I-digest (**B**). SM, non-transformed 'Sei-Marine'.

7. Reduced Leaf Senescence (yellowing) in Soil-grown Mature Plants after Exposure to Exogenous Ethylene

In the following experiments, the non-transformed (NT) control 'Sei-Marine' (SM) and seven transgenic lines (No 10, 14, 17, 19, 33, 39 and 41) were used. These transformants were preliminary selected from *in vitro* plantlets by observing reduced sensitivity to exogenous ethylene as compared with SM. When *in vitro* plantlets, which had been grown aseptically in a plastic culture vessel, were transplanted to soil in a plastic container, all plants of SM and transgenic lines grew vigorously and flowered almost normally.

'Sei-Marine' No. 8 No. 18 No. 41

Figure 12. Leaf yellowing profiles of the control and putative transgenic lines of chrysanthemum after treatment with ethylene. In vitro plantlets of the control non-transformed 'Sei-Marine' and the lines of the mDG-ERS1(etr1-4) transformant were enclosed in a tight plastic container with ethylene at 10 ml l^{-1} and left for 6 days. According to the magnitude of leaf yellowing, the lines were classified into three groups of ethylene sensitivity; the lines with ethylene sensitivity similar to that of 'Sei-Marine' (ES), the lines with reduced ethylene sensitivity (RES1 lines showing yellowing in one or two of the lowermost leaves) and the lines with more reduced ethylene sensitivity (RES2 lines showing no leaf yellowing). In the photographs are shown the control 'Sei-Marine' and representatives of the ES, RES1 and RES2 lines, i.e., No. 8, No. 18 and No. 41 in this order.

Four or 5 shoots, 20 cm high with 8-10 leaves, were cut from soil-grown plants maintained as perennial plants in plastic containers, and treated with ethylene at 10 ml l^{-1} for 9 days. The shoots of some transgenic lines showed reduced leaf yellowing as compared with those of Sei-Marine (SM). A typical profile of leaf yellowing is shown in Fig. 13, in which a shoot of SM showed severe yellowing (almost all the leaves showed yellowing), but a shoot of No.

Sei-Marine No. 33

Figure 13. Leaf yellowing profiles of the non-transformed 'Sei-Marine' control and a transgenic line No. 33 after treatment with ethylene. Shoots of 20-cm long were detached from soil-grown plants, enclosed in a glass container with 10 ml l^{-1} ethylene, and left for 9 days at 25°C under continuous white fluorescent light.

33 line did not. Fig. 14 shows the degree of leaf yellowing caused by exposure to exogenous ethylene in SM and seven transgenic lines. The leaf yellowing was significantly suppressed in shoots of transgenic lines, No. 10, 17, 19 and 33, as compared with SM, whereas, the reduction in leaf yellowing was not significant in shoots of the transgenic lines No. 14, 39 and 41. These results suggested that the sensitivity to ethylene, shown by leaf yellowing was reduced in the transgenic line No. 10, 17, 19 and 33.

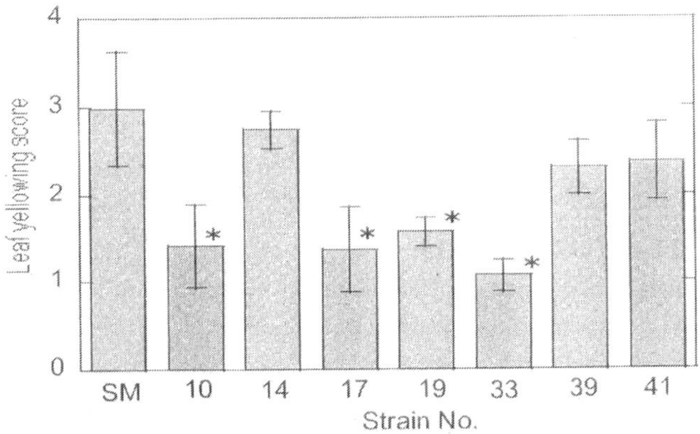

Figure 14. Leaf yellowing after treatment with ethylene in the 'Sei-Marine' and transgenic lines. Shoots of 20-cm long were detached from soil-grown plants of the 'Sei-Marine' and transgenic lines, enclosed in a glass container with 10 ml l^{-1} ethylene, and left for 9 days at 25°C under continuous white fluorescent light. Leaf yellowing score was determined as 0, 1 , 2, 3 and 4 when leaf yellowing covered 0, 1-10, 11-40, 41-70 and 71-100% from the bottom of the shoot, respectively. Leaf yellowing score was shown as the average of 4 or 5 shoots. Data were shown by the mean ± SE of 4 experiments. * shows significant difference from the value in 'Sei-Marine' at p=0.05 by Dunnett's multiple test.

8. Reduced Leaf Deterioration During Storage in Darkness without Ethylene Treatment

We examined the leaf deterioration (yellowing, browning and rotting) of cut shoots kept in darkness without ethylene treatment, in SM and the transgenic lines. Shoots (20-cm long with 8-12 leaves) cut from the soil-grown plants (about 40 cm in height) were enclosed in a plastic bag with minimum air space and left in the dark for one week at 25°C. This treatment caused leaf deterioration characterized by yellowing, browning and rotting in the shoots of SM. In some transgenic lines, however, leaf deterioration was alleviated showing less yellowing, deterioration and rotting of the leaves than in SM. Fig. 15 shows typical profiles of leaf deterioration. The shoot of SM showed severe leaf deterioration, yellowing, browning and rotting, but the shoots of transgenic lines No. 17 and 33 showed little or no leaf deterioration. Reduced rotting probably resulted from a significant reduction in susceptibility to bacteria or fungi, because some pathogens often infect senescent tissue.

These results indicated that the integration of the transgene *mDG-ERS1(etr1-4)* suppressed the senescence of leaves in the transgenic chrysanthemum plants, irrespective of ethylene-induced or dark-induced senescence. Chrysanthemum plants belong to the family Compositae. Compositae plants include leafy vegetables such as lettuce, garland chrysanthemum, chicory and butterbur. These vegetables are vulnerable to senescence, resulting in yellowing, browning and rotting of leaves. Usually, to prevent senescence, harvested vegetables are handled and stored under cool temperature or controlled atmosphere conditions. However, the Compositae vegetables cannot be stored for a long time in a cool temperature because of cold injury, which is possibly accelerated by the action of ethylene. Thus, the senescence of these vegetables could be preventable by the generation of transgenic plants with reduced leaf senescence. The present study suggests that *mDG-ERS1(etr1-4)* could be used to generate transgenic Compositae leafy vegetables with retarded leaf senescence resulting in healthy green leaves and a greatly extended shelf life.

Sei-Marine No. 17 No. 33

Figure 15. Leaf senescence profiles of the non-transformed 'Sei-Marine'control and transgenic lines No.17 and No. 33 of chrysanthemum after natural senescence. Shoots of 20-cm long were detached from soil-grown plants, enclosed in a plastic bag and left for 1 week in the dark at room temperature.

Table 1 : Transformation of 'Lillipot' carnation with constructs harboring *DC-ACO1* cDNA in antisense or sense orientation (*a/sACO* transgenes)

Transformation vector	No. of explants	No. of lines containing *HPT* gene	No. of lines containing *a/sACO* transgenes	*a/sACO* transgene-containing plants/explants (%)
pIG121/*sACO*	200	4	3	1.5
pIG121/*aACO*	200	10	10	5.0
pMLH2113/*sACO*	200	16	14	7.0
Total	600	30	27	4.5

Table 2 : Transformation of 'Sei-Marine' chrysanthemum with pBIEF1a/*mDG-ERS1 (etr1-4)*

No. of explants	No. of regenerants	No. of Paromomycin*-resistant lines	No. of *EF1a/mDG-ERS1(etr1-4)* containing lines	*EF1a/mDG-ERS1(etr1-4)* containing lines/ explants (%)
1600	62	45	42	2.6

* Paromomycin is an analog of the antibiotic Kanamycin.

4. Conclusion

In our studies, carnation plants, irrespective of their use for cut flowers and potted plants, were successfully transformed with *sACO* or *aACO* transgenes. Cut flowers of the resultant transformants had reduced ethylene production and showed extended vase life. The results suggest that these transgenic techniques could be used to lengthen the vase life of carnation flowers instead of currently-used chemical preservatives of the flowers. In addition, chrysanthemum plant was transformed with a mutated ethylene receptor cDNA, *mDG-ERS1(etr1-4)*. The resultant transformants had reduced sensitivity to exogenous ethylene, exhibiting less yellowing in their leaves after ethylene treatment. Moreover, the transformants showed reduced deterioration during storage without ethylene. This result suggests that this transgenic technique is useful to retard leaf deterioration and improve postharvest performance in Compositae leafy vegetables. Since ornamentals like carnations and chrysanthemums are exclusively cultivated by vegetative propagation, T0 plants that have gained useful traits will be practically useful without the establishment of homozygous descendent lines.

References

Abeles FB, Morgan PW, and Saltveit ME, Jr 1992. Ethylene in Plant Biology. 2nd ed. Academic Press, San Diego.

Aida R, Yoshida T, Ichimura K, Goto R, and Shibata M 1998. Extension of flower longevity in transgenic torenia plants incorporating ACC oxidase transgene. *Plant Science* 138, 91-101.

Aida R, Ohira K, TanakaY, Yoshida K, Kishimoto S, Shibata M, and Ohmiya A 2004. Efficient transgene expression in chrysanthemum, *Dendranthema grandiflorum* (Ramat.) Kitamura, by using the promoter of a gene for chrysanthemum chlorophyll-a/b-binding protein. *Breeding Science* 54, 51-58

Borochov A, and Woodson WR 1989. Physiology and biochemistry of flower petal senescence. *Horticultural Review* 11, 15-43

Bovy AG, Angenent GC, Dons HJM, and van Altvorst A-C 1999. Heterologous expression of the *Arabidopsis etr1-1* allele inhibits the senescence of carnation flowers. *Molecular Breeding* 5, 301-308

Ciardi JA, Tieman DM, Lund ST, Jones JB, Stall RE, and Klee HJ 2000. Response to *Xanthomonas campestris* pv. *vesicatoria* in tomato involves regulation of ethylene receptor gene expression. *Plant Physiology* 123, 81-92

Chang C, Kwok SF, Bleecker AB, and Meyerowitz EM 1993. *Arabidopsis* ethylene-response gene *ETR1*: similarity of product to two-component regulators. *Science* 262, 539-544

Doi M, Nakagawa Y, Watanabe S, Aoe K, Inamoto K, and Imanishi H 2003. Ethylene-induced leaf yellowing in cut chrysanthemum (*Dendranthema grandiflora* Kitamura). *Journal of the Japanese Society for Horticultural Science* 72, 533-535.

Doi M, Aoe K, Watabe S, Inamoto K, and Imanishi H 2004. Leaf yellowing of cut standard chrysanthemum (*Dendranthema grandiflora* Kitamura) 'Shuho-no-chikara' induced by ethylene and postharvest increase in ethylene sensitivity. *Journal of the Japanese Society for Horticultural Science* 73, 229-234.

Ferrante A, Tognoni A, Mensuali-Sodi A, and Serra G 2003. Treatment with thidiazuron for preventing leaf yellowing in cut tulips and chrysanthemum. *Acta Horticulturae* 624, 357-363.

Firoozabady E, Moy Y, Tucker W, Robinson K, and Gutterson N 1995. Efficient transfomation and regeneration of carnation cultivars using *Agrobacterium*. *Molecular Breeding* 1, 283-293.

Harada T, Murakoshi Y, Tanase K, Onozaki T, and Satoh S 2009. Structure of genomic DNA of *DcACS1*, a 1-aminocyclopropane-1-carboxylate synthase gene expressed in senescing carnation (*Dianthus caryophyllus*) petals. *Horticultural Research (Japan)* 8 (Suppl.1), 257

Henskens JAM, Rouwendal GJA, ten Have A, and Woltering EJ 1994. Molecular cloning of two different ACC synthase PCR fragments in carnation flowers and organ-specific expression of the corresponding genes. *Plant Molecular Biology* 26, 453-458

Hood EE, Gelvin SB, Melchers LS, and Hoekemma A 1993. New *Agrobacterium* helper plasmids for gene transfer to plants. *Transgenic Research* 2, 208-218

Hua J, and Meyerowitz EM 1998. Ethylene responses are negatively regulated by a receptor gene family in *Arabidopsis thaliana*. *Cell* 94, 261-271

Iwazaki Y, Kosugi Y, Waki K, Yoshioka T, and Satoh S 2004. Generation and ethylene production of transgenic carnations harboring ACC synthase cDNA in sense or antisense orientation. *Journal of Applied Horticulture* 6, 67-71

Jones ML, Larsen PB, and Woodson,WR 1995. Ethylene-regulated expression of a carnation cysteine proteinase during flower petal senescence. *Plant Molecular Biology* 28, 505-512

Jones ML, and Woodson WR 1999. Differential expression of three members of the 1-aminocyclopropane-1-carboxylate synthase gene family in carnation. *Plant Physiology* 119, 755-764

Kende H 1993. Ethylene biosynthesis. *Annual Review of Plant Physiology and Plant Molecular Biology* 44, 283-307

Kinouchi T, Endo R, Yamashita A, and Satoh S 2006. Transformation of carnation with genes related to ethylene production and perception: Towards generation of potted carnations with longer display time. *Plant Cell, Tissue and Organ Culture* 85, 27-35

Kosugi Y, Shibuya K, Tsuruno N, Iwazaki Y, Mochizuki A, Yoshioka T, Hashiba T, and Satoh S 2000. Expression of genes responsible for ethylene production and wilting are differently regulated in carnation (*Dianthus caryophyllus* L.) petals. *Plant Science* 158, 139-145

Kosugi Y, Waki K, Iwazaki Y, Mochizuki A, Tsuruno N, Mochizuki A, Yoshioka T, Hashiba T, and Satoh S 2002. Senescence and gene expression of transgenic non-ethylene-producing carnation flowers. *Journal of the Japanese Society for Horticultural Science* 71, 638-642.

Mitsuhara I, Ugaki M, Hirochika H, Ohshima M, Murakami T, Gotoh Y, Katayose Y, Nakamura S, Honkura R, Nishiyama S, Ueno K, Mochizuki A, Tanimoto H, Tsugawa H, Otsuki Y, and Ohashi Y 1996. Efficient promoter cassettes for enhanced expression of foreign genes in dicotyledonous and monocotyledonous plants. *Plant and Cell Physiology* 37, 49-59

Narumi T, KannoY, Suzuki M, Kishimoto S, Ohmiya A, and Satoh S 2005a.Cloning of a cDNA encoding an ethylene receptor (DG-ERS1) from chrysanthemum and comparison of its mRNA level in ethylene-sensitive and -insensitive cultivars. *Postharvest Biology and Technology* 36, 21-30

Narumi T, Aida R, Ohmiya A, and Satoh S 2005b. Transformation of chrysanthemum with mutated ethylene receptor genes: *mDG-ERS1* transgenes conferring reduced ethylene sensitivity and characterization of the transformants. *Postharvest Biology and Technology* 37, 101-110

Norikoshi R, Niki T, and Ichimura K 2008. A new ACO gene associated with ACO activity in carnation gynoecium. *Horticultural Research (Japan)* 7(Suppl 1), 233

Ohta S, Mita S, Hattori T, and Nakamura K 1990. Construction and expression in tobacco of b-glucronidase (GUS) reporter gene containing an intron within the coding sequence. *Plant and Cell Physiology* 31, 805-813.

Panavas T, Pikula A, Reid PD, Rubinstein B, and Walker EL 1998. Programmed cell death of daylily petals: activities of wall-based enzymes and effects of heat shock. *Plant Physiology and Biochemistry* 36, 379-388

Panavas T, Pikula A, Reid PD, Rubinstein B & Walker EL 1999. Identification of senescence-associated genes from daylily petals. *Plant Molecular Biology* 40,237-248

Park KY, Drory A, and Woodson WR 1992. Molecular cloning of an 1-aminocyclopropane-1- carboxylate synthase from senescing carnation flower petals. *Plant Molecular Biology* 18, 377-386

Reid MS, and Wu M-J 1992. Ethylene and flower senescence. *Plant Growth Regulation* 11, 37-43

Reyes-Arribas T, Barrett JE, Huber DJ, Nell TA, and Clark DG 2001. Leaf senescence in a non-yellowing cultivar of chrysanthemum (*Dendranthema grandiflora*). *Physiologia Plantarum* 111, 540-544

Satoh S, Kosugi Y, Iwazaki Y, Shibuya K, and Waki K 2000. Two ethylene signaling pathway in senescing carnation petals: exogenous ethylene-induced expression of genes for 1-aminocyclopropane-1-carboxylate (ACC) synthase and ACC oxidase is different from that of the gene for cysteine proteinase. *Journal of Plant Biotechnology* 2, 83-87

Savin KW, Baudinette SC, Graham MW, Michael MZ, Nugent GD, Lu C-Y, Chandler SF, and Cornish D 1995. Antisense ACC oxidase RNA delays carnation petal senescence. *HortScience* 30, 970-972

Shibuya K, Yoshioka T, Hashiba T, and Satoh S 2000. Role of the gynoecium in natural senescence of carnation (*Dianthus caryophyllus* L.) flowers. *Journal of Experimental Botany* 51, 2067-2073.

Tieman DM, Taylor MG, Ciardi JA, and Klee HJ 2000. The tomato ethylene receptors NR and LeETR4 are negative regulators of ethylene response and exhibit functional compensation within a multigene family. *Proceedings of National Academy of Science USA* 97, 5663-5668

Veen H., 1979. Effects of silver on ethylene synthesis and action in cut carnations. *Planta* 145, 467-470

Wilkinson JQ, Lanahan MB, Clark DG, Bleecker A, Chang C, Meyerowitz EM, and Klee HJ 1997.A dominant mutant receptor from *Arabidopsis* confers ethylene insensitivity in heterologous plants. *Nature Biotechnolgy* 15, 444-447

Wang H, and Woodson WR 1991. A flower senescence-related mRNA from carnation shares sequence similarity with fruit ripening-related mRNAs involved in ethylene biosynthesis. *Plant Physiology* 96, 1000-1001

Yang SF, and Hoffman NE 1984. Ethylene biosynthesis and its regulation in higher plants. *Annual Review of Plant Physiology* 35, 155-189.

The Science of Horticulture : Vol. 1, 2011
© *K.V. Peter (ed.), pp. 347-378*
New India Publishing Agency, New Delhi (India)
E-mail : newindiapublishingagency@gmail.com
Web: www.bookfactoryindia.com

Chapter 15 Tuber Crops Biotechnology

SHEELA, M.N. AND RAJMOHAN, K.

Root and Tuber crops are the most important food crops of man after cereals. Tuber crops are efficient producers of carbohydrates especially under suboptimal conditions of rainfall, fertility and low input conditions encountered in tropics. The group contributes about 6 % of the world's dietary calories. In fact, one out of 6 kg of human food comes from roots and tubers. Tuber crops find an important place in the dietary habits of small and marginal farmers especially contributing to food security of tribal populations. Tuber crops grown in India include Cassava, Sweet potato, *Dioscorea, Colocasia, Amorphophallus* and minor tuber crops like Chinese potato, Arrowroot, Canna, Yam bean etc. The biotechnology can contribute to the solution of the major problems and realize greater benefits for tuber crops. Traditional breeding in tuber crops is difficult due to their heterozygous nature, low fertility and out crossing coupled with strong inbreeding depression. Many of the agronomic traits are polygenic and/or recessive which make breeding for such characteristics complicated. Biotechnology is a powerful tool to complement traditional breeding. The developments in the field of tuber crops biotechnology are discussed below.

Cassava

Cassava (*Manihot esculenta* Crantz) which comes under the family Euphorbiaceae, has many common names like Manihot, Manioc, Tapioca, Mandioca etc. Cassava is the most important staple crop in the humid tropics and an important food source for more than 800 million people (FAO, 2002). It is a reliable food security crop for rural house holds, an effective feedstuff for farm animals, and a low cost versatile raw material for agroindustrial development. Cassava starch forms the basis of textiles, pharmaceuticals, paper, glues, alcohol etc. Unlike many other crops, cassava can be grown with minimal inputs and it is able to produce reasonably well under unfavorable conditions like low soil fertility, acidic soil or drought. In cassava, great advances are made in the field of *in vitro* conservation, characterization of genome, gene isolation, development of molecular maps, map based cloning and genetic engineering.

1. Meristem Culture

In vitro propagation and storage techniques are well established in cassava making it possible to multiply and store a large number of germplasm accessions. *In vitro* techniques were developed for collection of wild and cultivated genotypes. Meristem culture is reported to be the most effective method in tissue culture technique, to develop virus free plants *in vitro.* In cassava, meristem culture is used widely to clean plants infected by the cassava mosaic disease (CMD) which is caused by the Gemini group of virus and spread by the white fly, *Bemisia tabaci.* Elimination of viruses through shoot apical meristem culture in cassava was first reported by Kartha *et al.* (1974). Meristem cultures for elimination of Cassava Mosaic Disease was done in the MS medium in the presence of BA ($5x10^{-7}$) and NAA (10^{-6} M) and GA_3 (10^{-7}M) and elimination of virus was reported (Kartha *et al,* 1974). Meristem culture combined with thermotherapy is being used for large-scale virus-free conservation of germplasm at Centro International de Agricultura Tropical (CIAT),Cali, Columbia. At CTCRI, Indian cassava mosaic disease (ICMD) free plantlets are obtained by culturing of meristems tips of about 0.4 mm size on Murashige and Skoog (MS) (Murashige and Skoog 1962) medium supplemented with 0.02 mgl^{-1} NAA, 0.1 mgl^{-1} BAP and 0.03 mgl^{-1} GA_3 (Unnikrishnan and Sheela, 2000). Nodal explants as well as apical buds also could be used for initiation of cultures. BA was the most effective cytokinin in cassava. Multiple shoot cultures could be produced from meristems by increasing concentration of BA from 2.5 to 5.0 μM (Roca *et al,* 1984). Response variations among accessions are documented.

2. *Slow Growth Culture*

Cassava cultures are amenable to *in vitro* conservation under normal conditions as well as reduced growth rates. *In vitro* genebanks are set up in cassava using slow-growth cultures at various cassava genetic resource centers in Africa, South America and Asia. The various conservation methods viz. low temperature incubation, use of low light intensity, use of osmotically active agents or growth retardants, manipulations in the concentration of the growth hormones, macro and micro-nutrients and sucrose are used individually or in combination for the medium term storage of the cassava germplasm. Experiments with minimal and mannitol media indicated that combination of both media was effective to induce slow growth. Osmotic retardants like mannitol or sorbitol (3%) along with sucrose (3%) in the medium was effective in extending the subculture period from 10 to 12 months. Sorbitol was better in imparting slow and healthy growth than mannitol (Unnikrishnan *et al.* 2002).Silver nitrate (1mM) and activated charcoal (1g/l) in MS medium with sucrose (3%) and growth regulators (0.05 μM NAA, 0.1μM BA and 0.3μM GA3) helped in maintaining the cultures upto 18 months. Addenda like silver nitrate and activated charcoal was effective in reducing leaf chlorosis and preventing leaf shedding, thereby contributing to culture life and regeneration capacity. Throughout the storage period, the cultures continued to produce axillary buds, indicating viability of the cultures.

3. *Cryopreservation*

Initially, Bajaj (1983) reported cryopreservation technique for long-term conservation of cassava using shoot tips. He found that isolated meristems precultured on a medium containing DMSO and treated with a mixture of DMSO, sucrose and glycerol, could be frozen in liquid nitrogen using rapid freezing and slow-freezing techniques. By following the method of preculture followed by slow freezing using programmable freezer, the protocol for cryopreservation, using shoot-tips was developed and tested for 15 genotypes of cassava at CIAT. In some cases, regeneration up to 70% was achieved.

4. *Somatic Embryogenesis in Cassava*

Somatic embryogenesis forms the basis of all current transformation methods of cassava. Cyclic embryogenesis is the most routinely used method of *de novo* plant regeneration of cassava *in vitro* and provides constant source of material for transformation. Plant regeneration from somatic embryos can be established via either direct germination or shoot organogenesis. Currently, shoot organogenesis from cotyledons of somatic embryos and embryogenic suspensions derived from friable embryogenic callus are the two most frequently used systems for transformed cassava plants.

To date, all reported cases of somatic embryogenesis in cassava have relied on inducing organized embryogenic structures from immature leaf lobes or cotyledons (Stamp and Henshaw, 1982; Szabados *et al.*, 1987). These tissues are exposed to Murashige and Skoog media (1962) with various types and concentrations of auxin. Similar media are then used to promote subsequent multiplication of these tissues *via* a secondary embryogenesis process (Raemakers *et al.*,1993a; Stamp and Henshaw, 1987).

In the organogenesis system, development of shoot primordia is induced directly on cytokinin containing media from explants of germinating somatic embryos. A cyclic system in which secondary somatic embryos are induced on cotyledon explants from maturing somatic embryos is used to provide a constant source of regeneration competent explant material. In some cultivars, a fraction of cyclic somatic embryos maintained on MS medium supplemented with picloram (12mg/l) is able to produce a new tissue type which consists mainly of small globular embryo like structures. When isolated, it produces a highly friable embryogenic callus (Taylor *et al.*, 1996). The frequency of friable embryogenic callus (FEC) production can be increased by substituting the MS medium with GD medium (Gresshoff and Doy, 1974). Once pure FEC is obtained, it can be transferred easily to liquid medium to establish a rapidly proliferating embryogenic suspension in SH medium (Schenk and Hildebrandt, 1972), medium supplemented with 6% sucrose and picloram (12mgl^{-1}). The conversion of globular embryos is still the limiting step of the regeneration protocol, as only a small fraction of the embryos is able to develop into torpedo stage embryos, and the genotypic variation is high. At CTCRI, somatic embryogenesis occured in five out of eleven Indian varieties viz., Sree Vijaya, Sree Harsha, Sree Sahya, Sree Jaya and H165 which showed embryogenesis in liquid medium with 2,4-D (8mg/l) (Sree Rekha *et al.*, 2005).

5. Molecular Markers in Cassava

5.1. Diversity Analysis

A number of works are reported on molecular finger printing of cassava. Starch and polyacrilamide gel electrophoretic analyses were carried out for determining isozyme patterns of 16 enzymes of cassava varieties as potential genotype markers. Varieties could be distinguished by sequential classification on the basis of isozyme patterns of acid phosphatase, esterase, glutamate oxalaoacetate, transaminase and phosphoglucoisomerase. Of these, esterase was the most stable isozyme system for cassava (Ramirez *et al.* 1987).

Later DNA fingerprinting using RAPD markers was reported by Yu and Nyugen 1994. Variable number of Tandom Report loci was used by Bertram (1993) to examine individual variations with in an accession . Report on cassava germplasm characterization is also available using RAPD and SSR primed PCR markers for assessing genetic diversity (Carvalho and Schaal, 2000). RFLP technique was successfully employed in the genetic diversity studies of an *in vitro* germplasm collection of African cassava clones (Beeching *et al*, 1995). DNA markers are also employed in cassava for multilocus finger printing by Bertram (1993).

5.2. Phylogenetic Studies

Recently genetic variation at microsatellite loci was used for the investigation on evolutionary and geographical origins of cassava from its wild relatives (Olsen and Schaal, 2001). RFLP analysis of chloroplast DNA and nuclear ribosomal (rDNA) sequences using heterologous probes were also applied to assess phylogenetical relationship among *Manihot* species. The evidences from molecular analysis point towards the possible origin of cassava from Brazilian wild species *viz. M.tristis* and *M.esculenta* sub sp *flabellifolia.* Conclusive evidence of the origin of cassava from *M.esculenta* sub sp *flabellifolia* came from phylogenetic study based on sequencing the single copy nuclear gene glyceraldehydes 3 phosphate dehydrogenase, *G3pdh.* DNA fingerprinting was carried out using M 13 (cla 1-B sml fragment) and cassava genomic probes, improving the discriminatory resolution among suspected duplicates (Ocampo *et al.* 1994). Amplified Length Fragment Polymorphism (AFLP) was used to study the phylogeny of cassava and its taxonomic relationship with other species of the genus *Manihot.* AFLP evaluation was done with six representative samples of the crop's diversity ('core collection') and six wild taxa to estimate genetic relatedness within the genus. The studies helped in uncovering the unique genetic diversity in *M. esculenta* subsp. *flabellifolia* and *M. esculenta* subsp. *peruviana* and their genetic similarity to cassava supporting the hypothesis that these two species might be the ancestors of cassava.

5.3.Marker Assisted Breeding

Over 500 microsatellite (SSR) markers were developed in cassava so far. Many of these were placed on the framework map. Markers were used to obtain quantitative assessment of genetic similarity in cassava (Beeching *et al.*, 1995, Second *et al.*, 1997, Elias *et al.*, 2000) and to study the genetic nature of germplasm resistant to diseases (Fregene *et al.*, 2000).

Resistance Breeding Using Molecular Marker

The predominant source of resistance to the cassava mosaic disease is known to be polygenic requiring evaluation in multiple environments to characterize resistant genotypes, which makes the detection of genes for resistance using segregation analysis inefficient. Bulked segregant analysis (BSA) was used to identify quickly a simple sequence repeat (SSR) marker linked to CMD resistance gene. The marker, SSRY28, identified is located on linkage group R of the male parent derived molecular genetic map. The gene designated as *CMD2*, is flanked by the SSR, RFLP marker GY1 at 9 and 8 cM, respectively. This is the first report of qualitative virus resistance in cassava and of molecular markers that tag CMD resistance in cassava (Akano *et al.*, 2002). Recently, some Nigerian landraces were identified which exhibited high levels of resistance to CMD. Based on the work of Akano *et al.*, (2002), the SSR markers SSRY 28 and NS158 are the closest markers to the gene *CMD2*, which confer a high level of resistance to CMD. The fine-mapping was carried out using a population of 1690 individuals from a cross between TME3 and TMS30572. The population was evaluated with the above mentioned SSR markers along with other AFLP and RAPD markers. Two polymorphic RAPD markers were identified. One, RME-1, is 1cM from the gene, it was cloned and converted into a sequence characterized amplified region (SCAR) marker. A bacterial artificial chromosome (BAC) library with more than 10x coverage for the cassava genome was constructed from the cassava variety TME3, donor parent of *CMD2*. Successive hybridizations of the BAC end markers to the BAC library will be carried out until markers which flank the gene on both sides are identified.

Lokko *et al.* (2005) studied molecular markers associated with resistance to CMD in a resistant Nigerian landrace, using F_1 progenies derived from a cross between TME7 x TMS30555. Quantitative Trait Loci (QTLs) associated with resistance to CMD were identified using F_1 progeny derived from a cross between the CMD resistant (TMS30572) and the susceptible landrace (TME117). Seven highly significant marker-associated (SSRY6, 7, 21^a, 21^b, 42, 77, 324) with QTL effects, explaining between 7.31 and 12.15% of the total phenotypic variation, were detected by regression and Kruskal Wallis analysis. Significant marker trait associations were due to markers donated by both parents, which confirm the polygenic and recessive nature of this source of resistance (Lokko *et al.*, 2004). The analysis further showed that another SSR marker, SSRY106 and an AFLP marker, E-ACC/M-CTC-225, accounted for 35.59% and 22.5% of the total phenotypic variation for resistance respectively.

Productivity and Plant Architecture

Okogbenin and Fregene (2003) attempted to identify quantitative trait loci for several productivity and plant architecture traits in a full sib progeny of 144 individuals from two non-inbred parents in cassava. A molecular genetic map of this cross constructed previously with over 250 markers was the source of molecular markers. A total of 30 primary QTLs and 84 secondary QTLs were detected. All QTLs identified for fresh shoot yield were associated with either component traits of productivity or architecture traits. Hence QTLs for plant architecture can be used to improve productivity. However an exhaustive search and analysis of QTLs controlling architecture are required for initiating marker assisted selection for productivity in cassava.

Early bulking is another important trait evaluated in the F_1 mapping progeny. QTL analysis was done for traits linked significantly to early bulking. Three QTLs were found for dry foliage weight that explains 25-33% of the phenotypic variance while five QTLs were found for harvest index which explain 18-27% of the phenotypic variance. Based on marker analysis of the early bulking lines, a selection criteria involving molecular markers for fresh root yield and harvest will be useful.

Apomixis

Microsatellites (simple sequence repeats) are powerful molecular markers used to understand detailed patterns of parentage composition and for individual discrimination in clone identification in cassava. Using microsatellite markers, the genetic facts with respect of apomixis inheritance in cassava was studied by Nassar and Collevatti (2005). The results revealed that apomixis is determined by recessive alleles and also it is controlled by more than one pair of genes, probably located in different chromosomes. One of the most important consequences of this result is the possibility of mapping and characterising the genes of apomixis by molecular markers and isolating them in the future.

Transgenics in Cassava

Transgenic technology offers an excellent tool for transferring a single or even quantitative traits without the problems of linkage encountered in the traditional breeding programmes. The transformation technologies are utilized to produce genetically transformed cassava expressing traits of agronomical importance. The important areas for cassava genetic engineering include yield and root quality improvement, pest and insect resistance and production of noble compounds for value added products in cassava.

A range of selectable and visual markers are tested and developed for use in cassava transformation systems. Several laboratories use the *npt II* gene that imparts resistance to amyloglucoside antibiotics, routinely with selection for successful transformation events possible by culturing tissues on medium containing kanamycin, paramomycin or geneticin. The efficacy of the hygromycin resistance gene (*hpt*) was also studied, but is reported to suppress the frequency of plant recovery compared to transgenic tissues selected by expression of *nptII*. Use of *bar* gene to impart resistance to the herbicide phosphinothricin acetyltransferase is utilized both as a gene of agronomic interest and as a selection system. The use of the *E.coli phosphomannose isomerase (pmi)* gene was used to transform FEC and cotyledon explants and procedures for the effective recovery of transgenic tissues and plants using positive selection are developed. In addition, the different visual marker systems like *GUS, luciferase* and *GFP* (Green fluorescent protein) are used routinely for the production of transgenic cassava (Taylor, *et al.*, 2004). The summary of transgenic research in cassava is given elsewhere.

Stable integration of foreign genes in transformed cassava tissues and plants

Target tissue	Transfer method	Gene(s) inserted	Transgenic product	Reference
Leaf, stem, somatic embryos	A	*npt*II, *bar*, *uid*A	Callus	Calderon (1988)
Somatic embryos	A	*npt*II, *uid*A	Chimeric embryos	Chavarriaga *et al.* (1993)
Leaves	A	*npt*II, *bar*, *uid*A, ACMV.CP, CsMV-CP	Callus	Schopke *et al.* (1993)
Embryogenic suspensions	P	*npt*II, *uid*A	Transgenic Plants	Schopke *et al.* (1996)
Embryogenic suspensions	P	*bar*, *luc*, *npt*II,	Transgenic Plants	Raemakers *et al.* (1996)
Cotyledons from somatic embryos	A	*npt*II, *uid*A, *hpt*	Transgenic Plants	Li *et al.* (1996)
Embryogenic suspensions	A	*npt*II, *uid*A	Transgenic Plants	Gonzalez *et al.* (1998)
Embryogenic suspensions	P	*Luc, par, cassava AGPase B antisense gene*	Transgenic Plants	Munyikwa *et al.* (1998)
Cotyledons from somatic embryos	A	*npt*II, *bar*, *uid*A	Transgenic Plants	Sarria *et al.* (2000)
Embryogenic suspensions	P	*npt*II, CsMV-CP	Transgenic Plants	Schopke *et al.* (2000)

Contd..

Contd..

Embryogenic suspensions protoplasts	E	*Pat, luc*	Transgenic Plants	Raemakers *et al.* (2000)
Embryogenic suspensions	A	*ASP1,hpt, uidA*	Transgenic Plants	Zhang *et al.*, (2003)
Germinated somatic embryos	A	*CYP79D1/ CYP79D2, nptII*	Transgenic Plants	Siritunga and Sayre, (2003)

A: *Agrobacterium*-mediated transformation, E: electrophorosis, P: Particle bombardment, *ACMV-CP* : African cassava mosaic virus – coat protein, *AGPase B* : ADP glucose phosphorylase B, *bar, pat*: Phosphinothricin acetyl transferase, *CsMV-CP* : Cassava common vein mosaic virus coat protein, *hpt* : *hygromycin phosphotransferase*, *luc* : luciferase, *nptII* : neomycin phospho transferase, *uidA*: â-glucuronidase

Cassava Mosaic Resistance

The most important disease in cassava is cassava mosaic disease (CMD) caused by the Gemini group of virus and spread by the white fly, *Bemisia tabaci*. The CMD is a potential threat to cassava production causing yield loss ranging from 20 to 80%. The virus has two covalently closed circular single stranded DNA (ssDNA) molecules known as DNA A and DNA B (Stanley, 1983). DNA A is required for replication of both stands and proliferation and together with DNA B for vector transmission and spread. Both the components are needed for the systemic infection of susceptible plants. DNA B is involved in cell to cell and long distance viral spread. For the control of Cassava mosaic disease, the integration of a defective interfering (DL) sequence derived from a deleted B component of a Kenyan strain of African cassava mosaic virus (ACMV) could impart elevated resistance to ACMV (Frischumuth and Stanley, 1991). Transgenic plants were recovered from the West Africa cv. 60444 containing integrated copies of a complete DI dimer from ACMV - Kenya (Taylor *et al.*, 2003). However, the efficacy of resistance induced by integration of the AC1 gene from ACMV-Kenya was significantly more effective (Verdageuur *et al.*, 1996). Increased ACMV resistance in transgenic cassava is also developed using antisense RNA technology at ETH, Zurich. Schopke *et al.* (2000) used different constructs carrying AC1 gene for the genetic modification of cassava. Three constructs were designed for the expression of viral antisense RNA's as 3'untranslated regions (3'UTR) of a selectable hygromycin phosphotransferase (*hpt*) gene driven by the *CaMV 35S* promoter. In another approach, a hypersensitive reaction is mimicked by using the bacterial barnase and barstar genes from *Bacillus amyloliquefaciens* under the control of the ACMV DNA, a bi-directional promoter. The results show that it is possible to reduce viral replication by 37-99% through transgenic approach and hence it offers genetic modification of popular varieties into resistant ones.

A programme for the transgenic control of stem borer and horn worm was initiated at CIAT and trasgenics were produced in which *cry 1Ab* genes under control of CaMV 35S promoter were integrated into economically important Columbian cassava cultivars.

Reduced Cyanogens Content

Two approaches are taken to produce cassava with reduced cyanogen content. In the first approach, genes encoding a small (CYPDI and CYP79D2) family of cytochrome P450s that catalyze the first dedicated step in linamarin and lotaustralin synthesis were expressed in an antisense orientation in transgenic cassava lines of cultivar Col215 which showed reduced cyanogen content (34-94%)(Siritunga and Sayre,2003). The second approach is by over expressing HNL (Hydroxy nitrile lyase) that helps in the detoxification (Siritunga *et al.*, 2003).

Herbicide Resistance

Transgenic plants of cassava cultivar Per183, expressing the *bar* gene that encodes phosphinothricin (*ppt*) and provides resistance to the herbicide basta are regenerated at CIAT (Sarria *et al.*, 2000).

Starch Quality

Work is also being carried out to produce transgenic cassava with altered starch quantity and quality. cDNAs encoding ADP glucose pyrophosphorylase, branching enzymes (BEI, BEII) and granule bound starch synthase (*GBSS*) involved in starch biosynthetic pathway are cloned from cassava. The researchers at Wageningen University, Netherlands cloned the *gbss* gene from cassava in the antisense orientation under the control of the *CaMV35S* promoter and the *gbss* promoter from potato. A total of 560 transgenic plants with the *gbss* antisense gene were recovered, two of which subsequently produced and stored amylase free starch that enabled the production of transparent starch with stable viscosity for industrial application (Raemakers *et al.*, 2003). The same work is being carried out successfully at CIAT also. The researchers at Ohio University were successful in up regulating overall starch biosynthesis by expressing a modified version of the *E.coli glgC* gene encodes *AGPase* which is the rate limiting step in the cassava starch biosynthesis, under the control of the tuber specific patatin gene from potato (Uzoma *et al.*, 2003). Sixty five % increase in *AGPase* activity was observed in transgenics as compared to control.

Protein Content

Researchers at ETH, Switzerland produced transgenic plants of cv.60444 with an artificial storage protein *ASP1* gene designed to be rich in essential amino acids (Zhang *et al.*, 2003). Research is also being undertaken to produce transgenics having high leaf retention by expressing the cytokinin regulation ipt gene under control of antisense controlling promoter. Research is also started to produce transgenic cassava resistant to post harvest deterioration. Longer term goals of transgenic improvement of cassava include induction of apomixis, bioplastic production , modified plant architecture and rapid bulking of storage roots.

Sweet Potato

Sweet potato is grown in more than 100 countries as a valuable source of food, animal feed and industrial raw material. It is a staple crop in many Southeast Asian and African countries. It is a major crop that feeds millions of people in the developing world. It is especially popular among farmers with limited resources, and produces more biomass and nutrients /ha than any other food crop in the world. New biotechnological approaches enable scientists to rapidly develop superior, disease and pest resistant cultivars in this crop. Because of the biological complexities of sweet potato, sexual hybridization strategies are not very effective in developing improved cultivars. Therefore, biotechnological tools, such as gene transfer, are very attractive in sweet potato improvement, as they enable direct introduction of desirable genes from other sources into preadapted cultivars. (Prakash, 1994).

1. In vitro Conservation

In vitro methods in sweet potato were used earlier at several international centers like Centro International de la Papa, (CIP), Peru, the International Institute of Tropical Agriculture (IITA), Nigeria and AVRDC, Taiwan for conservation of sweet potato collections. Besides, several national programmes in USA, Costa Rica and India are maintaining a large number of *in vitro* collections of sweet potato for the last over 10 years. Meristem cultures in conjunction with thermotherapy and use of antiviral drugs are effective in virus elimination in sweet potato (Ashmore, 1997). Meristem culture technique coupled with indexing by ELISA and grafting on indicator plants was used to establish healthy cultures, at NBPGR (Mandal and Chandel 1996). *In vitro* propagation technique is refined at NBPGR, to develop a single medium capable of eliciting desirable response from all varieties and related species. MS medium supplemented with 4 % sucrose, 0.2 mg l^{-1} kinetin and 0.1 mg l^{-1} IAA is used to establish cultures of 260 sweet potato accessions maintained at NBPGR. For the medium term storage of sweet potato, various methods are used successfully. These include incubation at low temperature (16-20°C) (Ng and Hahn, 1985), addition of mannitol to the medium coupled with reduction in incubation temperature and use of sucrose (3%) with 3% mannitol (Dodds and Roberts, 1985). Though low temperature incubation (18°C) is effective in reducing growth rate, maintenance of large collections at this temperature during summer when the ambient temperature exceeds 40°C is expensive and risky in view of frequent power breakdowns or equipment failure Among the several modifications tested, supplementation of the medium with 1-2% mannitol was the most effective (Mandal and Chandel, 1991). Using this method, 260 accessions are maintained at culture room temperature of 25±2°C for ten years with an annual subculture. While genotypic variation in response to mannitol does exist, the concentration of 1-2 % was suitable for all accessions.

Application of cryopreservation technology for sweet potato conservation has not been satisfactory. The use of somatic embryos or embryogenic tissues seems to hold more promise since recent studies (Bhatti *et al.* 1997; Blackesley *et al.* 1997) demonstrated the recovery of complete plantlets from Liquid Nitrogen exposed embrygenic tissues.

2. Molecular Fingerprinting

Sweet potato cultivars are identified using starch gel electrograms of nine enzymes (Kennedy and Thompson 1991).Genetic polymorphism was identified using DNA amplification fingerprinting (DAF) technique. Seven, out of the twenty-eight octamer, screened, were found suitable and detected a high degree of variability in sweet potato genotypes. The DAF profiles of 70 accessions collected from around the world showed very high variability (Prakash *et al.* 1996). Lakhanpaul *et al.* (1990) also studied the polymorphism using Random Amplified Polymorphic DNA (RAPD) analysis.

Dhillon and Ishiki (1999) used random amplified polymorphic DNA (RAPD) markers for fingerprinting and analyzing genetic relationship among 29 *Ipomoea* accessions from different geographical locations around the world including unique wild species. Reproducible profiles were obtained for all accessions using random decamer primer. The primers generated 46 polymorphic markers , one primer alone having 10 products enabling the discrimination of all the 29 accessions studied. Based on RAPD profile, *Ipomoea trifida* was identified as the probable progenitor of *Ipomoea batatus*. RAPDs were proved useful for sweet potato systematics and should be valuable for germplasm management, gene tagging, and for identifying diverse parents for breeding programmes.

Molecular characterization and expression of starch granule bound starch synthase in the sink and a source tissue of sweet potato were studied by Shu-Jen Wang *et al.*, (1999). A cDNA clone for starch granule bound starch synthase (*GBSSI*) which encodes a 67 kDa protein was isolated from a cDNA library prepared from tuberous roots of sweet potato.

3. Transgenics

Biotechnology is not well developed for sweet potato because of the difficulty of establishing an efficient and reproducible system for plant regeneration from cultured tissues or cells. In 1996, Otani and Shimada developed an efficient and variety independent method for the production of embryogenic callus from meristem tissues using a medium containing altered plant growth regulators, picloram, dicamba, or 4-flourophenoxyacetic acid (4FA). Otani *et al.*(1998) reported an efficient method for the production of transgenic sweet potato plants via *Agrobacterium tumefaciens* mediated transformation using embryogenic calli.The summary of the reports on transformation is given elsewhere.

One of the most serious diseases of sweet potato is russet crack disease caused by Sweet potato feathery mottle virus (SPFMV). Okado *et al.* (2001) reported development of transgenic plant having resistance to this dreadful disease. They developed an expression vector carrying the coat protein (CP) and hygromycin phosphotransferase (*hpt*) genes driven by cauliflower mosaic virus 35S promoters. The transgenics developed through electroporation had not only primary but also secondary resistance to SPFMV strain.

Among the pests, sweet potato weevil (*Cylas formicarius* ssp) is the major pest worldwide. The expression of foreign genes in plants to confer resistance is an appropriate alternative to reduce crop losses. Moran *et al.*(1998) transformed sweet potato plants of the variety Jewel cultivar with the *cry III A* delta-endotoxin gene from *Bacillus thuringiensis* var. *tenebrionis* using *A.tumfaciens* mediated transformation.

The development of new cultivars which have a novel physico-chemical property of starch is an important objective of sweet potato breeding. It is thought that amylase -free starch with very low amylase content will have a major role in new industrial developments. The full length cDNA and the gene for sweet potato granule bound starch synthase I (GBSSI) were isolated and characterized (Wang *et al.*,1999; Kimura *et al.*, 2001). This full length cDNA for sweet potato *GBSS I* driven by the CaMV 35S promoter was introduced into the sweet potato by *Agrobacterium tumefaciens* mediated transformation . The transgenic plant obtained showed the absence of amylase in the tuberous root as determined by iodine calorific method.

Summary of published studies on transformation of Sweet potato

Target tissue	Transfer method	Gene(s) inserted	Transgenic product	Reference
In vitro whole plant	*A.rhizogenus*	Synthetic genes	Transgenic plants	Dodds *et al.*(1991)
Leaves and petioles	Particle bombardment	*Gus* A, *npt* II	Transformed calli	Prakash and Varadarajan (1992)
Leaves	*A.rhizogenus15384*	*Gus* A, *npt* II	Transgenic plants	Otani *et al.*(1993)
Storage roots	*A.tumefaciens* LBA 4404	*Gus* A, *npt* II, cowpea trypsin inhibitor, snowdrop lectin	Transgenic plants	Newell *et al.*, (1995)
Embryogenic callus	*A.tumefaciens* EHA 101	*Gus* A, *npt* II	Transgenic plants	Gama *et al.* (1996)
Embryogenic callus	*A.tumefaciens* EHA 101	*Gus* A, *hpt*	Transgenic plants	Otani *et al.* (1998)
Leaves	*A.tumefaciens* C58C1	*cry*IIIA, *npt* II	Transgenic plants	Moran *et al.* (1998)
Embryogenic callus	*A.tumefaciens* EHA 101	*GBSS, hpt*	Transgenic plants	Kimura *et al.* (1999)
Mesophyll protoplasts	Electroporation	*hpt*, SPFMV-S coat protein gene	Transgenic plants	Okado *et al.*,2001

YAMS

Yam (*Dioscorea* spp.) is an important crop plant for sustainable food production in tropical and subtropical regions. *Dioscorea* species probably originated from tropical areas of three separate continents: Africa (the *D. cayenensis–D. rotundata* complex), South-East Asia and the South Pacific (*D. alata*) and South America (*D. trifida*) (Coursey 1976). Of the 600 known species of yams, only six are edible or food yams. *Dioscorea rotundata* and *D. cayenensis* (both known as Guinea yam) are the most popular and economically important yams in West and Central Africa where they are indigenous, while *D. alata* (referred to as water or greater yam) is the most widely distributed species globally. Greater Yam is the main *Dioscorea* species grown in India. Yam production has increased steadily in the last decade, from 18 million tonnes tonnes in 1990 to recent estimates of over 39 million tonnes (FAO, 2002).

Yams are characterized by vegetative propagation, dioecious nature, rare and erratic flowering leading to non synchrony of the sexes and prevalence of polyploidy. The early testing of progenies from hybridization was complicated by yam's slow multiplication rate and the juvenile phase during which it is unreliable to use conventional screens for some economic traits. Progress is made in recent years in germplasm characterization and development of molecular markers for genome analysis. Tissue culture protocols are being employed for the medium and long term conservation of germplasm and for elimination of pathogen from planting material.

1. In vitro Conservation

Being a vegetatively propagated crop, yam is seriously affected by an accumulation of pathogens. To overcome the problems of field maintenance, IITA, and IPGRI recommended use of *in vitro* techniques for yam conservation (Hanson, 1986). In yams, two kinds of *in vitro* germplasm preservation were considered : slow growth condition for medium term preservation and cryopreservation using the encapsulation/dehydration technique for long term conservation.

Many species of yams are successfully established under *in vitro* conditions. Axillary buds and nodal cultures are used most frequently. In India, the work related to the micropropagation of yams was carried out mainly at NBPGR and CTCRI (Mandal and Chandel 1996; Nair and Chandrababu 1996).

Using nodal cuttings as explants, *in vitro* tuber development was recorded in several species (Ng and Ng, 1997; Mandal, 1999). Nodal segments with one or two axillary buds were cultured on MS medium alone or supplemented with kinetin or NAA @ 0.25 mg l^{-1} each. However, for routine sub culturing, a single medium *i.e.* MS supplemented with 0.15 mg l^{-1}NAA, is applicable to all the accessions maintained in the repository.

Incubation temperatures play a significant role in growth rate of yam plantlets. Plantlets storage period could be extended to 1-2 years by incubating at 18-20°C. Use of mannitol and high level of sucrose either alone or in combination, has significant effect on plantlet growth rate at the culture-room temperature. Addition of mannitol to the medium can prolong storage at normal culture room temperature of 25-30°C for a period up to 12 months with high recovery rates (80%). Use of 2.0-mg l^{-1} BAP or kinetin helped in maintaining cultures of *D.alata* for two years. Induction of micro tubers increased shelf life by an additional 3-6 months. At CTCRI, *D. alata, D.*

rotundata and *D. esculenta* nodal cultures could be conserved up to 12 months without subculture on MS basal medium containing 3 % mannitol and no growth regulators.

Cryopreservation of yam shoot tips of two cultivated edible species (*Dioscorea alata, D. bulbifera*), a medicinal yam (*D. floribunda*) and one wild edible type (*D.wallicii*) were successfully carried out using the technique of encapsulation-dehydration (Mandal and Chandel 1996). Though, survival and growth initiation from cryopreserved shoot-tips were recorded at a frequency of 28-71 %, regeneration of complete plantlets were recorded at a frequency of 21-37 % in *D. alata* and *D. wallichii* (Mandal *et al.* 1996). Malaurie *et al.* (1998) reported successful cryopreservation of shoot tips *of D. alata* and *D. bulbifera* using the technique of encapsulation - dehydration. Very recently, success of cryopreservation of shoot-tips/somatic embryos of several edible and medicinal yams using both vitrification and encapsulation-dehydration was registered. These include, shoot-tips of *D. rotundata,* by vitrification (Ahuja *et al.,* 2000;), *D. floribunda* (Mandal, 2000), *D. pentaphylla* (Ahuja *et al.* 2000), *D. deltoidea* (Mandal and Dixit, 2000) and somatic embryos of *D.bulbifera* (Dixit *et al.* 2000). The regenerated plants obtained from cryopreserved shoot-tips/somatic embryos were transferred to field with 90-100 % establishment. The field grown plants obtained from cryopreserved explants were morphologically similar to those of their mother stocks maintained *in vitro* conditions and transferred to the field.

2. Molecular Breeding

The advent of molecular markers, genome studies, and plant genetic transformation has opened avenues for circumventing breeding obstacles in yams. A better understanding of the available genetic diversity and the breeding potential of specific accessions is important for the choice of parents for use in breeding programs. Molecular markers linked to genes controlling these economic traits would be useful in selection at an early stage of the plant's growth, thereby enhancing the speed and efficiency of selection.

Polymorphism was studied in different species of yams using the isoelectric focussing of soluble peroxidase and acid phosphatase isozymes (Twyford, 1990). Isozyme pattern studies were useful in the characterization of varieties of the *D. cayenesis-rotundata* complex (Hamon and Toure, 1990). Zoundjihekpon *et al.* (1994) used isozyme markers *viz.* esterase, isocitrate dehydrogenase, malate dehydrogenase, 6-phosphogluconate dehydrogenase, shikimate dehydrogenase, and phosphoglucoisomerase to check the progenies and detect outbreeding in yams. While isozymes have proven to be valuable tools for genetic discrimination of yam cultivars (Dansi *et al.,* 2000a), polymorphic isozyme systems are often inadequate for the unequivocal identification of individual cultivars within broad cultivar groups or morphotypes in yams. Co-dominant molecular markers were used (i) to study relationship between the African cultivated species *D. rotundata* and its assumed ancestors, *Dioscorea praehensilis* and *Dioscorea abyssinica,* (ii) to study relationship between the cultivated species *D. alata* and the related species *Dioscorea nummularia* from South–East Asia and Oceania, and (iii) to genetically identify the cultivars in three diversification areas.

Several microsatellite markers were identified and characterized from wild yams also. Terauchi and Konuma(1994) characterized six microsatellite loci in *Dioscorea tokoro*, a wild yam species in East Asia.

Hochu *et al.*(2006) developed microsatellite markers in American yam (*Dioscorea trifida*). A microsatellite sequence-enriched genomic library was screened, and after sequencing, primers were designed for 20 microsatellites. Among these, eight primer pairs yielded amplification products which were both interpretable and polymorphic in 24 yam cultivars. The number of alleles per locus ranged from two to 13 and the overall expected heterozygosity was around 0.5. Six of the eight *Dioscorea trifida* microsatellite loci gave amplification products in other *Dioscorea* species. Large cross-species applicability indicates that the primers will be useful for other studies within the *Dioscorea* genus in future.

Tostain *et al.* (2006) developed new simple sequence repeat (SSR) markers in different species of yam (*Dioscorea* sp.). A microsatellite-enriched bank was created from *Dioscorea alata, Dioscorea abyssinica* and *Dioscorea praehensilis.* Sixteen polymorphic loci were characterized. Several of these markers are transferable to species of other *Dioscorea* sections.

3. Molecular Marker Assessment of Genetic Diversity

Molecular markers are now being used to elucidate the genetic architecture of yams. Earlier the cultivated yam species white yam, *D.rotundata*(2n=40) was considered by most author as tetraploid species with a basic chromosome number of ten. Using two isozyme loci and six microsatellite markers Scarcelli *et al.* (2005) conducted genetic analysis of progeny derived from a self fertilizing monoecious plant of *D.rotundata*. They found that white yam is a diploid species with 2n=20.

Similarly based on Microsatellite segregation analysis and cytogenetic evidence Mustapha, *et al.* (2006) reported tetrasomic inheritance in the American yam *(Dioscorea trifida)* and a new basic chromosome number 10 or 9) in the *Dioscoreae*.

The impact of traditional farmers' management on genetic diversity of vegetatively propagated crops was also studied using microsatellites (Scarcelli *et al.*, 2006). They used 11 microsatellite markers on yam tubers and demonstrated that wild × cultivated hybrids are spontaneously formed.

The relative efficiency of AFLP, RAPD and SSR markers for the assessment of genetic relationships, as well as for cultivar identification and discrimination was studied by Mignouna *et al.*, (2003) among 45 West and Central African white yam cultivars belonging to 22 morphotypes/cultivar groups. The results showed that each of the three marker-systems could unequivocally identify each cultivar, but that marker-systems differed in the mean number of profiles generated per primer (or primer combination) per cultivar, referred to as genotype index (GI). The order of merit based on this criterion was AFLP (GI =1.0), SSR (GI= 0.39) and RAPD (GI =0.35). Thus AFLP showed the highest efficiency in detecting polymorphism and revealed genetic relationships that most closely reflected morphological classification. Mignouna *et al.* (1998) previously showed that AFLP is a sensitive and robust technique for yam genome fingerprinting. In spite of their robustness, technical and financial burdens associated with AFLP and SSRs limit their availability to laboratories in developing countries where most of the research on yam is conducted. Thus cost considerations may make RAPD the method of choice for laboratories with rather limited resources.

4. Genome Mapping in Yams

The ability to use molecular marker biotechnology to exploit information at the gene level, based on genome mapping and marker-assisted genetic analysis and breeding, now offers a practical solution to many of the intractable problems in yam breeding. Molecular genetic maps and marker-aided analysis of complex traits can be used to elucidate the genetic control of yield potential and tuber quality, and locate genes of pest and disease resistance, nutrient use efficiency, and flowering (Mignouna,2003).Yam genome mapping was initiated in the mid-1990s. However, the lack of mapping populations and adequate polymorphic molecular markers hampered rapid progress. By the end of 2001, considerable progress had been made in developing appropriate molecular tools and mapping populations for developing linkage maps of the two most important species, *D. alata* and *D. rotundata*

Mignouna *et al*, (2002) also constructed a genetic linkage map of white yam (*D. rotundata* Poir). Based on 341 co-dominantly scored amplified fragment length polymorphism (AFLP) markers were found segregating in an intraspecific F1 cross. The mapping population was produced by crossing a landrace TDr 93-1 as female parent to a breeding line TDr 87/00211 as male parent. The marker segregation data were spilt into maternal and paternal data sets and separate genetic linkage maps were constructed since the mapping population was an F1 cross between two presumed heterozygous parents. The markers segregate like a diploid cross pollinated population suggesting that *D. rotundata* genome is an allotetraploid (2n=4x= 40). They also constructed a genetic linkage map of greater yam (D. *alata* L). Based on 469 co-dominantly scored amplified fragment length polymorphism (AFLP) markers segregating in an intraspecific F1 cross. The mapping population was produced by crossing two improved breeding lines TDa 95/00328 as female parent to a breeding line TDa 87/01091 as male parent. The 469 markers were mapped on 20 linkage groups with a total map length of 1.233 cM and a mean marker spacing of 2.62cM. In greater yam also, the markers segregated like a diploid cross pollinated population suggesting that *D.alata* genome is an allotetraploid (2n=4x= 40). This map covered 65% of the yam genome and was the first linkage map reported in *D.alata*. The map provided a tool for further genetic analysis of traits of agronomic importance and for using marker assisted selection in *D. alata* breeding programmes. QTL mapping opened new avenues for accumulating anthracnose resistance genes in preferred *D.alata* cultivars.

5. Markers for Agronomic Traits

Selective genotyping coupled with DNA sample pooling is very efficient for initial identification of disease resistance-linked markers. The approach is successfully applied in yams for identification of YMV and anthracnose resistance genes (Mignouna *et al.*, 2002 a,b).

Yam Mosaic Virus is a major limiting factor for stable production of yams, and *D. rotundata* is particularly susceptible to the virus. Genetic resistance is the only economically viable method of control. A study of the genetic control of YMV resistance in three *D. rotundata* cultivars to a Nigerian isolate of YMV showed that resistance is manifested differentially as the action of a single dominant gene in simplex condition or a major recessive gene in duplex condition (Mignouna *et al.*, 2001b). The dominant locus that contributed to YMV resistance was tentatively named Ymv-1, until tests of allelism were conducted. Two RAPD markers OPW18850 and OPX15850 closely linked in coupling phase with the dominant YMV-resistance locus Ymv-1 were identified. These markers successfully identified the resistance gene in resistant genotypes among a sample of 12 *D. rotundata* varieties. Based on the map constructed in white yam, Mignouna *et al.*(2001b) also identified three QTL's on maternal linkage map and one on paternal linkage map were linked to yam mosaic virus resistance

Anthracnose disease, caused by *Colletotrichum gloeosporioides* is a major constraint to the production of yam worldwide with *D. alata*, the most widely distributed species, being particularly susceptible to the disease. Resistant yam cultivars could form the basis of sustainable management strategies for anthracnose. Initial genetic inheritance studies showed that resistance to yam anthracnose in *D. alata* is dominantly but quantitatively inherited and controlled by the gene designated Dcg-1, (Mignouna *et al.*, 2001a). Two RAPD markers, OPI171700 and OPE6950, closely linked in coupling phase with Dcg-1 were identified. Both markers successfully identified *Dcg1* in resistant *D. alata* genotypes among 34 breeding lines, indicating their potential use in marker-assisted selection. Recently QTL mapping revealed one AFLP marker E-14/M52-307 located on linkage group 2 was associated with anthracnose disease resistance explaining 100% of the phenotypic variance (Mignouna, 2006). Further AFLP mapping is needed to identify additional QTLs and strengthen existing marker-QTL linkages. Candidate gene analyses are yet to be employed to investigate a variety of traits. Initial QTL studies indicated that the pyramiding of major resistance genes may be a valuable strategy for YMV and anthracnose resistance breeding .

Aroids

Aroids like taro (*Colocasia esculenta*) and elephant foot yam (*Amorphophallus campanulatus*) are used as vegetables in India and south east Asian countries. Taro starch has very small granule size and is ideal for the preparation of baby food and cosmetics. *Amorphophallus* spp are traditionally used in ayurvedic preparations recommended for piles and dysentery.

In vitro technology in aroids is well developed and is being used routinely in national and international centers for conservation of these crops (Staritsky *et al.* 1986; Bessembinder *et al.* 1993). There are several reports on production of virus-free taro. Shoot tips (meristems dome plus one or two-leaf primordial) 0.2 to 0.3 mm in size were cultured on MS medium with 0.1 mg l^{-1} NAA at 25°C under 12 h day length (Morishita and Yamada 1978). These meristems regenerated to shoots and produced roots after four months of culture. Successful cormel tip culture of *Colocasia esculenta* leading to the elimination of Dasheen mosaic virus was done at CTCRI (Nair *et al.* 1996). In *Xanthosoma sagittifolium*, addition of spermine, arginine and ornithine enhanced plant regeneration from shoot tip cultures (Sabapathy and Nair 1995). In India, work on medium term *in vitro* conservation was carried out at CTCRI and NBPGR. Taro cultures could be stored upto ten months under light (3000 lux) on MS basal medium containing sucrose and mannitol (3%) without vitamins or growth regulators. Half strength basal medium also induced slow growth (Unnikrishnan and Sheela, 2000). *Colocasia* and *Alocasia* cultures could be conserved at normal culture room temperature of 25°C ± 2°C upto 10 months without subculturing (Mandal, 1999). For long–term conservation, slow freezing of embryogenic callus prior to cryopreservation resulted in 70% survival, with no loss of regenerability (Takagi *et al.* 1997). Vitrification of axillary buds, which initially met with only partial success, was modified for cryopreservatioon of *in vitro* shoot tips resulting in almost 80 % shoot recovery (Takagi *et al.,* 1997; Thinh *et al.,* 2000). Shoot tips were precultured for 16 h on 0.3 M sucrose-containing medium loaded with mixture of 2M glycerol and 0.4 M sucrose for 20 min. at 25°C and treated with PVS_2 (Plant Vitrification solution-2) for 20 min at 25°C prior to immersion in LN.

Limitations in flowering, lack of synchrony, pollen sterility and protogyny are the main constraints in utilization of taro in breeding program. Molecular techniques are used routinely for fingerprinting of taro germplasm . Also efforts are being made to characterize genes controlling important traits especially protein.

Electrophoretic patterns of peroxidase and esterase were studied for identifying genetic diversity in different varieties of *Colocasia esculenta* and *C. gigantia*. Cluster analysis, based on 17 morphological characters and isoenzyme patterns, was done besides studying the chromosome number. Large variation was noticed in morphological characters and isozyme composition among *C. esculenta* strains but not among strains. Genetic diversity was analysed in taro germplasm using isozyme and RAPD markers. Moderate to high polymorphism was observed in the isozyme patterns of enzymes, acid phosphatase, peroxidase and SOD, while monomorphism with shikimate dehydrogenase and MDH. Amplification of genomic DNA using random decamer primers revealed high polymorphism as only one of the 22 primers analyzed, gave monomorphic patterns. Cluster analysis of the allozymes and RAPD data also revealed prevalence of high genetic diversity among the morphotypes studied (Lakhanpaul *et al.,* 1996).

Limited studies were made to unravel the expression of genes associated with protein and corm development in Taro. Hirai *et al* (1993) isolatd cDNAs encoding for storage proteins in the tubers of taro (*Colocasia esculenta* Schott). Two major protein groups of taro (*Colocasia esculenta*) tuber were purified, and their antisera used for the screening of the cDNA library constructed from poly(A)+ RNA of taro tuber. The amino acid sequence deduced from the nucleotide sequence of this clone indicated that the 25 kD proteins were homologous to the trypsin inhibitors of soybean and winged bean as well as sporamins, the storage proteins of sweet potato.

Bezerra *et al.* (1995) isolated a gene encoding a globulin from a major taro corm protein family, tarin (G1, ca. 28 kDa) was isolated from a lambda

Charon 35 library, using a cDNA derived from a highly abundant corm-specific mRNA, as probe. The gene, named tar1, and the corresponding cDNA were characterized and it has an open reading frame (ORF) of 765 bp, and the deduced amino acid sequence indicated a precursor polypeptide of 255 residues which were post-translationally processed into two subunits of about 12.5 kDa each. Significant similarity was also found at the nucleic acid sequence level with genes encoding lectins from plant species of the Amaryllidaceae and Lilliaceae families. Luiz *et al.* (1992) studied the gene expression pattern that occurs during taro corm development . They found that corm development is associated with the differentiation of specialized cells and tissues and that these differentiation events are coupled with the temporal and spatial expression of corm specific genes.

Minor Tuber Crops

There are many minor tubers cultivated as vegetable or medicinal crop. Arrow root is grown for the extraction of starch which has medicinal value. Research on improvement of minor tubers using biotechnological interventions is meagre and restricted mainly to standardization of *in vitro* regeneration protocols. In Chinese potato (*Solenostemon rotundiifolius*) widely grown vegetable crop, standardized protocols for *in vitro* conservation and micropropagation are available. Direct regeneration was obtained from lamina explants and internodal cuttings at higher levels of BAP (1 to 10mM). Higher levels of NAA (1 to 10 mM) produced callus and indirect regeneration (Nair, 1992). Nodal cultures grew on hormones free MS medium. Nodal explants developed shoot and root directly in MS medium.

The progress achieved so far in the fields of transgenic technology and molecular breeding with limited resources will help to increase the productivity and utility of tropical tuber crops. However, there is a need to develop/consider approaches such as the use of microarrays technology to study the expression of genes, development of a large number of ESTs ,isolation and cloning of novel genes, etc. Also the crops like sweet potato can be used for molecular farming.

References

Ahuja Sangeeta, Mandal,B.B. and. Srivastava,P.S. 2000. Vitrification: A potential cryopreservation technique for long-term conservation of yam germplasm. Abst. *National symposium on Prospects and Potentials of Plant Biotechnology in India in the 21 st Century. Jodhpur,* Oct. 18th -21st, 2000. p. 159

Akano, A.O., Dixon, A.G.O., Mba, C. , Barrera, E. and M.Fregene.2002.Molecular genetic mapping of resistance to the African cassava mosaic disease. *Theoret. Applied Genetics,* 105 :521-525.

Ashmore, S .E.1997. *Status report on the development and application of in vitro techniques for the conservation and use of plant genetic resources.* IPGRI, Rome

Bhatti, M. H., Percival, T, Davey, C. D. M, Henshaw, G.G. and Blakesley, D. 1997. Cryopreservation of embryogenic tissue of a range of genotypes of sweet potato (*Ipomoea batatas* (L.) Lam.) using an encapsulation protocol. *Pl. Cell Rep.* 16: 802-806

Beeching ,J.R., Marmey, P., Hughes .M.A and Charier ,A. 1995 . Evaluation of Molecular Approaches for determining Genetic diversity in cassava Germplasm,*Proceeding of the second International Scientific meeting of the cassava Biotechnology Network,* Boyor, Indonesia (working document 150) 2: 80-89.

Bertram , R.B.1993. Application of Molecular Technique to genetic resources of cassava. *(Manihot esculenta* Crantz)(*Euphorbiaceae*) inter specific evolutionary relationship and intra specific characterization PL.D dissertation University of Meryland, USA 465.

Bessembinde, J. J. E., Stritsky, G. and Zandvoort, E .A.1993. Long term *in vitro* storage of *Colocasia esculenta* (L.) Schott under minimal growth conditions. *Pl. Cell Tiss. Org. Cult.* 33: 121-127.

Bezerra ,I.C., Castro, L.A., Neshich,G., de Almeida ER, de Sa MF, Mello LV and Monte-Neshich, D.C. 1995. A corm-specific gene encodes tarin, a major globulin of taro (*Colocasia esculenta* L. Schott). *Plant Mol Biol.* 28(1):137-44.

Blackesley, D., Percival, T., Bhatti , M. H. and Henshaw, G. G. 1997. A simplified protocol for cryopreservation of embryogenic tissue of sweet potato (*Ipomoea batatas* (L.) Lam.) utilizing sucrose preculture only. *Cryo-lett.* 18: 77-80.

Calderon, A. 1988. Transformation of *Manihot esculenta* Crantz (cassava) using *Agrobacterium tumefaciens* and expression of the introduced genes in transformed cell lines. M.Sc. Thesis Vrije Universiteit Brussel, Belgium, 37p.

Carvalho, L.J.C.B Schaal B.A., and Fukuda,W.M.G.2000.Agronomic DNA (RAPD) marker were used to assess the genetic diversity of cassava *(Manihot esculenta* Crantz) in Carlvalho J.C.B, Thro . Thro A.M. and IV Intl sci Meet – CBN. EMBR. APA – Genetic Res. Biotech. Brasilia.- DF Brazil PP – 51- 64.

Chaavariaga P., Scopke, C., Sangare A., Fauquet,C.M. and Beachy, R.N. 1993. Transformation of transgenic cassava (*Manihot esculenta* Crantz) embryogenic tissues using *Agrobacterium tumefaciens* In : Roca, W.M., Thro, A.M.(eds), *Proc. I*[st] *Int. Sci. Meet Cassava Biotechno. Network,* Cartagena, Colombia, 25-28 August 1992. *Centro Internacional de Agricultura Tropical,* Cali, Colombia, pp : 222-228.

Coursey, D.G.1976. The origins and domestication of yams in Africa. *In: Origin of African Plant Domestication* (eds Harlan JR, De Wet JMJ, Stemler Mouton ABL), p. 383–408. Mouton, Hague, the Netherlands.

Dansi, A., Mignouna, H.D., Zoundjihékpon, J., Sangaré, A., Asiedu, R. and Ahoussou, N . 2000. Using isozyme polymorphism to assess genetic variation within cultivated yams (*Dioscorea cayenensis/ Dioscorea rotundata* complex) of the Benin Republic. *Genet. Resourc. Crop Evol.* 47:371-383.

Dhillon, N.P.S. and Ishiki,K. 1999. genomic variation and genetic relationships in Ipomoea spp. *Plant Breeding* 118 : 161-165.

Dixit Sonali, Mandal B B, and Srivastava P S. 2000. Cryopreservation of somatic embryos/ embryogenic tissues: Application in Conservation and Biotechnology. Abst. In: *National symposium on Prospects and Potentials of Plant Biotechnology in India in the 21 st Century*. Jodhpur, Oct. 18th -21st 2000. pp. 161.

Dodds, J. H. and Roberts, L.W. 1985. *Experiments in Plant Tissue Culture*, 2nd edn., Cambridge, NY: Cambridge University Press

Dodds,J.H., Merzdorf, C., Zambrano, V., Siguenas, C . and Jaynes, J.1991. Potential use of *Agrobacterium* mediated gene transfer to confer insect resistance in sweet potato. In: R K Jansson , K V Raman eds. *Sweet potato Pest Management: A Global Perspective*. Boulder, CO: Westview, 1991, pp 203-219.

Elias,M., Panaud,O. and Robert 2000. Assessment of genetic variability in a traditional cassava (*Manihot esculenta* Crantz) farming system using AFLP markers. *Heredity*, 85 : 219-230.

FAO 2002. FAOSTAT Agriculture data. Food and Agriculture Organisation of the United Nations. http://apps.fao.org/collections

Frischmuth,T. and Stanley,J. 1991African cassava mosaic virus DI DNA interferes with the replication of both genomic components. *Virology*, 183:539-534.

Fregene,M., Angel,F., Gomez,R., Rodríguez, F., Chavaariaga, P., Roca,W., Tohme,J. and Bonierbale, M. 1997. A molecular genetic map of cassava (*Manihot esculenta* Crantz.) . *Theor. Appl. Genet.*,95: 431-441.

Fregene,M., Bernal,A., Duque,M., Dixon,A. , Tohme,J. and Bonierbale, M. 2000. AFLP analysis of African cassava (*Manihot esculenta* Crantz) . germplasm resistant to the cassava mosaic disease(CMD). *Theor. Appl. Genet.*, 100: 678-685.

Fregene,M., Okogbenin,E., Mba,C., Angel,F., Suarez,M.B.,Guitierez,J., Chavaariaga, P., Roca,W., Bonierbale, M. and Tohme,J. 2001. Genome mapping in cassava improvement: challenges, achievements and opportunities. *Euphytica*, 120 : 159-165.

Gama, M.I.C.S., Leite Jr, R.P., Cordeiro, A.R. and Cantliffe, D.J.1996. Transgenic sweet potato plant obtained by *Agrobacterium tumefaciens* – mediated transformation. *Plant Cell Tissue Org .Cult* 46:237-244.

Gonzalez,A.E., C.Schopke, Taylor,N.J. Beachy R.N and Fauquet CM. 1998. Regeneration of transgenic cassava plants (*Manihot esculenta* Crantz) through *Agrobacterium* – mediated transformation of embryogenic suspension cultures. *Plant Cell Reports* , 17: 827-831.

Gresshoff,P. and Doy,C.1974. Derivation of a haploid cell line from Vitis vinifera and the importance of the stage of meiotic development of the anthers for haploid culture of this and other genera. *Zeitschrift fir Pflanzenphysiologie* 73, 132-141.

Hamon P and Toure B.1990.Charaterisation of traditional yam varieties belonging to the *Dioscorea cayenensis-rotundata* complex by their isozymic pattern. *Euphytica* 46: 101-107.

Hanson, J.1986.Methods of storing tropical root crop germplasm with special reference to yam. *FAO/IBPGR Pl. Genet. Res. Newsl.* 64: 24-32.

Hirai,M., Nakamura,K., Imai,T,and Sato,T., 1993. cDNAs encoding for storage proteins in the tubers of taro (*Colocasia esculenta* Schott). *Jpn.J. Genet.* 68 (3) : 229-36.

Hochu, I., Santoni, S. and Bousalem, M. 2006.Isolation, characterization and cross-species amplification of microsatellite DNA loci in the tropical American yam *Dioscorea trifida* . *Molecular Ecology Notes.*, 6 (1) : 137 - 140.

Kartha, K.K., Gamborg, O.L., Constenbel,F., and Shyluk, J.D . (1974) Regeneration of cassava plant from Apical meristem plant *Sci.Lett.*2:107-113.

Kennedy, L. S. and Thompson, P.G.1991.Identification of sweet potato cultivars using isozyme analysis. *Hort. Sci.* 26: 300-302.

Kimura,T., Otani,M.,Noda.T., Ideta, O., Shimada, T., and Saito, A. 1999. Decrease of amylose content in transgenic sweet potato. *Breed.Sci.* 1: 142.

Lakhanpaul, S., Mandal,B.B. and Chandel, K.P.S.1990. Isozyme studies in the *in vitro* regenerated plants of *Ipomoea batatas* (L.) (Lamarck). *J.Root Crops,* ISRC Nat. Sym. 1990, Special: 305-310.

Li,H.Q., Sautter,C., Potrykus, I. and Pounti Kaerlas J.1996. Genetic transformation of cassava (*Manihot esculenta* Crantz) . *Nature Biotech.,* 14: 736-740.

Lokko, Y., Gedil, M. and Dixon, A. G. O. 2004. QTLs associated with resistance to the cassava mosaic disease: new directions for a diverse plant. In: Proceedings of the 4th International Crop Science Congress, 26th September-1stOctober, 2004, Brisbane,Australia. http://www.cropscience.org.au/icsc2004/poster/3/7/4/1154_lokko.htm.

Lokko, Y., Danquah, E. Y., Offei, S. K., Dixon A. G. O. and Gedil, M. A. 2005. Molecular markers associated with a new source of resistance to the cassava mosaic disease. *African J. Biotech.*, 4: 873-881.

Lopez, C.E., Zulugua, A.P., Cooke, R., Delseny,M., Tohme,J. and Verdier,V.2003. Isolation of resistance gene candidates (RGCs) and characterization of an RGC cluster in cassava. *Mol. Gen. Genomics,* 268 : 658-671.

Luiz,A.B. de Castro, Mauro Carneiro, Damares de C.M. Neshich and Genaro R. de Paiva. 1992. Spatial *and Temporal Gene Expression Pattern Occur during Corm development. The Plant Cell,* 4 : 1549-1559.

Malaurie,B., Trouslot,J., Berthaud,J., Bousalem,M., Pinel,A. and Dubern, J.. Medium term and Long term *in vitro* conservation and safe international exchange of yam *Dioscorea* germplasm. *Electronic Journal of Biotechnology,* 1(3) , ISSN: 0717-3458.

Mandal,B.B. and Chandel, K.P.S.1991.Utilization of tissue culture technique in preservation of sweet potato germplasm. *J. Root Crops* 17:291-295.

Mandal, B.B. and Chandel,K.P.S. 1996. Conservation of genetic diversity in sweet potato and yams using *in vitro* strategies. In G.T.Kurup, M.S. Palaniswami, V.P. Potty, G.Padmaja, S. Kabeerathumma and Santha V. Pillai, (*Eds.*) *Tropical Tuber Crops: Problems, Prospects and Future Strategies.* Oxford and IBH, New Delhi p.49-59.

Mandal, B.B .and Sonali Dixit. 2000. Cryopreservation of shoot-tips of *Dioscorea deltoidea* Wall. - An endangered medicinal yam, for long-term conservation. *IPGRI News Letter for Asia, the Pacific and Oceania.* No. 33. Sept.-Dec.2000. pp. 23.

Mandal B B, Chandel K P S and Dwivedi S. 1996. Cryopreservation of yams (*Dioscorea spp.*) shoot apices by encapsulation - dehydration. *Cryo-Letters* 17: 165-174.

Mandal,B.B. 1999. Conservation biotechnology of endemic and other economically important plant species of India. In: E.E. Benson (ed.) *Plant Conservation Biotechnology*, Taylor and Francis, London. pp. 211-225.

Mandal, . B. 2000. Cryopreservation of yam apices: comparative study with three different techniques. *In*: F. Engelmann and H. Takagi (*eds.*) *Cryopreservation of tropical plant germplasm: Current research progress and application, JIRCAS proceedings of International workshop* 20th -23rd oct, Japan/ IPGRI, Rome Italy. pp. 233-237.

Mba,R.E.C., Stephenson,P., Edwards,K.,Melzer,S., Kumbira,J.N., Gulberg,U., Apel,K., Gale, M., Tohme,J and M.Fregene. 2001. Simple sequence Repeat (SSR) marker suevey of the cassava genome towards an SSR based molecular genetic map of cassava. *Theoret. Applied Genetics*, 102 :21-31.

Mignouna, H.D., Abang, M.M., and Fagbemi, S.A. 2003. A comparative assessment of molecular marker assays (AFLP, RAPD and SSR) for white yam (*Dioscorea rotundata* Poir.) germplasm characterisation. *Ann. Appl. Biol.* 142:269-276.

Mignouna, H.D., Abang, M.M., Onasanya, A., Agindotan, B., and Asiedu, R . 2002b. Identification and potential use of RAPD markers linked to *Yam mosaic virus* resistance in white yam (*Dioscorea rotundata* Poir.). *Ann. Appl. Biol.* 140:163-169.

Mignouna, H.D., Njukeng,P., Abang,M.M, and Asiedu, R. 2001b. Inheritance of resistance to Yam Mosaic Virus, genus Potyvirus, in white yam (*Dioscorea rotundata*). *Theor. Appl. Genet.* 103:1196-1200.

Mignouna, H.D., Ellis, T.H.N., Knox, M., Asiedu, R.and Ng, S.Y.C. 1998.Analysis of genetic diversity in Guinea yams (*Dioscorea* sp) using AFLP fingerprinting. *Tropical Agriculture* 75 : 224 – 229.

Mignouna, .H.D., Abang,M.M., Green, K.R., and Asiedu, R.2001a. Inheritance of resistance in water yam (*Dioscorea alata*) to anthracnose (*Colletotrichum gloeosporioides*). *Theor. Appl. Genet.* 103:52-55.

Mignouna, H.D., Mank, R.A. , Ellis, T.H.N., van den Bosch, Asiedu, R., Ng, S.Y.C., and Peleman, J. 2002. A genetic linkage map of Guinea yam (*Dioscorea rotundata* Poir.) based on AFLP markers. *Theor. Appl. Genet.*, 105 (3): 716-725.

Mignouna,H.D., Abang ,M.A. and Asiedu,R. 2003.Harnessing modern biotechnology for tropical tuber crop improvement: Yam (*Dioscorea* spp.) molecular breeding . *African Journal of Biotechnology*, 2(12) : 478-485.

Moran ,R., Garcia, R., Lopez, A., Zasldua, Z., Mena,J., Garcia, M., Armas ,R., Somonte,D., Rodriguez, J.,Gomez,M. and Primentel,E. 1998. Transgenic sweet potato plant carrying the delta endotoxin gene from *Bacillus thuringiensis var. tenebrionis. Plant Sci.*,139:175-184.

Morishita, M. and Yamada, K.1978. Studies on tissue culture of taro (*Colocasia esculenta* Schott) (1) Inducing plantlets from apical meristem. *Bull. Osaka Agri. Res. Cent.* 15:9-12.

Munyikwa, T.R.I., Raemakers, C.J.J.M., Schreuder,M., Kok, R., Rozeboom,M., Jacobsen, E., and Visser, R.G.F. 1998. Pinpointing towards improved transformation and regeneration of cassava. *Plant Sci.*135:87-101.

Murashige, T. and Skoog, F. 1962. A revised medium for rapid growth and bio-assays with tobacco tissue cultures. *Physiologia Plantarum*, 15: 473–97.

Murata,T., Okada, Y., Fukuoka, H., Saito, A., Kumura,T., Mori, M., Nishiguchi, M., Handa, K., Sakai, J. and Fukuoka.H.1997. Transformation by directly gene transfer in sweet potato [*Ipomoea batatus* (L.) Lam.]. Proceedings of International workshop on sweet potato production system towards the 21 st century. Kyusyu National Agricultural Experimental Station. Miyakonojo, Japan, pp 159-180.

Mustapha Bousalem, Gemma Arnau, Isabelle Hochu, Richard Arnolin, Vaeronique Viader, Sylvain Santoni, and Jacques David (2009) Microsatellite segregation analysis and cytogenetic evidence for tetrasomic inheritance in the American yam *Dioscorea trifida* and a new basic chromosome number in the *Dioscoreae, Theor. Appl. Genet.*, 113 (3): 439-451.

Nagib M. A. Nassar and Rosane G. Collevatti. 2005.Microsatellite markers confirm high apomixis level in cassava bred clones. *Hereditas*, 142: 33-37.

Nair,N.G. and Chandra Babu, S.1996. *In vitro* production and micropropagation of three species of edible yams.In: Kurup GT, Palaniswami MS, Potty VP, Padamaja G, Kabeerathumma S and Pillai SV(Eds.). *Tropical Tuber Crops: Problems, prospects and future strategies*, pp.55-60, Oxford and IBH Publishing Co. Pvt. Ltd., New Delhi.

Nair N G.1992. Growth and morphogenesis of Chinese potato (*Coleus parviflorus* Benth) in culture. InSubha Rao, N.S., Balagopalan, C., Ramakrishna,S.V. (Eds). New trends in Biotechnology, pp 57-62. Oxford & IBF publishing Co. Pvt.Ltd., New Delhi., India.

Newell, C.A., Lowe, J., Merryweather, M.A ., Rooke, L.M. and Hamilton,W.D.O.1995. Transformation of sweet potato [*Ipomoea batatus* (L.) Lam.] with *Agrobacterium tumefaciens* and regeneration of plants expressing cowpea trypsin inhibitor and snowdrop lectin. *Plant Sci* 107:215-227.

Ng,S.Y.C and Hahn, S. K.1985. Application of tissue culture to tuber crops at IITA.(*In*): *Biotechnology in International Agricultural Research*, pp. 27-40, International Rice Research Institute, Manila, Phillipines.

Ng,S.Y.C. and Ng, N.Q.1997. Germplasm conservation in food yams (*Dioscorea* spp.):constarints, applications and future prospects. (*In*:) Razdan,M.K and Cocking,

E.C.(Eds.) *Conservation of Plant Genetic Resources in vitro. Vol. 1: General Aspects,* pp. 257-286, Scientific Publishers Inc.,USA.

Noyer, J.L., Billot,C., Weber, A., Brottier,P., Guero-Garcia and V.Lebot. 2003. Genetic Diversity of Taro, assessed by SSR markers. *Proc. of Taro Genetic Symposium*, Nadi, Fiji, May 22-24, 2003.

Nweke,F.I.,Spencer,D.D.C. and Lynam,J.K., 2002. The Cassava transformation.pp.101-114. Michigan State University Press, East Lansing.

Ocampo,C., Angel,F., Jim Crez,A., Jaramillo, G., Hershey,C., Granados,E., and Iglesias,C., 1994. DNA fingerprinting to confirm posible genetic duplicates in cassava germplasm. *Proc.* 2nd Intern. Sci. Meeting CBN II. 22-26 , August 1994, Bogor, Indonesia, Abst. p.44

Okado,Y., Saito,M., Nishiguchi,M., Kimura,T., Mori, M., Hanada,K., Sakai,J., Miyazaki, Matsudo,Y and Murata,T. 2001. Virus resistance in transgnic sweetpotato (*Ipomoea batatus* (L.)Lam) expressing the coat protein gene of sweet potato feathery mottle virus. *Theor. Appl. Genet*. 103:743-451.

Okogbenin,E. and Fregene,M. 2003. Genetic mapping of QTLs affecting productivity and plant architecture in a full sib cross from non-inbred parents in cassava (Manihot esculenta Crantz). *Theor. Appl. Genet*. 107:1452-1462.

Olsen,K.M., and Schaal,B.A. 1999. Evidence on the origin of cassava : phylogeography of *Manihot esculenta. Proceedings of the National academy of Sci.*,96: 5586-5591.

Otani, M. and Shimada, T.1996. Efficient embryogenic callus formation in sweet potato [*Ipomoea batatus* (L.) Lam.]. *Breed. Sci.* , 46 : 257-260.

Otani, M., Shimada, T., Kimura,T., and Saito,A.,1998. Transgenic plant production from embryogenic callus of sweet potato (*Ipomoea batatus* (L.) Lam.) using *Agrobacterium tumefaciens. Plant Biotechnol,* 15:11-16, 1998.

Otani,M., Mii, M., Handa, T., Kamada, H. and Shimada T.1993. Transformation of sweet potato [*Ipomoea batatus* (L.) Lam.] by *Agrobacterium rhizogenes. Plant Sci* 94:151-159.

Prakash, C. S, Desai, A. P, Ramana Murthy, G., Korsi Dumenyo, C, He G, Zheng, Q, Egnin, M .and Kanyand, M.1996. Biotechnological approaches for improving sweet potato. In: Kurup G T, Palaniswami M S, Potty V P, Padmaja G, Kabeerathumma S and Pillai S V (Eds.). *Tropical tuber crops: Problems, Prospects and Future Strategies*, pp. 27-36, Oxford and IBH publishing Co. Pvt. Ltd. New Delhi

Prakash, C.S. 1994.Sweet Potato Biotechnology: Progress and potential. *Biotechnology and Development Monitor*, No. 18, : 1819-22.

Prakash,C.S. and Varadarajan,U.1992. Genetic transformation of sweet potato by particle bombardment. *Plant Cell Rep* 11:53-57.

Raemakers, C.J.J.M., Amati, M., Staritsky, G., Jacobsen, E. and Visser, R.G.F. 1993a. Cyclic somatic embryogenesis and plant regeneration in Cassava. *Annals of Botany*, 71: 289–294

Raemakers, C.J.J.M., Schreuder, M., Munyikwa,T., Jacobsen, E. and Visser,R.G.F.2000. Towards a routine transformation procedure for cassava. (In): Carvalho, L., Thro, A.M., and Vilarinhos, A.D. (Eds) Proc. 4th Int. Sci. Meet Cassava Biotechno. Network, Cartagena, Colombia, 3-7 November 1998, Salvador, de Bahia, Brazil, Brazilian Agricultural Research Corporation - EMBRAPA, Brazil, pp 236-243. August 1992. Centro Internacional de Agricultura Tropical, Cali, Colombia, pp : 250-266.

Raemakers, C.J.J.M., Schreuder, M., Pereira,I., Munyikwa,T., Jacobsen, E. and Visser,R.G.F. 2001. Progress made in FEC transformation of cassava. *Euphytica* 120:15-24.

Raemakers, C.J.J.M., Schreuder,M., Pereira,I., Munyikwa,T., Jacobsen, E. and Visser,R.G.F. 1996. Production of transgenic cassava plants by particle bombardment using luciferase activity as the selection marker. Mol. Breeding, 2: 339-349.

Raemakers, C.J.J.M., Sofiari,E., Jacobson, E., and Visser, R.G.F. 1997. Regeneration and transformation of cassava. Euphytica *96: 153-161.*

Ramirez H, Hussain A, Roca W and Bushuk W.1987.Isozyme electrophoregrms of sixteen enzymes in five tissues of cassava (*Manihot esculenta* Crantz.) varieties. *Euphytica* 36: 39-48.

Roca ,W.M,1984.Cassava (In) Sharp ,W.R.,Evans , D.A.,Ammirato and P.V., Yumada, Y.(eds).*Handbook of plant cell culture* .Vol 2. Crop species Mac Millan New York , 269-301.

Roca, A.G., Maya, M.M., Daque ,M.C., Tohme,J., Allem , A.C., and Bonier bala ,H.W.2000.AFLP analysis of relationship among cassava and other Manihot species.*Theor Applic.Genetic.*95.741-750.

Sabapathy, S. and Nair, H.1995. *In vitro* propagation of taro, with spermine, arginine and ornthine: Plantlet regeneration via callus. *Pl. Cell Rep.* 14: 520-524.

Sarria, R., Torres . E., Angel,F., Chavarriaga, P. and Roca, W.M. 2000. Transgenic plants of cassava (*Manihot esculenta*) with resistance to Basta obtained by *Agrobacterium* mediated transformation. *Plant Cell Rep.* , 19: 339-344.

Scarcelli, N. , Tostain, S., Vigouroux, Y. , Agbangla, C. , Daïnou , O. and Pham., J.L. 2006. Farmers' use of wild relative and sexual reproduction in a vegetatively propagated crop. The case of yam in Benin. *Molecular Ecology*, 15 (9) 2421 -2431

Scarcelli, N. , Daïnou , O. , Agbangla, C. Tostain, S.and Pham., J.L. 2006. 2006. Segregation patterns of isozyme loci and micro satellite markers show the diploidy of African Yam (Dioscorea rotundata (2n=40). *Theor. Appl. Genet.* 111 : 226-232.

Schenk,R.U., and Hildebrandt,A.C.1972. Medium and techniques for induction and growth of monocotyledonous and dicotyledonous plant cell cultures. *Canada J. Bot.* 50 : 199-204.

Schopke, C.,Chavarriaga, Fauquet, C.M. and Beachy, R.N., 1993. Cassava tissue culture and transformation : improvement of culture media and the effect of different antibiotics on cassava. In : Roca, W.M., Thro, A.M.(eds), *Proc. Ist Int. Sci. Meet Cassava Biotechno. Network*, Cartagena, Colombia, 25-28 August 1992. *Centro Internacional de Agricultura Tropical*, Cali, Colombia, pp : 140-145.

Schopke, C.,T MasonaM.V., Taylor,N.J. , Carcamo,R., Beachy, R.N. and Fauquet, C.M. 2000. Production and characterization of transgenic cassava plants expressing the coat protein gene of cassava common mosaic virus (In) : Carvalho, L., Thro, A.M., and Vilarinhos, A.D. (Eds) *Proc. 4th Int. Sci. Meet Cassava Biotechno.* Network, Cartagena, Colombia, 3-7 November 1998, Salvador, de Bahia, Brazil, Brazilian Agricultural Research Corporation - EMBRAPA, Brazil, pp 236-243. August 1992. *Centro Internacional de Agricultura Tropical*, Cali, Colombia, pp : 140-145.

Schopke, C.,Taylor,N.J. , Carcamo,R., Konan,N.K., Marmey,P., Henshaw, G.G., Beachy, R.N., and Fauquet, C.M. 1996. R.N.Beachy and CM.Fauquet. 1998. Regeneration of transgenic cassava plants (*Manihot esculenta*) from micro bombarded embryogenic suspension cultures. *Nature Biotech.* 14: 731-735.

Schreuder, M.M., Raemakers, C.J.J.M., Jacobsen, E. and Visser, R.G.F. 2001. Efficient production of transgenic cassava plants by Agrobacterium mediated transformation of cassava (*Manihot esculenta* Crantz) , *Euphytica,* 120: 35-42.

Second,G.,Allem,A.,Emperaire,L.,Ingram,C.,Colombo,C.,Mandez,R., and Carvalho,L., 1997. AFLP based *Manihot* and cassava numerical taxonomy and genetic structure analysis in progress: implications for dynamic conservation genetic mapping. *African Journal of Root and Tuber Crops*,2 : 196-200.

Shu-Jen Wang, Kai-Wun Yeh and Chia-Yin Tsai. 1999. Molecular characterization and expression of starch granule bound starch synthase in the sink and source tissues of sweet potato. *Physiologia Plantarum* , 106 : 253-261.

Siritunga,D. and Richard T. Sayre. 2003. Generation of cyanogen free transgenic cassava. *Planta,* 217:367-373.

Sreerekha, V.R., Asha, V., Unnikrishnan, M., Sheela, M.N., Subhra Chakraborty., Niranjan Chakraborty and Asis Datta, 2005. Response to callusing and somatic embryogenesis in some indigenous cassava varieties in 2,4 D Medium. *J. Root Crops.*

Stamp, A.A.and Henshaw, G.G.1982. Somatic embryogenesis and plant regeneration in cassava. *ZPflanzenphysiol.,* 105: 183-187.

Stamp,A.A., and Henshaw,G.G. 1987. Somatic embryogenesis from clonal leaf tissues of cassava. *Anna. Bot.* 59: 445-450.

Stanley,J. 1983. Infectivity of cloned Geminivirus genome requires sequences from both DNA s, ,*Nature*, 305 :643-645.

Staritsky, G., Dekkers, A. J., Lowwaars, N .P. and Zandvoort, E. A.1986. *In vitro* conservation of aroid germplasm at reduced temperature and under osmotic stress. In: Withers L A and Anderson P G (Eds.) *Plant Tissue Culture and its Agriculture applications.* Butterworth, London, pp. 277-283.

Szabados, L., Hoyos, R., and Roca, W.M. 1987. *In vitro* somatic embryogenesis and plant regeneration of cassava. Plant Cell Rep. 6: 248-251.

Takagi, H., Tien Tninh, N., Islam, O. M., Senboku, T .and Sakai, A.1997.Cryopreservation of *in vitro*- grown shoot tips of taro (*Colocasia esculenta* (L.)Schott) by vitrification. 1.

Investigation on basic conditions of the vitrification procedure. *Pl. Cell Rep.* 16:594-599.

Taylor, N.J., Edwards, M., Kieman , R.J., Davey,C.D.M.,Blakesley,D., and Henshaw, G.G. 1996. Development of friable embryogenic callus and embryogenic suspension systems in cassava (*Manihot esculenta* Crantz.), *Nature Biotechnol.* 14 : 726-730.

Taylor, N.J., Masona, M.V., Ogbe,F.and Fauquet, C.M. 2003. (In): C.M.Fauquet and N.J.Taylor., (Eds).Cassava: An ancient crop for modern times (Compact disc 3). *Proceedings of the 5th International Meeting of the Cassava Biotechnology Network*(4-9 November,2001,St.Louis,MO.

Taylor, N.J., Masona, M.V., Carcamo, R., Ho, T., Schopke, C. and Fauquet, C.M. 2001. Production of embryogenic tissues and regeneration of transgenic plants in cassava (*Manihot esculenta* Crantz) . *Euphytica,* 120: 25-34.

Taylor, N.J., Chaavariaga P., Raemakers, K., Siritunga, D., and Zhang,P. 2004. Development and application of transgenic technologies in cassava. Plant molecular Biology, 56:671-688.

Terauchi,R. and Konuma,A. 1994.Microsatellite polymorphism in *Dioscorea tokoro,* a wild yam species. *Genome* , 37(5) : 794-801.

Thinh, T. N., Takagi, H. and Sakai, A. 2000.Cryopreservation of *in vitro* grown shoot tips of five vegetatively propagated monocots by vitrification. *In*: Cryopreservation of tropical plant germplasm. *Current Research progress and application* (Engelmann, F. and Takagi, H. (*Eds.*) JIRCAS/IPGRI, pp.227-232.

Tostain, S., Scarcelli, N., Brottier, P. , Marchand, J.-L ., Pham, J.-L. and Noyer, J.-L. 2006. Development of DNA microsatellite markers in tropical yam (*Dioscorea* sp.) .*Molecular Ecology Notes*, 6(1) : 173-175.

Twyford C T.1990.Charaterization of species and vegetative clones of *Dioscorea* using isoelectric focussing of peroxidase and acid phosphatese isoenzymes. *Trop. Agri.* 67: 337-341.

Unnikrishnan M and Sheela M N.2000.Biotechnology in conservation and improvement of tuber crops.(In): Biotechnology in Horticultural and Plantation Crops:Chadha K L, P N Ravindran and Leela Sahijram(Eds.), Malhotra Publishing House, New Delhi, India.

Unnikrishnan,M., M.N., Sheela, ,V. Gayathri and J.R Nair,. 2002. Genotype response variation to micro propagation and effect of retardants on *in vitro* growth in cassava. CBN-V. *Fifth International Scientific meeting of the cassava Biotechnology Network,* Nov.4-9, 2001. Donald Danforth Plant Science centre, St.Louis – Missouri – USA. Abstract Book (Eds.) Taylor, N.J., Ogbe, F. and Fauquet, C.M. S7-33.

Uzoma,E.I., Arias-Garzon,D.I. and Sayre,R.T. 2003. Genetic modification of cassava for enhanced starch production . Harnessing Crop Technologies to Alleviate hunger in Africa. Proceedings of the 6th Biennial Conference of the African Crop Science society, 12-16th October, 2003, Nairobi, Kenya, p65.

Verdaguer, B., de kochko, A., Fux, C.I., Beachy, R.N. and Fauquet, C.M. 1998. *Functional organization of the cassava vein mosaic virus (CsMV) promoter. Plant Mol. Biol.* 37: 1055-1067.

Zhang,P., Jaynes,J., Potrykus,I., Gruissem,W. and Pounti Kaerlas, J. 2003. Transfer and expression of an artificial storage protein (ASP1) gene in cassava (*Manihot esculenta* Crantz). *Transgenic Res.* 12: 243-250.

Zhang,P.,Phansiri, S. and Pounti Kaerlas, J. 2000. PIG mediated cassava transformation using positive and negative selection. *Plant Cell Rep.* 19: 1041-1048.

Zhang,P.,Potrykus, I. and Pounti Kaerlas, J. 2000. Efficient production of transgenic cassava using negative and positive selection. *Plant Cell Rep.*,9: 405-415.

Zoundjihekpon J., Hamon, S. , Tio-Touré B. and Hamon , P. 1994.First controlled progenies checked by isozyme markers in cultivated yams *Dioscorea cayenensis-rotundata. Theor. Appl. Genet.*, 88 (8): 1011-1016.

Fruit Science

The Science of Horticulture : Vol. 1, 2011
© K.V. Peter (ed.), pp. 379-406
New India Publishing Agency, New Delhi (India)
E-mail : newindiapublishingagency@gmail.com
Web: www.bookfactoryindia.com

Chapter 16 Viticulture in the Napa Valley : A Farmer's Perspective

RICHARD MENDELSON AND ROBERT STEINHAUER

California's Napa Valley, situated at the northern end of the San Francisco Bay 30 miles (48 kilometers [km]) from the Pacific Ocean, is the most famous wine appellation in the United States. This small valley, 56 km long and 8 km wide at its broadest point, has produced some of North America's finest wine grapes from the 1850s to the present. Today it contains approximately 18,616 hectares [ha] of vineyards along the valley floor, in the foothills and mountains flanking the main valley and in the smaller valleys in the eastern part of Napa County [45].

---

1 Richard Mendelson is a lawyer specializing in vineyard and wine law with the law firm of Dickenson, Peatman and Fogarty in Napa, California. He directs the Wine Law and Policy Program at the University of California, Berkeley School of Law, and is past president of the International Wine Law Association.

2 Robert Steinhauer managed vineyard operations at Beckstoffer Vineyards (1971-1979) and Beringer Vineyards (1979-2004) and is now a viticultural consultant with and co-owner of Wineland Consulting, LLC. He is a past president of Napa Valley Vintners and in 2008 received the Merit Award from The American Society for Enology and Viticulture.

3 The authors wish to acknowledge and thank Hal Huffsmith and Tucker Catlin for their peer review of this article.

The uniqueness of this vineyard landscape is determined, in large part, by the great richness and variety of its geology, soil types, microclimates and topography, all of which affect viticulture [19]. The recognition of Napa Valley as an American viticultural area in 1981 and the subsequent establishment of 12 sub-appellations wholly within Napa Valley[4] are evidence of this diversity.

The Valley's vineyards and wineries have maintained a remarkable continuity in the face of cyclical market forces, recurrent threats from vine pests and diseases and the unique American experience of Prohibition. These cycles have directly affected viticulture, often adversely, but they also have provided an opportunity for innovation and improvement which are hallmarks of American viticulture generally and Napa Valley viticulture in particular.

Maps : courtesy of Napa Valley Vintners

4 In 1978, the U.S. government began the process of recognizing American viticultural areas ("AVAs"), defined as delimited grape growing regions with geographical features (soil, climate, elevation, physical features, etc.) that distinguish the area viticulturally from surrounding areas [10]. An AVA can be used on a wine label if, *inter alia*, at least 85% of the grapes in that wine were grown inside the AVA. Two AVAs, Carneros and Wild Horse Valley, are located partly in Napa County and partly in neighboring counties (Sonoma County and Solano County, respectively). As of June, 2009, the proposed Calistoga AVA is pending before the federal Alcohol and Tobacco Tax and Trade Bureau.

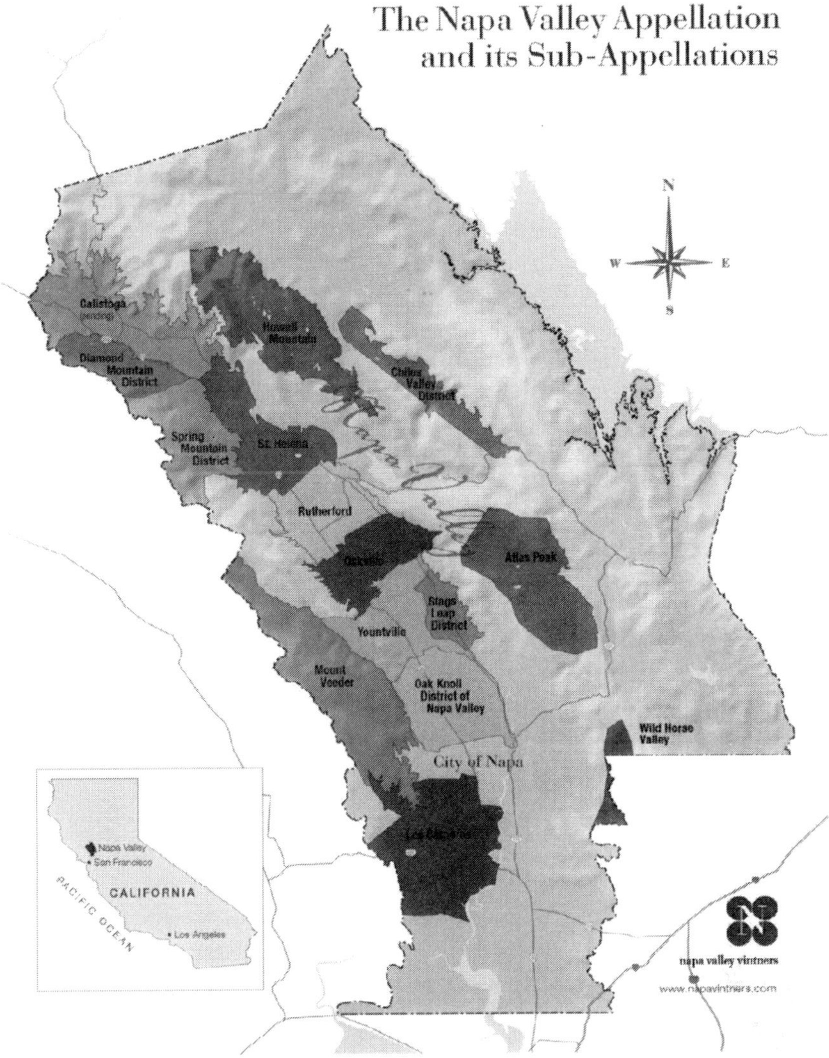

Courtesy of Napa Valley Vintners

In this article, we examine modern day viticulture in Napa Valley. Following a general introduction to the Valley's natural conditions, we address the human factors – site selection, planting material, vineyard layout, infrastructure and cultural practices. Although none of these practices is legally prescribed as part of the American appellation system, there is widespread agreement today in the viticultural community of Napa Valley about best farming practices which are chronicled below.

A Century and a Half of Grape Growing and Wine Making

Vineyards were introduced to Napa Valley with the earliest European settlement and within a half century came to dominate the Valley's landscape, as they do to this day. The first vines were planted in Napa Valley sometime between 1838 and 1845 by settler George Yount [29]. The discovery of gold in the Sierra foothills in 1848 brought a large number of immigrants to California, including many from the wine regions of Europe who were familiar with grape growing and winemaking. Thirsty miners packed into the new city of San Francisco and provided a ready market for domestic wines. By the mid-1850s, there were regular, commercial shipments of wine from Napa to San Francisco [29]. The completion of a railroad along the length of the Valley, from the city of Napa to the town of Calistoga, in October, 1868, made Napa Valley wines even more accessible to the growing city of San Francisco.

Vineyard and winery development increased during the 1870s and early 1880s until, in 1884, the relatively small County of Napa surpassed Los Angeles County as the state's leading producer of dry table wines [29]. By 1886, the Valley had an estimated 175 wineries. The vast majority of the early vineyards in the County were planted on the valley floor, but as early as the 1870s, some vineyardists planted hillside vineyards in the quest for higher grape quality and to avoid spring frosts. The Valley's reputation for fine wines spread around the country and the world. At the 1889 Paris World's Fair, the Valley's wines captured 20 of 34 medals awarded to California wines [28].

Phylloxera ravaged the Valley's vineyards in the 1880s and 1890s. Although Phylloxera resistant rootstocks were identified and vineyards were replanted beginning around the turn of the century, the specter of Prohibition presented another threat to the wine economy. In 1919, the Eighteenth Amendment to the United States Constitution was passed forbidding the manufacture, sale or transportation of alcoholic beverages. Interestingly, viticulture continued during Prohibition with grapes sold and shipped by rail to home winemakers; each family was allowed to make 200 gallons for its own use.

The Twenty-first Amendment to the United States Constitution ended Prohibition in 1933. Yet the revival of the United States wine industry was slowed by World War II. Only a handful of new wineries opened their doors in the Napa Valley between the end of Prohibition and 1966. Since that time, however, the Valley's wine economy has burgeoned. Today the Valley is home to around 400 wineries.

Geographic Specificity and Viticultural Uniqueness

The complex geology of the Napa Valley is determined, in large part, by its location on the seismically unstable western edge of the North American continent. The Franciscan Formation, the Great Valley Sequence and the Sonoma Volcanics make up most of the bedrock elements which determine the chemical and mineral elements of the Valley's soils. Seismic forces along the San Andreas Fault brought these three elements together and formed the Mayacamas and Vaca Mountains that border the main valley on, respectively, its western and eastern sides. The chemical and mineral contents of the bedrock affect not only the nutrient status of soils but also their ability to retain water and the chemical content of the groundwater upon which most of the Valley's vineyards depend for irrigation [67].

The extreme diversity of the 33 distinct soils series which have evolved from these bedrock components is attributable to the complex interplay of the Valley's varied topography, climate and biological resources [59]. The small size of the Valley accentuates the effect of its topography in relationship to the climate.

Climate in the Valley varies along a complex set of variables. These include the marine influences of the San Francisco Bay and Pacific Ocean to the Southwest of the Valley, the most apparent in the Carneros region, and elevation variations from the Valley floor to the crests of its mountain ranges. In general, warm dry summer daytime conditions are regularly followed by the influx of a cooling marine fog layer in the evening. A dry season typically extending from April to November allows vineyardists to carefully calibrate the water stress of the vines through irrigation and dry farming, adding another tool to determine grape quality [60].

Because of the wide range of viticultural conditions and the lack of any governmental controls over grape varietal selection, Napa Valley is home to a wide array of grape varietals, including principally Cabernet Sauvignon (19,460 planted acres [7,875 ha] in 2008) and Chardonnay (7,199 planted acres [2,914 ha] in 2008).

Legal Protections

Napa Valley is directly impacted by its proximity to a major urban center, the San Francisco Bay metropolitan area. It is likely that only the highest quality grapes, capable of making wines of exceptional quality, had any chance of inspiring the political actions necessary to protect this irreplaceable vineyard resource from unchecked suburban development and rampant tourism.[5] The Valley's vineyard and wine industry benefit from some of the United States' earliest agricultural land protection laws and regulations.

5 California has lost some of its historically important coastal vineyards to urban and suburban expansion, among them the vineyards of Santa Clara, Alameda and Los Angeles counties.

In 1965, a California law known as the Williamson Act created a mechanism for individual counties to protect agricultural land through controls on local property taxation [7]. In order to enact these protections, counties were required to establish "Agricultural Preserve" districts, which Napa County did in 1968 [47]. The Agricultural Preserve still covers most of the Valley floor, preventing individual parcels of less than 16 ha from being created and allowing property owners to pay taxes based upon the value of the land as farmland, rather than as land available for subdivision and urban development. In addition, most of the foothills and mountain lands of the county are zoned Agricultural Watershed and are subject to a minimum parcel size of 65 ha, which further restricts residential development potential in such areas [50].

In 1983, Napa County adopted urban growth boundaries which limit the development of urban infrastructure outside established municipal boundaries [46]. In 1990, the voters of the county adopted, by initiative, restrictions on the conversion of agricultural land to urban uses, requiring until the year 2020 a county-wide vote on every proposal to redesignate agricultural lands to non-agricultural purposes or to change the building intensity of such lands [43]. That law has since been extended to 2053 [44]. Also in 1990, County government began to regulate more closely the establishment of new wineries and the expansion of existing wineries, restricting their commercial and industrial operations on agriculturally zoned lands, further protecting the Valley from encroaching urbanization and from the pressures of tourism [48].

In 1991, agriculturalists and environmentalists worked together to put in place regulations designed to minimize the environmental impacts of vineyard development. Inspired by the continuing development of hillside vineyards, these local regulations supplement state and federal environmental laws and mandate the inclusion of erosion control measures, riparian and wildlife corridors, protections for streams and wetlands and other environmental controls in vineyard development and operations [49]. The so-called Conservation Regulations apply to any new or redeveloped site with a slope of more than five percent.

Developing a Vineyard: *Site Selection, Infrastructure and Plant Material*

The diverse and complex soils of Napa Valley, combined with the wide range of climatic conditions, demand challenging viticultural decisions in the development of new sites and the redevelopment of existing vineyards. Experienced growers recognize that physical features and biological factors dramatically influence the wine's structure and flavors so they strive to fully understand all the characteristics of the site to make the best decisions in planting a new vineyard.

In terms of climate, the breezes from San Francisco Bay make the southern end of the valley the coolest, with a warming gradient to the northwest. However, other climatic conditions also are important viticulturally, including wind direction and speed, slope, aspect, solar radiation, elevation, humidity and rainfall which vary from less than .508 meters [m] per year in Carneros to more than 1.016m per year in Calistoga. Fortunately, there is sophisticated weather monitoring equipment throughout Napa Valley today that helps the grower to delineate weather conditions.

The diverse soils of the Napa Valley are the result of the interaction over time of climate, geology (specifically, tectonic or earthquake moving and shaking), topography and biology. Growers and their viticultural consultants typically conduct inspections to determine the soil characteristics of a particular development or redevelopment site. Soil pit analysis allows them to determine soil texture, physical restrictions, soil structure, mottling (anaerobic conditions indicating poor drainage), effective rooting depth, water holding capacity (including the need for subsurface drainage), the presence of pests such as parasitic nematodes and various chemical parameters. Figures 1 and 2. The chemical measurements generally include nutrient status, salinity and other chemical concentrations which affect soil structure or that may be toxic to vines.

Water availability is a critical site development parameter, particularly in water deficient areas. Both sources (stream flow,[6] run-off, drain tile and/ or well) and chemical composition of the source water are important. For example, some wells are very high in iron which may precipitate and plug drip emitters; other isolated wells have high boron concentrations that can be toxic to vines.

Napa Valley grape growers have long understood that terroir, which is the synergy of the biophysical components of soil, geology, topography and climate, and the human factors of rootstock and varietal selection and cultural practices, give a sense of origin and uniqueness to the resulting wine. With this knowledge and appreciation, the winegrower (a term of art that recognizes that wine is "grown" in the vineyard) uses his intuition, technical education and historical experiences to make the key decisions for a specific site development. These decisions include the need for subsurface drainage, whether and how to irrigate and frost protect, whether to add pre-plant nutrients like gypsum or lime and the depth and method of ripping (deep

6 The California State Water Resources Control Board under the authority of the *California Water Code* Sections 100 and 275 (2008) controls and issues the permits for water diversions from Napa streams and for storage of that water.

cultivation) and soil preparation. The grower also must decide on rootstock, grape variety (also known as scion or cultivar) and clone, row direction, vine spacing, trellis type and block layout to match soil types, including the location of roadways and avenues.

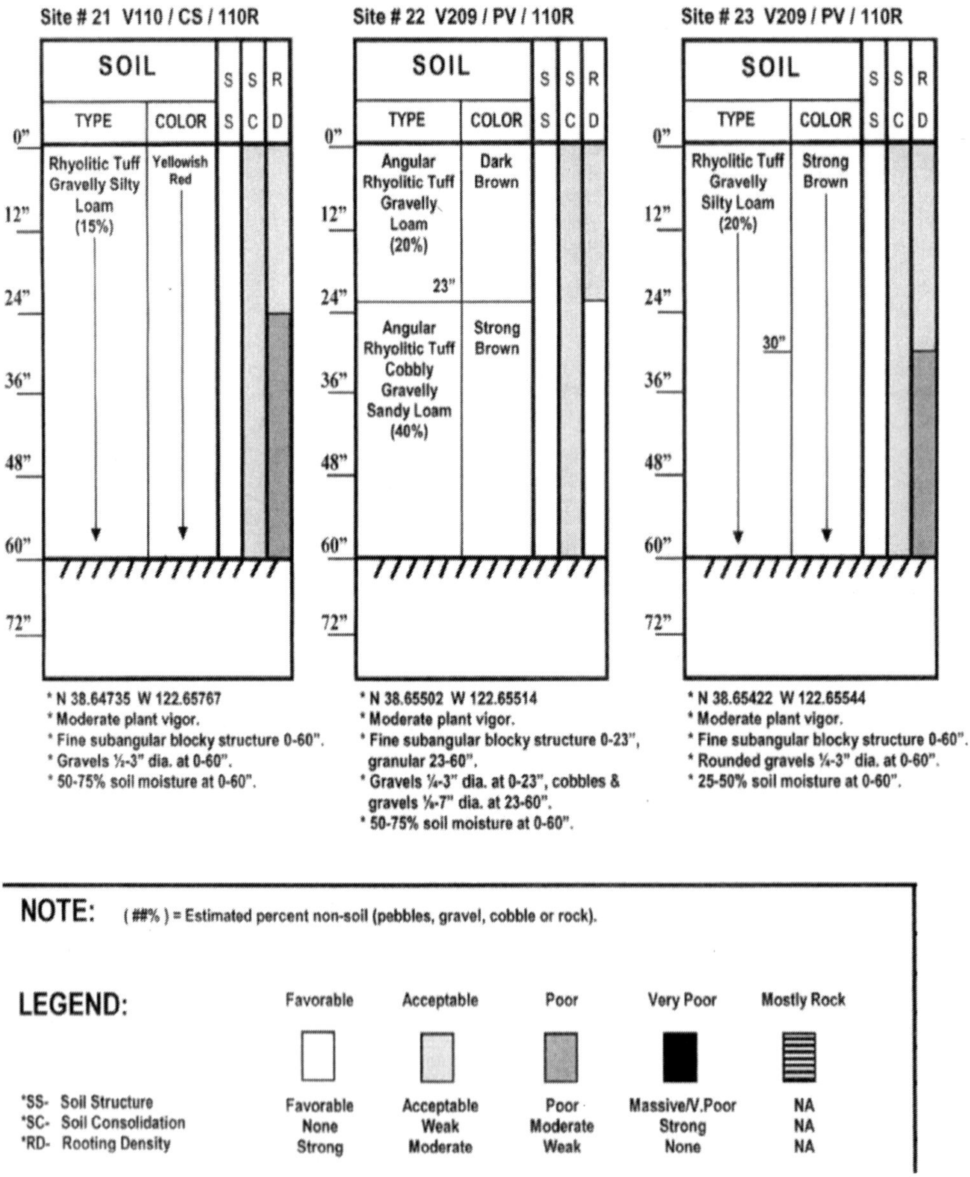

Figure 1. Sample Soil Physical Profile

SOIL ANALYSIS REPORT

Laboratory No.
For: *Project No.*
Client:

Reported: 5/1/2009
Received: 5/1/2009

Site	Sample Depth (in)		Texture	Salinity & Permeability								Nutrients					Extractable Cations					Ratios		T/ac-6"	%
					mmho/cm	meq/l	meq/l	meq/l		ppm		ppm	ppm	ppm	ppm	meq/100g	Percentage of CEC								
			Sat%	pH	ECe	Ca	Mg	Na	SAR	B	Free Lime	NO_3-N	P	K	Zn	CEC	Ca	Mg	K	Na	H	Ca/Mg	K/Mg	Lime Reqmt	Organic Matter
V110/CS/110R SITE 21	18	36	57	6.1	0.1	0.1	0.1	0.6	2.4	0.04	0	0.5	2	327	0.1	26.2	63	24	3	0	10	2.7	0.14	0.0	0.2
"	36	60	58	5.5	0.1	0.1	0.2	0.6	1.7	0.03	0	0.5	2	209	0.1	26.9	61	25	2	1	11	2.4	0.08	0.9	0.1
V209/PV/110R SITE 22	0	18	37	6.5	0.3	0.9	0.3	0.5	0.7	0.03	0	3.6	2	203	1.1	8.9	80	13	6	1	0	6.2	0.45	0.0	1.6
"	23	47	36	5.8	0.2	0.4	0.2	0.4	0.8	0.03	0	2.0	1	118	1.9	8.1	54	22	4	1	19	2.4	0.17	0.0	0.5
V209/PV/110R SITE 23	0	18	39	6.6	0.5	2.3	0.4	0.7	0.6	0.06	0	1.9	3	222	0.9	8.4	81	11	7	1	0	7.2	0.61	0.0	1.9
"	18	36	39	5.9	0.6	2.5	1.1	0.7	0.5	0.04	0	2.6	2	227	0.8	9.5	54	16	6	1	23	3.4	0.38	0.8	1.8
"	36	60	42	5.7	0.8	3.4	1.9	0.9	0.5	0.04	0	1.2	1	249	0.5	9.1	52	18	7	1	22	2.8	0.38	0.5	1.1
Desired Levels for Grapes				6.5 to7.5	Below 1.5				Below 5	0.2 to1.0		2 to 20	8+	100+	1.0+		60 to80	12 to20	3 to7+	Below 6	Below 8	2 to 10	0.1 to 0.4		

Figure 2. Sample Chemical Soil Analysis

While all of these viticultural practices have been modified over time, the selections of rootstock, variety and clone have changed most dramatically in the past 20 years. Prior to 1989, most of the Valley's vineyards were planted on the rootstocks St. George, 5C and AXR#1. Nearly 70% of Napa Valley vineyards were at one time planted on AXR#1 rootstock which eventually proved to be susceptible to biotype B Phylloxera infestation. Phylloxera root damage ultimately led to a decline in vine vigor and required major vineyard redevelopment [23].

The replanting of the AXR#1 sites started around 1987 [72]. A group of winegrowers hired Lucie Morton, a viticultural consultant, to review the French literature on twelve rootstocks. Although this information was never published, it was very useful in the selection of replacement rootstock. Dr. Jim Wolpert of the University of California, Davis, prepared a bibliography containing 200 technical papers on rootstocks from all over the world [74]. With such extensive technical information, supplemented by reports from European and Australian growers on the performance of the various rootstocks, the experienced winegrower today, assisted by trained scientists and consultants, is able to make informed judgments on rootstock selection [37, 41, 53].

Generally, rootstock is determined based on soil type, soil chemistry and potential irrigation requirements along with the growth characteristics of the scion. The growers seek a balance between capacity (fruit yields) and vigor (vegetative growth). Another criterion for rootstock selection is resistance to disease and pests. For example, in sites where Fanleaf Degeneration, a virus disease, is carried in the soil, Dagger Nematode, *Xiphinema index,* acquires the virus while feeding on diseased roots and can transmit the virus to new healthy roots [21]. Fortunately, the rootstock 039-16, developed by Dr. Harold Olmo at the University of California, Davis, and released in 1988, has a high resistance to Dagger Nematode feeding and to date prevented the expression of Fanleaf virus symptoms in the scion.

Other vineyard sites may have populations of other plant parasitic nematodes. These include Root Knot Nematode (*Meloidogyne incognita* and other species), Dagger Nematode (*Xiphinema americanum*), Root Lesion Nematode (*Pratylienchus spp.*), Citrus Nematode (*Tylenchulus semipenetrans*) and Ring Nematode (*Criconemella xenoplax*) [21]. Considerable research is available to guide the grower in the selection of Phylloxera resistant rootstock that also will provide some resistance to these nematodes [11]. For example, Dr. Andy Walker of the University of California, Davis, and Dr. Howard Ferris developed new rootstocks with broad and durable nematode and Phylloxera resistance. Five of these stocks are available to nurseries for commercial distribution [70].

The grower's selection of variety and clone is partially driven by market demand for various wine types.[7] The selection of the best variety for the site is also a critical factor because it affects potential wine quality. Of the total Napa Valley grape acreage 18,616 ha in 2008, Cabernet Sauvignon is the most widely planted varietal, comprising 46% of the total planted acres in the Valley [45]. This figure has risen dramatically since 1997, as the market demand for Cabernet Sauvignon wines has soared. Presently, there are only six red varieties and two white varieties of the 35 varieties planted in Napa Valley that each total over 405 ha, Cabernet Sauvignon, Merlot, Pinot Noir, Zinfandel, Cabernet Franc, Syrah, Chardonnay and Sauvignon Blanc [45].

Most winegrowers are careful to select varieties and clones that are disease free. Growers have known for many years that viral diseases can be spread by propagation materials so it is imperative that they select plant material that is disease free [9, 33, 66]. Winegrowers frequently consult the National Grape Registry, (http://ngr.ucdavis.edu), for the current status of viruses and the history of particular varieties and clones.

In addition, the winegrower must select a trellis system to support the vines and distribute the foliage and fruit so as to expose the leaves to sunlight and provide filtered shade the fruit. The trellis design also facilitates the harvesting operation, whether by hand or machine.

The most common trellis system in the Napa Valley is Vertical Shoot Position (VSP). Picture 1. This system, which is used at many different spacings, can facilitate both cane and cordon-spur pruning. Recently, the VSP system was modified by some growers to include small cross arms for shade. Picture 2. Other common trellis systems include Geneva Double Curtain (GDC), Picture 3; modified lyre, Picture 4; and different forms of a "T" trellis, Picture 5. The GDC and modified lyre trellis types typically are found where the soil is deep and fertile; some growers prefer these trellis systems for vigorous cultivars like Sauvignon Blanc.

Head trained (Goblet) vines, shown in Picture 6, still exist in the Napa Valley. In fact, a few viticulturists prefer the head trained system for varieties like Zinfandel and Grenache.

---

7 If market conditions change or if the grower's initial selection of variety or clone was inadvisable, the grower has the option, in a viable vineyard, to bud or graft on a new variety or clone.

Trellis Systems in Napa Valley

Picture 1. VSP

A VSP, spaced six feet between rows and four feet between vines, known as a 6' x 4' (1.8m x 1.2m), in a Cabernet Sauvignon vineyard.

Picture 2. VSP with Cross Arms

A modified VSP with cross arms in a Cabernet Sauvignon vineyard spaced 8' x 6' (2.4m x 1.8m). Note the wind machine for frost protection. The vineyard has alternating cultivated and sod rows and is drip irrigated.

Picture 3. GDC

A GDC trellis in a drip-irrigated Cabernet Sauvignon vineyard spaced 12' x 6' (3.65m x 1.8 m).

Picture 4. Modified Lyre

A Lyre system, using movable foliage "catch" wires, in a Sauvignon Blanc vineyard spaced 12' x 7' (3.65m x 2.1m). Note overhead sprinklers for frost protection. The vineyard is also drip irrigated.

Picture 5. "T" Trellis

A "T" trellis in a Cabernet Sauvignon vineyard spaced 10' x 7' (3m x 2.1m). This vineyard is drip irrigated.

Picture 6. Goblet

Goblet trained Zinfandel vineyard spaced 10' x 8' (3m x 2.4m). This vineyard is drip irrigated, although the drip line is not shown in this photograph.

Vine spacing in Napa Valley, both between rows and vines, has evolved to much higher densities than in years past. Row widths from six feet (1.8m) to eight feet (2.44m) and distances between vines from three feet (.91m) to six feet (1.8m) are quite common. Due to excess shading, trellis height typically does not exceed row width. Planting densities range from 907 vines /acre (2,240 vines/ha) to 2,420 vines /acre (5,978 vines/ha).

A commonly used row direction, depending on site configuration and slope, is northeast by southwest. This row direction provides for moderate sun exposure on each side of the canopy and reduces the potential for sun burn and high fruit temperatures [14, 65, 73].

Sustainability

The Napa Valley wine industry adheres to and engages in sustainable and environmentally sound farming practices. The Code of Sustainable Winegrowing Workbook presents hundreds of sustainable practices and provides a self-assessment tool that allows winegrowers to review their current practices, identify areas where improvement can be made and develop action plans to increase the sustainability of their operations [13].

There are different but complementary methods of implementing sustainable farming. Many growers have become certified organic or biodynamic. The organic farms in Napa County in 2008 covered some 2,490 acres (1,008 ha), representing 6.25% of the total 46,000 planted acres [45].

Napa Green Certified Land program[8] is a voluntary, third party certified program for Napa Valley growers and vintners who seek to restore, protect and enhance the regional watershed. The participants develop and certify plans for their respective properties using Fish Friendly Farming program guidelines[9] that include wildlife habitat restoration, healthy riparian corridors and sustainable agricultural practices. The program covers not only vineyard land but also wild land, roadways and stream banks within the property. The Napa Green program is recognized and certified by National Marine Fisheries Service, the Napa County Agricultural Commissioner's Pesticide Regulatory Program and the Regional Water Quality Control Board. Additionally, the United States National Resources Conservation Service is involved with the program through the development of conservation plans. Approximately 33,150 acres (13,416 ha), comprising 72% of total planted

---

8 http://www.napavintners.com/napa_green/index.aspx.

9 Fish Friendly Farming is a certification program for vineyards that are managed to restore fish and wildlife habitat and improve water quality. *See* www.fishfriendlyfarming.org.

acreage, are currently enrolled in the program, 16,900 (6,840 ha) of which are now certified, comprising 37% of total planted acreage.

River restoration projects are common in Napa Valley as a way to enhance the environment for fish and wildlife. These projects generally follow the Fish Friendly Farming guidelines and are undertaken by individual winegrowers, often under the auspices of public agencies.

Most winegrowers have adopted integrated pest management (IPM) practices which integrate chemical, cultural and biological controls.[10] The IPM approach involves basal leaf removal, the use of fungicides and pesticides that are effective but more environmentally friendly, increased farm biodiversity to favor beneficial organisms, vineyard floor management practices, insect mating disruption and biological insect controls such as the release of parasitoids and beneficial insects [6, 12, 18, 21, 27, 40, 52]. These practices are employed for a wide range of pest control.

The grower in Napa Valley, as in many other agricultural communities, must deal with a large number of pests and diseases. A complete summary of these insects and diseases is found in *Grape Pest Management*, Second Edition [21]. Support for the suppression and control of pests in California comes from licensed pest control advisors, ongoing research and the dissemination of technical findings by country agriculture extension personnel, universities and local colleges.

New invasive pests and diseases are a recurring problem in the Napa Valley. In the current decade, three insects have become a major concern – the Vine Mealybug, the Light Brown Apple Moth and the Glassy-winged Sharpshooter. A recently introduced and highly invasive Vine Mealybug, *Plannococcus ficus* (VMB) is requiring special control efforts. The vine and grape clusters are damaged by a build up of sooty mold on the bugs' excreted honeydew which can damage grape dusters and berries. The insect is also a vector for grape vine leafroll viruses. The introduction of Vine Mealybug has added additional pressures on the Napa Valley winegrower. It is now well documented that leafroll virus disease is rapidly spreading in the Napa Valley and that mealybugs, which entered Napa Valley on propagation materials, i.e., green growing bench grafts, from nurseries in the San Joaquin Valley of California, have become an increasingly important vineyard pest. Reports in Australia, Argentina, Europe and South Africa substantiate the transmission role of mealybugs [12, 22, 40, 54, 66, 68].

---

10 The California Department of Pest Regulation (DPR) protects human health and the environment by regulating pesticide sales and use. Before a pesticide can be sold or used in the state, it must be evaluated and registered by DPR.

Light Brown Apple Moth (LBAM), *Epiphyas postvittana,* is a native of Australia and is found in isolated vineyards in Napa, Sonoma and other California counties. The California Department of Food and Agriculture has imposed a quarantine that prohibits the movement of plants and plant parts unless the shipment is certified as "free from" the presence of LBAM. If LBAM is not eradicated and becomes established in the vineyards, it will increase control pressures for the growers. The quarantine is designed to prevent the establishment of LBAM as a resident pest in Napa Valley vineyards.

The Glassy-winged Sharpshooter (GWSS), *Homalodisca corqulata,* was first reported in Southern California in 1994. It is a native insect of the southeastern United States and Mexico that feeds on xylem fluids of a large number of plants. By 1999 it had become established in the vineyards of Temecula, California, where it caused substantial damage.

GWSS can cause serious damage by feeding, but it also is a vector insect for Pierce's Disease (PD), *Xylella fastidiosa*. PD destroyed the vineyard industry in Anaheim, California, in the early 1900s and has been a challenge to winegrowers for a number of years. Historically, the vector for PD in the Napa Valley has been the Blue Green Sharpshooter (BGSS), *Graphocephala atropunctata,* which resides in riparian areas and on ornamental plants used for landscaping. The GWSS is much larger, a much more aggressive feeder and can fly further than the BGSS, thus creating the potential for widespread dissemination of PD. Fortunately, the entire California grape industry, working cooperatively with the United States government, California Department of Food and Agriculture, and County Agricultural Commissioners, have prevented the entry of GWSS into the north coast grape districts, including Napa Valley. To pay for the tasks of survey and detection, containment, rapid response and treatment of GWSS infestations, there is a statewide grower assessment, a local Napa assessment and federal and state funding for various GWSS/PD programs. Some of these funds support major research projects on the control of GWSS and PD and its vectors. The success of these efforts to date provides the Napa Valley winegrowers with a model for future invasive pests.

Cultural Practices

Cultural practices that are integrated with the particular vintage along with a vineyard design that augments the natural features of the particular site determine the ultimate wine quality in any given harvest year. The most important cultural practices are pruning, canopy management, irrigation, fertilization, vineyard floor management, yield management and harvesting. Pruning is a critical factor in managing yield and wine quality. The Napa Valley winegrower strives for a balance between yield and vegetative vine growth on a consistent, sustainable basis. Optimal values include 0.66 pounds (0.3 kilogram [kg]) to 2.2 pounds (1 kg) of pruning weight per meter of canopy length, and between five and seven pounds of fruit (2.27 to 3.18 kg) for every pound (.45 kg) of dormant vegetative growth[11] [32, 55]. Some growers relate pounds (kgs) of prunings to retained bud count. For example, an average of seven buds retained per pound (15 buds/kg) of prunings will provide for a balanced vine in most Cabernet Sauvignon sites. The experienced winegrower also will take into account the historical yields at the specific site, vine vigor, wine quality and the amount of summer trimming.

There is no clear preference for cane pruning versus spur pruning in Napa Valley. Both methods are used depending on the growers' preference.

The Napa Valley is subject to spring frosts. Once bud break occurs, the green shoots can be damaged if temperatures fall below 32° Fahrenheit [F] (0° Celsius [C]). The most common type of frost event is a radiational freeze.[12] On some very rare occasions, an advective freeze[13] may occur.

Most vineyards which are vulnerable to frost are protected in some manner.[14] The use of wind machines (one per 10 ± acres [4 ± ha]) mixes the warmer, inverted air with the colder air on the vineyard floor. During a radiational freeze, a wind machine will raise temperatures by two to three degrees Fahrenheit (0.8 to 1.2°C). The air movement caused by the wind machine of two to ten miles (3 km to 16 km) per hour also results in a shallower boundary layer, a larger heat gradient, enhanced heat transfer, and a warmer leaf [63]. As a result of these combined effects, wind machines are observed to protect vines against temperatures as low as 26° F (-3.3°C). It should be noted that wind machines will not protect the vine, and may cause greater damage, during an advective freeze.

---

11 A pruning trial conducted in Napa Valley utilizing Cabernet Sauvignon grafted to 110R rootstock resulted in significant differences in sensory attributes. That seemed to defy the conventional wisdom. The low yields of 1.74 tons / acre (4.3t/ha.) resulted in "veggie aroma and flavors, bell pepper aroma, bitterness and astringency," while the high yield wines of 8.98 tons / acre (22.2t/ha.) were "higher in red/black aroma, jam aroma and fresh fruit aroma" [8]. Other research has pointed out the need to achieve balance between capacity (yields) and vine vigor [1, 56].

12 A radiational freeze occurs when a cold but dry (low dew point) air mass moves into the valley. Temperatures can be in the 60s in degrees Fahrenheit in the afternoon but after sunset radiational cooling of the soil surface causes temperatures in the lower atmosphere to fall, sometimes by 30° F or more. Temperatures a few hundred feet above the ground remain fairly constant, forming an inversion (temperature increases with elevation).

13 An advective freeze occurs when a cold air mass from the arctic regions migrates into the Napa Valley. The temperatures at the vineyard are below freezing and become colder at higher elevations.

14 There are many sites, especially in the hills, that are generally frost free.

Another commonly used method for frost protection is a properly designed overhead sprinkler system. Heat released by the freezing of the applied water, which forms an ice film around the green leaves, clusters and shoots, protects the vine by keeping temperatures near the 32°F (0°C) mark. This water application can protect to approximately 24°F (-4.4°C) [57, 64], although co-author Robert Steinhauer has observed protection down to 21°F (-6°C). Picture 7.

Picture 7. Vines at 21°F (-6°C) protected by overhead sprinklers.

The overhead sprinkler system consumes large volumes of water, 52 gallons / minute / acre (490 liters / minute / ha). For most properties, large reservoirs and large pump capacity are required. In return for this investment, the sprinkler system will protect against much lower temperatures than will wind machines. It also will provide protection in most advective conditions, and the same system is sometimes used for irrigation and cooling during summer heat spike events.

Part of the trend to sustainable practices in Napa Valley is the management of the vineyard floor, both in row and under vine. Best farming practices generally have moved away from pre-emergent herbicides under the vine row to mechanical tilling or applications of glyphosate for weed control. Generally, between rows are permanent sod cover crops that are mowed or cultivated row centers. Currently, many sites alternate every other year between sod cover crop and cultivated row middles. Some growers plant a winter cover crop which is incorporated into the soil as a source of nutrition as well as for erosion control. Other growers utilize a compost for the same purpose. Typically, these management practices are site specific.

The benefits of cover crops are reduced tractor wheel compaction, improved physical properties of the soil, increased organic matter, reduced chemical use, habitat for beneficial insects, dust control and reduction in water run-off. However, permanent sod cover crop is found to compete for water and nutrients which potentially reduces, over time, vine vigor [4, 24, 38, 39, 42, 62].

Vine nutrition is an important component of a successful vineyard, both for the particular vintage and for the long term health of the vineyard. Most growers rely on field observations to determine nutrient status, along with soil chemistry analysis, bloom time petiole nutrient status and leaf blade nutrient analysis. The Napa Valley winegrower has at his disposal many advanced nutrient and soil amendment application capabilities such as fertilization by drip irrigation, mechanical soil incorporation and materials that can be sprayed on grapevine leaves.

The ultimate goal of the winegrower is to produce the largest quantity of the very highest quality grapes which are in the greatest demand in Napa Valley. To achieve this goal, the viticulturist must have a solid scientific understanding of the effects of cultural practices, the vintage and terroir plus the historical perspective of farming on the particular site. The grower must be flexible in adjusting cultural practices throughout the growing season from site to site, block to block and even within a block.

While many factors influence the selection of grape variety and the ultimate wine character and quality, site (terroir) and season (vintage) are perhaps the most dramatic. In terms of farming practices, canopy modification and irrigation strategies are demonstrated mechanisms for manipulating wine quality. The response to these and other management practices depends upon variety, row orientation and trellising [2, 16, 31, 55].

Canopy management is important because filtered light has a positive influence on fruit composition [26]. However, too much light and high fruit temperatures are deleterious to wine quality; therefore, the foliage must protect the fruit from sunburn and dehydration [15, 65, 76]. Additionally, sunlight on the shoots will improve bud fruitfulness the following year [5, 17].

Most Napa Valley winegrowers retain about five shoots / foot of canopy (15 shoots / meter) [61, 77]. The spacing of the shoots generally allows for optimum sunlight on the leaves. The shoot length on a balanced vine should be uniform and grow to between 3.5 feet (1m) and 5 feet (1.5m) by veraison (the point at which the grape berries turn from green color to their final color, representing the change from berry growth to berry ripening). The shoots should stop growing about two weeks before veraison and retain about 12 to 15 active leaves that remain green and healthy to harvest. The grape clusters typically are protected by filtered light through one to two layers of leaves. Growers may decide to remove basal laterals or laterals plus a few leaves based on observations of the site and leaf cover. Row direction and leaf size (Merlot and Zinfandel are examples of large sized leaves) will influence the decision about leaf removal to obtain the desired amount of fruit exposure. North by south row direction may require leaf removal on the east (morning sun) side and no leaves on the west side of the canopy to shade the fruit.

The water status of the vine has a major effect on the canopy and ultimate fruit and wine quality. As such, irrigation management is one of the most important viticultural tools. The Napa Valley generally receives enough rainfall from November through April to fill the soil profile. Normally, very little rain occurs during the summer and fall months. Almost all Napa vineyards are either dry farmed or are equipped with drip irrigation facilities. Following the advent of drip irrigation, water use by Napa Valley growers has become very efficient and prescribed. A tremendous amount of research on irrigation treatments and their effects on vine physiology and wine quality have been completed [3, 20, 30, 36, 58, 69, 71, 75].

There are sites with deep loam or clay loam soils which require no irrigation or are intentionally dry-farmed. Many growers advocate deficit irrigation strategies. These irrigation strategies maintain the vine water status post-veraison at between 70 to 80% of the evapo-transpiration rate. Most growers understand spatial variations in the vineyard and adjust irrigation within blocks or sub-blocks of the vineyard. Along with constant visual inspection of the water status of the vines, many growers rely on soil moisture measurements and determine the midday leaf water potential utilizing a pressure chamber. Many growers will not start irrigating until the vines get to -10 bars of midday leaf water potential. Along with the soil and plant water measurements, the field observation of vine status, especially observing shoot tips, is critical.

Cluster removal, known as thinning, occurs just prior to veraison or at veraison in almost every ultra-premium vineyard in Napa Valley. Typically, the grower removes the clusters which are touching or are compacted together in order to provide light to the berries and also reduce the potential for fungal disease.

Many growers, especially the producers of red grape varieties, "color thin" at 80 to 95% veraison by removing any green or partially green clusters. This process is undertaken in an attempt to achieve uniform maturity at harvest. Some growers will remove clusters in excess of one cluster / shoot. This process is gaining favor among some ultra-premium producers of Cabernet Franc, Cabernet Sauvignon, Pinot Noir and certain other varieties.

Growers understand that thinning early, while the berries are in the growth phase, will increase the berry size of remaining berries. For this reason, almost all growers wait to thin until veraison.

The winemaker's demands for concentrated flavor and supple tannins at harvest have lead to the practice of leaving grapes on the vine for days or weeks after the traditional harvest date. This practice is commonly referred to as "hang time." This requires continued maintenance irrigation to keep the leaves green to just before harvest.

Viticulturists in Napa Valley are divided on the effects of yield, expressed as tons / acre (hectoliters [hl] / hectare), on wine quality. Undercropping (too few clusters) can lead a vine to excessive vegetative growth, vegetative wine flavors and high total pH. At the opposite extreme, overcropping (too many clusters) a vine can lead to low alcohol and light, poor quality wines. A majority of growers believes that precision farming for small yields produces the highest quality ultra-premium wines. The success of these growers makes it difficult to argue with this position. Another group, however, believes that precision farming for balanced capacity and vine vigor is the best avenue to produce ultra-premium wines. Whichever approach is correct depends ultimately on a multitude of factors. Dr. Mark Matthews reports, "Prior research has shown no effect of yield on wine quality;" he also states that "quality cannot be measured by a scientist because it is so subjective" [34]. Most viticulturists agree that for the ultra-premium wine market, small yields that are balanced with the vine's vigor result in superior wines [1, 8, 51, 56, 61, 69]. The Napa Valley average yield / acre from 1986 to 2004 was 3.65 tons / acre (54.75 hl/ha). By contrast, the California average for Cabernet Sauvignon for this same period was 5.55 tons / acre (83.2 hl/ha).

The Napa Valley climate allows for the full maturity of even late ripening varieties like Cabernet Sauvignon and Cabernet Franc. For many years, harvest date was determined by measuring sugar, acid and pH. These measurements are still made today, but the winemaker also tastes the fruit to determine when to harvest. For red varieties, most winemakers look for ripe, intense flavors with supple tannin structure. They also want the grape seeds to be brown and mature. Bibiana Guerra describes an Australian research project on Syrah confirming that seed color is a good indicator of physiologically ripe berries, ready for harvest [25]. A given vineyard may be harvested block by block or even sub-block by sub-block on different days based upon the flavor profile of the fruit.

Most ultra-premium fruit is harvested by hand. However, the mechanical harvester technology is so advanced that a substantial amount of Napa Valley fruit is now machine harvested.

Harvest is the culmination of the growers' efforts. Before and during harvest, the winegrower and the winemaker (sometimes one and the same) observe the conditions of the vines and determine how best to meet the winemaker's parameters for the fruit. They observe conditions down to the level of the blocks and sub-blocks (precision farming), and the grower then adjusts the cultural practices accordingly, including irrigation, fertilization and canopy management.

When harvest concludes, the winegrower and winemaker focus on the wine to determine if modifications to the cultural practices would improve yields and quality. The grower then considers whether to adjust those practices in future vintages, cognizant of the fact that each vintage year will present its own unique conditions and challenges.

Conclusion

The winegrowers and winemakers of Napa Valley understand that they are fortunate to grow grapes and make wine in such a favorable terroir. Winegrowers and winemakers work collaboratively to ensure that their grapes meet the wine quality parameters and will vinify high quality wines. This strategic alliance of winegrower and winemaker is a key component of the success or the failure of the vineyard.

The grower and winemaker also constantly work together to improve their know how and stay at the forefront of technology. They recognize, respect and appreciate that each wine growing region of the world has its own valuable history, tradition, science and practices which may differ from those of the Napa Valley, and they remain open to learn and adapt their skills and practices to improve the quality of Napa Valley wines. This commitment to excellence is evident in the Napa Valley for many generations, resulting in the well-deserved international reputation that Napa Valley has long enjoyed.

References

Archer, Eben and Hunter, J.J. 2006. "Vine balance drives grape quality." *Practical Winery & Vineyard* (May/June 2005): 35-50.

Arnold, R.A. and Bledsoe, A.M. 1990. "The Effect of Various Leaf Removal Treatments on the Aroma and Flavor of Sauvignon blanc Wine." *Am. J. Enol. Vitic.* 41 (1) (1990): 74-76.

Battany, Mark, 2006. "Fine-tuning irrigation management with a post-season evaluation." *University of California Cooperative Extension, Grape Notes* (December 2006).

Bettiga, L., Smith, R., Cahn, M. and Tourte, L. 2007. "Evaluation of the impact of in-row practices on soil/water erosion, vine growth and productivity, weed control, economics of wine grape production." *Practical Winery & Vineyard* (September/ October 2007): 42-45.

Buttrose, M.S. 1970. "Fruitfulness in grapevines: The response of different cultivars to light, temperature and day-length." *Vitis.* 9 (1970): 121-125.

California Department of Food & Agriculture. 2006. "Pierce's Disease Control Program: 2006 Annual Report to the Legislature." (2006).

California Land Conservation Act of 1965, Statutes 1965, Chapter 1443. Codified at *California Government Code* § 51200 *et seq.*

Chapman, Dawn M., Mark Matthews, and Jean-Xavier Guinard. 2004. "Sensory Attributes of Cabernet Sauvignon Wines Made from Vines with Different Crop Yields." *Am. J. Enol. Vitic.* 55 (4) (2004): 325-334.

Charles, J.G., Cohen, D., Walker, J.T.S., Forgie, S.A., Bell, V.A. and Breen, K.C. 2009. "A review of the ecology of grapevine leafroll associated virus type 3 (GLRaV-3)." *New Zealand Plant Protection* 59 (2006): 330-337. Available at http://www.nzpps.org/ journal/59/nzpp59_330.php (accessed June 15, 2009).

Code of Federal Regulations 27 CFR §§ 4.25(e)(1) and 9.3, *et seq.*

Coombe, B.G. and Dry, P.R. (Eds.) *Viticulture. Volume 1, Resources in Australia.* Adelaide: Australian Industry Publishers, 1988.

Daane, Kent M., Monica Cooper, Serquei V. Triapitsyn, Vaughn M. Walton, Glenn Y. Yokota, David R. Haviland, Walter Joseph Bentley, Kris Godfrey and Lynn R. Wunderlich. 2009. "Vineyard managers and researchers seek sustainable solutions for mealybugs, a changing pest complex." *Cal Ag* 62 (4) (October-December 2008): 167. Available at http://repositories.cdlib.org/anrcs/californiaagriculture/v62/ n4/p167 (accessed June 15, 2009).

Dlott, J., Cliff Ohmart, John Garn, Kari Birdseye and Karen Ross, (Eds.) "The Code of Sustainable Winegrowing Workbook," a project of the Wine Institute and the California Association of Winegrape Growers.

Dokoozlian, N.K. and Kliewer, W.M. 1995. "The Light Environment Within Grapevine Canopies. I. Description and Seasonal Changes During Fruit Development." *Am. J. Enol. Vitic.* 46 (2) (1995): 209-218.

Dokoozlian, N.K. and Kliewer, W.M. 1996. "Influence on Light on Grape Berry Growth and Composition Varies during Fruit Development." *J. Am. Soc. Hort. Sci.* 121 (5) (1996): 869-874.

Downey, Mark O., Dokoozlian Nick K., and Mark P. Krstic. 2006. "Cultural Practice and Environmental Impacts on the Flavonoid Composition of Grapes and Wine: A Review of Recent Research." *Am. J. Enol. Vitic.* 57 (3) (2006): 257-268.

Dry, P.R. 2000. "Canopy management for fruitfulness." *Aust. J. Grape Wine Res.* 6 (2) (2000): 109-115.

Dufour, Rex. 2009. "Farmscaping to Enhance Biological Control; Pest Management Systems Guide." *ATTRA* CT065 (December 2000). Available at www.attra.ncat.org/attra-pub/farmscape.html (accessed June 15, 2009).

Elliott-Fisk, and Deborah L. 1993. "Viticultural Soils of California, with Special Reference to the Napa Valley." *Journal of Wine Research*, 4 (2) (1993): 67-77.

Ezzahouani, Abdelazia and Larry E. Williams. 2007. "Effect of Irrigation Amount and Preharvest Irrigation Cutoff Date on Vine Water Status and Productivity of Danlas Grapevines." *Am. J. Enol. Vitic.* 58 (3) (2007): 333-340.

Flaherty, Donald L., Peter Christensen, L., Thomas Lanini, W., James J. Marois, Phil A. Phillps, and Lloyd T. Wilson, (Eds.) 1992. *Grape Pest Management,* Second Edition. Davis: University of California, Division of Agriculture and Natural Resources, 1992.

Golino, Deborah A., Weber, Ed, Susan Sim, and Adib Rowhani. 2008. "Leafroll disease is spreading rapidly in a Napa Valley vineyard." *Cal Ag* 62 (4) (October-December 2008): 156-160.

Granett, Jeffrey, John A. De Benedictis, James A. Wolpert, Edward Weber and Austin C. Goheen. 1991. "Phylloxera on rise...: Deadly insect pest poses increased risk to north coast vineyards" *Cal Ag* 45 (2) (March-April 1991): 30-32.

Greenspan, Mark. 2009. "Weed and Vineyard Floor Management Practices." *Wine Business Monthly* (March 2009): 50-54.

Guerra, Bibiana. 2009. "The Seed of the Problem. Using seed color to determine when berries are ready to be picked." *Wine Business Monthly* (May 2009): 44-47.

Hashizume, Katsumi and Takashi Samuta. 1999. "Grape Maturity and Light Exposure Affect Berry Methoxypyrazine Concentration." *Am. J. Enol. Vitic.* 50 (2) (1999): 194-198.

Heeger, Jack. 2008. "UC Berkeley introduces beneficial insects to control vine mealybugs." *Practical Winery & Vineyard* (May/June 2008): 66-71.

Heintz, William F. 1999. *California's Napa Valley: One Hundred Sixty Years of Wine Making,* San Francisco: Scottwall Associates, 1999, 149.

Heintz, William F. 1990. *Wine Country: A History of Napa Valley.* Santa Barbara: Capra Press, 1990, 25, 29, 45, 176.

Hepner, Y., and Bravdo, B. 1995. "Effect of Crop Level and Drip Irrigation Scheduling on the Potassium Status of Cabernet Sauvignon and Carignane Vines and Its Influence on Must and Wine Composition and Quality." *Am. J. Enol. Vitic.* 36 (2) (1985): 140-147.

Jackson, D.I. and Lombard, P.B. 1993. "Environmental and Management Practice Affecting Grape Composition and Wine Quality – A Review." *Am. J. Enol. Vitic.* 44 (4) (1993): 409-430.

Kliewer, W. Mark and Dokoozlian, Nick K. 2005. "Leaf Area/Crop Weight Ratios of Grapevines: Influence on Fruit Composition and Wine Quality." *Am. J. Enol. Vitic.* 56 (2) (2005): 170-181.

Martinelli, L. 2000. "Major Graft-transmissible diseases of grapevines: Nature, diagnosis and sanitation." Presented at the ASEV 50th Anniversary Annual Meeting, Seattle, Washington (June 19-23, 2000), 231-236.

Matthews, Mark. 2005. Verbal communication, *University of California, Davis, RAVE Symposium* (2005).

Matthews, M.A. and Michael M. Anderson. 1988. "Fruit Ripening in *Vitis vinifera* L.: Responses to Seasonal Water Deficits." *Am. J. Enol. Vitic.* 39 (4) (1988): 313-320.

Matthews, M.A., Rie Ishii, M.M. Anderson, Michael O'Mahony. 1990. "Dependence of wine sensory attributes on vine water status." *J. Sci. Food Agric.* 51 (3) (1990): 321-335.

May, Peter. 1994. *Using Grapevine Rootstocks: The Australian Perspective.* Adelaide: Winetitles, 1994.

McGourty, Glenn. 2008. "Choosing Cover Crops." *Vineyard & Winery Management* (September/October 2008): 71-75.

McGourty, Glenn, Steve Tylicki, Julie Price and Jim Nosera. 2006. "Performance of 18 crop species in high elevation North Coast vineyard." *Practical Winery & Vineyard* (September/October 2006): 9-19.

Monis, Judith. 2008. "Detecting leafroll disease: strategies for testing, management and control." *Wines & Vines* (February 2008): 64-67.

Morano, Lisa and Mark Kliewer, W. 1994. "Root Distribution of Three Grapevine Rootstocks Grafted to Cabernet Sauvignon Grown on Very Gravelly Clay Loam Soil in Oakville, California." *Am. J. Enol. Vitic.* 45 (3) (1994): 345-348.

Morlat, René and Alain Jacquet. 2003. "Grapevine Root System and Soil Characteristics in a Vineyard Maintained Long-term with or without Interrow Sward." *Am. J. Enol. Vitic.* 54 (1) (2003): 1-7.

Napa County Agricultural Lands Preservation Initiative (November 6, 1990).

Napa County Agricultural Lands Preservation Initiative (November 4, 2008).

Napa County, Department of Agriculture and Weights and Measures. "2008 Agricultural Crop Report" (2009). Available at http://www.co.napa.ca.us/GOV/Departments/26400/Forms/2008AgriculturalCropReport2 0090407.pdf (accessed June 15, 2009).

Napa County General Plan (adopted June 7, 1983), Policy 4.10 "Urban Expansion Limits" and Policy 4.11 "Urban Services Facilities."

Napa County Ord. 274 (April 9, 1968) (amending Ord. 186 of January 25, 1955, codified in *Napa County Code* ch. 18.16 (2009).

Napa County Ord. 947 (January 23, 1990), codified in *Napa County Code* ch. 18 (Zoning Ordinance) (2009).

Napa County Ord. 991 (1991) codified in *Napa County Code* ch. 18.108 (2009).

Napa County Ord. 1058 (1994) (increased minimum parcel size from 40 to 160 acres) codified in *Napa County Code* §18.101.010 (2009).

Noble, A.C., Deborah L. Elliott-Fisk. 1990. "Evaluation of the Effects of Soil and Other Geographical Parameters on Wine Composition and Flavor: Napa Valley, California." 37-45 in *Actualities Oenologiques* 89, 4 Symposium International d'Oenologie, Paris: Dunod (1990): 567.

Ohmart, Cliff. 2009. "Vineyard View: Elicitors Put Vines in Charge." *Wines & Vines* (March 2009): 36-39.

Ough, C.S. and Lider, L.A. 1968 et al. "Rootstock-Scion Interactions Concerning Wine Making. I. Juice Composition Changes and Effects on Fermentation Rates in the St. George and 99-R Rootstocks at Two Nitrogen Fertilizer Levels." *Am. J. Enol. Vitic.* 19 (4) (1968): 213-227.

Rieger, Ted. 2008. "Leafroll Virus Spreads in California Vineyards." *Vineyard & Winery Management* (September/October 2008): 63-70.

Rieger, Ted. 2005. "UCD Rave Looks at Crop Yields in Relation to Wine Composition." *Vineyard & Winery Management* (July/August 2005): 78-84.

Ross, Jordan. 1999. "Balancing Yield and Quality: Part I." *Practical Winery & Vineyard* (March/April 1999): 1-4.

Schultz, H.B. and Weaver, R.J. 1977. "Preventing frost damage on vineyards." *Ca. Agr. Expt. Sta. Ext. Serv.* Leaflet 2139 (January 1977).

Sipiora, M.J., Anderson, M.M. and Matthews, M.A. 2004. "A role of irrigation in managing vine potassium status on a clay soil." Edited by L.P. Christensen and D.R. Smart. Presented at Proceedings of the soil environment and vine mineral nutrition symposium, San Diego, California. (June 29-30, 2004).

Skinner, Paul W. 2008. *Soils and Wine Grapes in the Napa Valley*. St. Helena: Terra Spase, 2003.

Skinner, Paul W. 2008. *Weather and Wine Grapes in the Napa Valley AVA*. St. Helena: Terra Spase, 2003.

Smart, R.E., Joy V. Dick, Isabella M. Gravett and Fisher, B.M. 1990. "Canopy management to improve grape yield and wine quality – principles and practices." *S. Afr. J. Enol. Vitic.* 11 (1) (1990): 3-17.

Smith, Richard, Larry Bettiga, Michael Cahn, Kendra Baumgartner, Louise E. Jackson and Tiffany Bensen. 2008. "Vineyard floor management affects soil, plant nutrition, and grape yield and quality." *Cal Ag.* 62 (4) (October-December 2008): 184-190.

Snyder, R.L., and Thompson, J.F. 1992. "The role of air movement in frost protection. Proc. Frost Protection Strategies for Trees and Grapevines." *Ca. Agr. Technol. Inst. Publ.* 11104 (1991): 63-74.

Solomon, Kenneth H. 1991. "The role of water in frost protection. Proc. Frost Protection Strategies for Trees & Grapevines". *Ca. Agr. Technol. Inst. Publ.* 11104 (1991): 75-83.

Spayd, S.E., Tarara, J.M., Mee, D.L. and Ferguson, J.C. 2002. "Separation of Sunlight and Temperature Effects on Composition of *Vitis vinifera* c.v. Merlot Berries." *Am. J. Enol. Vitic.* 53 (3) (2002): 171-182.

Sweet, Nancy. 2008. "Leafroll Virus Threatens California Vineyards." *Foundation Plant Services Grape Program Newsletter* (October 2008): 34-40 and *Practical Winery & Vineyard* (September/October 2008): 5-82

Swinchatt, Jonathan, David G. Howell. 2004. *The Winemaker's Dance: Exploring Terroir in the Napa Valley.* Berkeley: UC Press, 2004, 32-41, 100-102.

Tsai, C.W., Chau, J., Fernandez, L., Bosco, D., Daane, K.M., and Almeida, R.P. 2008. "Transmission of grapevine leafroll-associated virus 3 by the vine mealybug (*Planococcus ficus*)." *Phytopathology* 98 (10) (2008): 1093-1098.

Vasconselos, M. Carmo and Steve Castagnoli. 2001. "Leaf canopy structure and vine performance." *Practical Winery & Vineyard* (September/October 2001): 5-17.

Walker, Andrew and Howard Ferris. 2009. "New nematode-resistant rootstocks released to nurseries." *Practical Winery & Vineyard* (January/February 2009): 106.

Walker, Larry. 2007. "Dry, Dry Again. Why a few California growers say no to irrigation." *Wines & Vines* (November 2007): 52-56.

Weber, Ed. "Vineyard Views: 1996 Phylloxera Survey Results." *University of California, Cooperative Extension, Napa County* (May 29, 1997).

Weiss, PhD., Stuart B., David C. Luth and Bibiana Guerra, 2003. "Potential solar radiation in VSP trellis at 38°N latitude." *Practical Winery & Vineyard* (May/June 2003): 16-27.

Wolpert, James A. 1990. "Rootstock Bibliography." *Unpublished,* University of California, Davis (July 1990).

Wong, Deborah Parker. 2008. "Remedy for Underperforming Cabernet Sauvignon." *Vineyard & Winery Management* (September/October 2008): 76-80.

Yamane, Takayoshi, Seok Tae Jeong, Nami Goto-Yamamoto, Yoshiko Koshita and Shozo Kobayashi. 2006. "Effects of Temperature on Anthocyanin Biosynthesis in Grape Berry Skins." *Am. J. Enol. Vitic.* 57 (1) (2006): 54-59.

Zoecklein, Bruce W., Tony K. Wolf, Lisa Pélanne, M. Kay Miller, and Sandy S. Birkenmaier. 2008. "Effect of Vertical Shoot-Positioned, Smart-Dyson and Geneva Double-Curtain Training Systems on Viognier Grape and Wine Composition." *Am. J. Enol. Vitic.* 59 (1) (2008): 11-21.

Photograph 1. Enriched cow dung applied to the de-navelled distal end of bunch of Robusta' banana

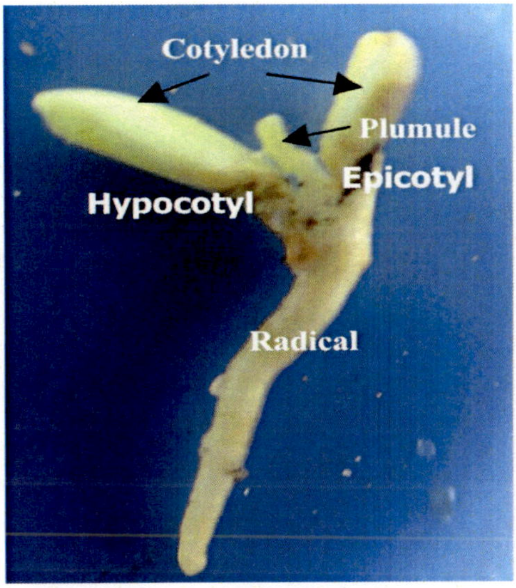

Figure 3. Organogenesis in *Citrus reticulata*

Figure 4. Citrus seed with two seedlings

Figure 5. Citrus seed with three seedlings

Figure 6. Citrus seed with four seedlings

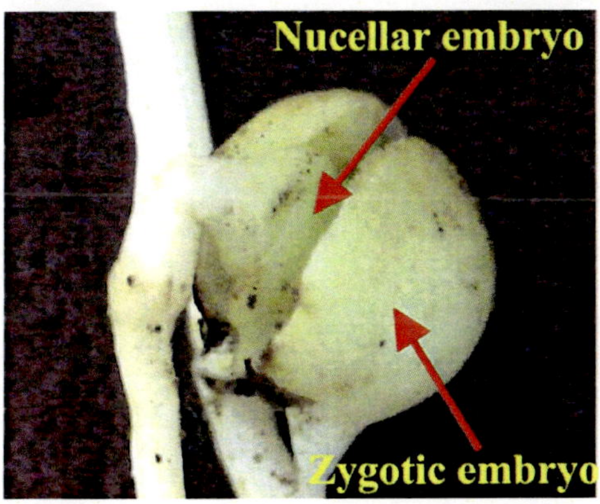

Figure 7. Zygotic and Nucelar embryos of *Citrus reticulata*

Figure 8. Zygotic and Nucelar seedlings of *Citrus reticulata*

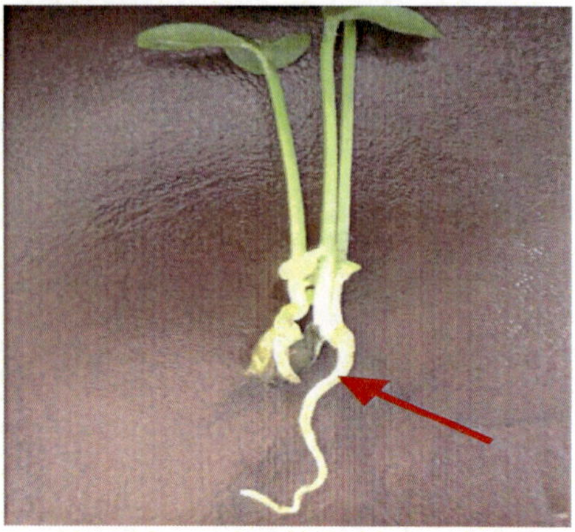

Figure 9. Zygotic twins in *Citrus reticulata*

Figure 10. Zygotic twin in *C. reticulate* having equal cotyledonsize

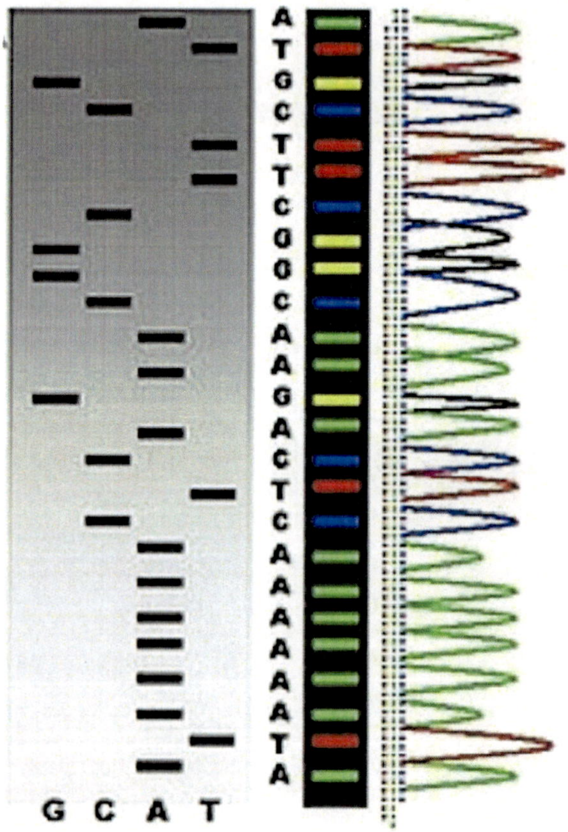

Sequence ladder by radioactive sequencing compared to fluorescent peaks

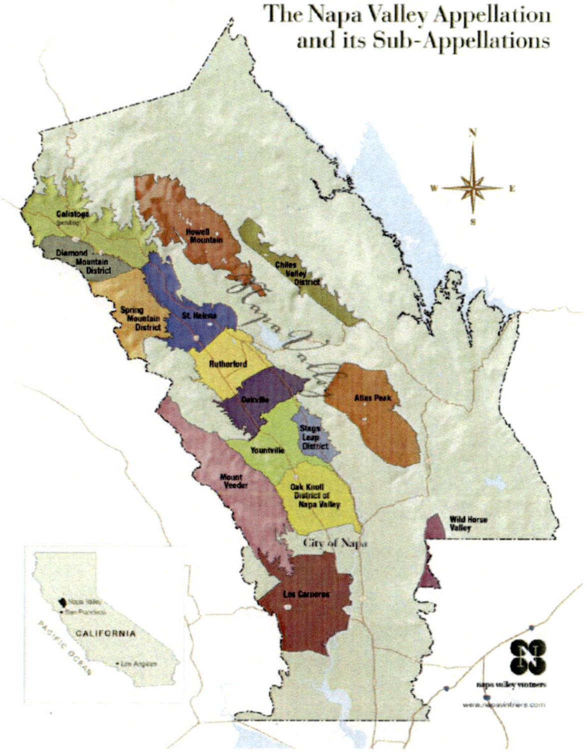

Courtesy of Napa Valley Vintners

Picture 1. VSP
A VSP, spaced six feet between rows and four feet between vines, known as a 6' x 4' (1.8m x 1.2m), in a Cabernet Sauvignon vineyard.

Picture 2. VSP with Cross Arms
A modified VSP with cross arms in a Cabernet Sauvignon vineyard spaced 8' x 6' (2.4m x 1.8m). Note the wind machine for frost protection. The vineyard has alternating cultivated and sod rows and is drip irrigated.

Picture 3. GDC
A GDC trellis in a drip-irrigated Cabernet Sauvignon vineyard spaced 12' x 6' (3.65m x 1.8 m).

Picture 4. Modified Lyre
A Lyre system, using movable foliage "catch" wires, in a Sauvignon Blanc vineyard spaced 12' x 7' (3.65m x 2.1m). Note overhead sprinklers for frost protection. The vineyard is also drip irrigated.

Picture 5. "T" Trellis
A "T" trellis in a Cabernet Sauvignon vineyard spaced 10' x 7' (3m x 2.1m). This vineyard is drip irrigated.

Picture 6. Goblet
Goblet trained Zinfandel vineyard spaced 10' x 8' (3m x 2.4m). This vineyard is drip irrigated, although the drip line is not shown in this photograph.

Picture 7. Vines at 21°F (-6°C) protected by overhead sprinklers.